Mobile Crane Training and Certification

Mobile Crane Training and Certification

Danny Thiemens

Industrial Press

Industrial Press, Inc.
1 Chestnut Street
South Norwalk, Connecticut 06854
Phone: 203-956-5593
Toll-Free in USA: 888-528-7852
Email: info@industrialpress.com

Author: Danny Thiemens
Title: Mobile Crane Training and Certification
Library of Congress Control Number: 2024950217

ISBN (print): 978-0-8311-3688-8
ISBN (ePUB): 978-0-8311-9635-6
ISBN (eMOBI): 978-0-8311-9636-3
ISBN (ePDF): 978-0-8311-9634-9

Publisher/Editorial Director: Judy Bass
Copyeditor: Judy Duguid
Compositor: Patricia Wallenburg, TypeWriting
Proofreader: David Johnstone
Cover Designer: Jeff Weeks

books.industrialpress.com
ebooks.industrialpress.com
1 2 3 4 5 6 7 8 9 10

Contents

19 The Effect of Sling Angles 311

20 Specialized Rigging, Slings, and Hardware Used in Craning 321

21 Specialized Crane Operations, Tools, and Equipment 383

Forewords

Every significant change has champions who lead the way. The crane industry is no different.

Up until the mid-1990s, the crane industry didn't really focus on safety, training, or certification. While the cranes themselves were undergoing substantial changes to incorporate load moment indicators and computer assisted design during this time, the operator was left behind. In fact, anyone entering crane operations actually had to complete less training and qualification than those who were going to cut hair.

Then the accidents started—one high-profile accident after another. One took place in the heart of a city, where the falling pieces of the crane crushed a school bus on a busy street, one occurred in a stadium that demolished it and killed several workers, one accident was in a city that a crane fell into a neighboring building and killed an individual in his apartment. As each accident occurred, the public became aware of the risks of these machines and began to demand answers. As the cries for public safety were elevated, legislature, Congress, OSHA, and regulators began to focus on the industry and recognized that there were major gaps in the requirements and qualifications for operators.

This is where Danny Thiemens, the author of this book, stepped up and became a champion in the crane industry. There weren't any existing training courses, exams, qualification requirements or mandates, or even industry standards. That didn't hold Danny back. He began compiling what information was out there and developed detailed training courses.

He was also an early adopter of the CCO (*Certification of Crane Operators*) certification and was a key leader in the development of crane operator certification for the country.

Then the changes came in the form of standards and regulations. States jumped ahead of OSHA to regulate cranes, crane operators, and crane operations. OSHA began its lengthy process, and the ASME standards started to define responsibilities and qualifications. Suddenly, the crane industry went from a regulation written in the 1960s to state, local, federal, and industry standard changes all at once. These changes culminated in the 2018 OSHA final rule. However, states, local and industry standards continue a rapid rate of change for the crane industry.

You will find that the information contained herein is thorough, complete, and up to date. I particularly appreciate Danny's perspective from the field and the additional information that is provided that serves to explain the thoughts and concepts behind the regulations and codes. He also does a remarkable job of giving practical advice on how to implement these important concepts in the field.

What drove Danny back in the 1990s when he first became a champion for the industry is exactly what drives him today. He wants to elevate safety for the crane industry so that workers can return home to their families, and the public isn't at risk. This is why he created these materials.

Thank you, Danny, for sharing your insights, wisdom, and passion for safety. It is an honor to serve the industry with you!

Thom Sicklesteel
CEO at CCO
National Commission for the Certification of Crane Operators

This book is a great resource for everyone involved in the crane industry. Danny has taken complex operational and engineering questions and answered them in a manner that all can understand. Anyone and everyone involved in or interested in cranes and how they work should put this book on their reading list. Congratulations, Danny, on a job well done.

Bill Lampson
President, Lampson International

A book written by a crane expert with such a vast knowledge of both the previous and current laws and regulations governing the crane industry has been long awaited. Finally, the wait is over, and Danny Thiemens has delivered in spectacular fashion.

This work is a comprehensive study of crane design, operation, and safety that brings to light the failures of the past and shines a new light on the successes that were borne of those failures. Danny's unique perspective as a crane operator and trainer keeps the reader's interest while offering sound advice on implementing programs that help ensure safe crane operations.

This book is a brilliant culmination of years of hard work, research, and hard-won experience by a talented technical author. Hats off to you, Danny, for your continued commitment to the safety in this industry and for the publication of this very informative book. I have been honored to work with you for many years and proud to call you a friend.

Bruce A. Stemp
Safety/Quality Assurance Director
Lampson International, LLC

Danny Thiemens was hired as Training Director in the spring of 1992 to head up the *International Union of Operating Engineers' (IUOE)* Local 370 training site in Boise, Idaho, for the Idaho operations. He came to the school after spending a decade working as supervisor for one of the nation's largest and most successful general contractors, Kiewit Pacific Company, from the northwest district. He was known then for his expertise in heavy highway construction finishing grading operations. He supervised highway, airport, and public service projects in Washington, Idaho, Oregon, and Hawaii. When he came to the training site in Boise, he brought with him other key Kiewit personnel to help him. They came on a mission, and in one short year they revised the property and building along with a fleet of tired heavy equipment. Danny and his team brought back much needed pride to a very underfunded training program. After a short stint managing the Idaho program, he was promoted and moved to Spokane, Washington, to administer both the Idaho and its sister program for Eastern Washington. In a short time, the two programs were merged into one. That's when Danny began his search for property large enough to host a training site for two states. Land was acquired near Spangle, Washington—a small town 25 miles south of Spokane. Danny had big hopes and dreams, and big goals for the 60-acre site. In only a few years, his team rebuilt an entire fleet of dirt-moving machinery and service equipment. The equipment and machinery were used to shape and mold the 60-acre property into a training site, offices, a classroom, a shop, dormitories, and large campus, all built by dislocated workers, apprentices, and journey operating engineers alike. Even union representatives (including myself) found themselves coming to Spangle to donate a helping hand. The new school created a lot of interest and pride while everything took shape.

The fleet of equipment had become very admirable, all painted and clean with our logos and equipped with all required safety equipment. As the training sessions advanced, the students with their fleet of equipment would mobilize into special projects they built for the community. Students would design, survey, and build baseball diamonds, skate parks, public parks, and church parking lots all over the area with the help of Danny's supervision and team of excellent instructors. The apprentices, journeymen, and instructors became quite respected in our area. I decided to give them the name, "Green Hands Incorporated." The program was underfunded, so Danny went to school and became a certified grant writer. With his newly found writing abilities, he was successful in winning many federal- and state-funded grants that helped bolster the school to a much higher level.

As time went on, one of the goals for the school was to build state-of-the-art classrooms, supported by a hand-on Mobile Crane Training area, all in the hopes of a new Mobile Crane Training program that was to be the best in the nation. That's a very bold undertaking, and Danny knew he needed some professional help. This is when he looked north to Toronto, Ontario, Canada, the home of the renowned Mobile Crane instructor and consultant Ron Cowper. Ron was very respected across North America for his knowledge and expertise on Mobile Cranes. At the time, Ron was helping to administer Canada's elite and highly respected Mobile Crane School near Toronto, the *Operating Engineers Training Institute of Ontario (OETIO)*. Ron was also known for his contributions as co-editor on the Institute's well-known book "Mobile Craning Today." When Danny called Ron and explained about the high number of high-profile crane accidents across the United States and Washington state in recent months, Ron wasn't surprised. He and was already aware of the problem. When Danny told him about his dreams of building a new school, Ron was very interested in lending a helping hand. Within only a few months, Ron was at our school in Spangle, Washington, working with us to start providing our first mobile crane classes.

Because of the seriousness of the accidents across the United States and Washington State, many of them with serious injuries and fatalities, I had a strong feeling that our state legislature would soon be introducing some sort of mobile crane licensing or certification mandate for our area. And sure enough, thanks to Danny Thiemens and Ron Cowper, we were ready. In fact, Danny rose to the occasion, scheduling a new mobile crane training program across the jurisdiction of the *International Union of Operating Engineers* (*IUOE*) Local 370. Ron Cowper had helped us solve a big problem, putting our school on the map almost instantaneously. Over the next several years, Danny continued his leadership at the school, heading up the normal activities. Along the way, Ron Cowper and Danny became very close—building something of a father-and-son relationship. Together they continued to develop many more crane related curricula for the school. They joined together to support the *National Commission for the Certification of Crane Operators* (NCCCO). Collectively, they helped the NCCCO develop their new Mobile and Tower Crane written examinations. Danny was successful in obtaining grant funding that provided for members of our sister IUOE locals 400 in Montana, 302 in Alaska and Washington, and 701 in Oregon, to send their members to attend crane programs for free.

Soon we were working closely with the NCCCO to provide practical examiner-accredited workshops, helping many of our members to earn their credentials to be NCCCO practical examiners, eligible to administer crane practical examinations on behalf of the NCCCO. Clearly teaming up with Danny Thiemens and Ron Cowper paid big dividends for our school and crane operators across our region and the country.

Ron Cowper and Danny Thiemens developed a strong bond, providing hundreds of crane programs at our school and across the country. Danny left our school in 2009 to accept a staff position with the NCCCO. He is still out there on the road, instructing crane programs for his company, Fulcrum Crane Training International. Our school in Spangle, Washington, is now going on 30 years, and continues to train operating engineer journeymen and apprentices. Danny and I also developed a great friendship that we still hold today.

No one will ever forget when it all started with a shared goal and a dream, in large part because of Danny and Ron. When I learned he had been working on a new crane book, I had to smile. It's wonderful to tell our story. As far as the new book goes, I have one on my book shelf, and like everything else those two have done and continue to do, it is excellent.

Curt Koegen
Business Manager
International Union of Operating Engineers (IUOE) Local 370
Retired

Introduction

Sadly, there's a high possibility that someone in the mobile crane industry isn't going to come home from work this evening because they were killed on the job today. Even graver, there's a very high likelihood that someone else will join the list of fatalities next week, or perhaps even more than one, in a bad week. The Bureau of Labor Statistics (BLS) reports there could be as many as four funeral services just this month and realistically, if this year is to be like so many others, there are likely to be wakes for as many as forty-two workers before the new year. Across the nation, crane disasters are widespread as the construction industry booms, developing urban infrastructure, high-rise buildings, factories, hydro-electric dams, green energy wind farms, bridges, and airports.

Companies of all sorts, large and small, are saturated with laborers ready to get to work. Some workers are highly skilled, thoroughly trained, are qualified, and have certifications earned within their craft. Other construction workers, eager and passionate, yet brand new to the industries, are creating a skill gap across all trades, especially those utilizing cranes. The Center for Construction and Training (CCT) began investigating crane-related fatalities in 2008 and provided a jaw-dropping report involving three major incidents occurring in a brief, ten-week period that year. One took place in Miami, Florida, and two other accidents in New York City, New York, collectively causing the deaths of ten construction workers, eleven first responders, and one bystander. CCT wrote in its findings that "ultimately a lack of worker and supervisor training" was to blame. With this discovery, CCT also found that injuries, lost time accidents, and their causes had failed to be reported in the past, skewing government accident reports from showing just how bad the situation really was. Crane inspections and scheduled repairs not being made on time also contributed heavily to the problem. The 1990s and 2000s brought plenty of bad reports like this, filled with horrible accident report data and statistics. Often these reports made recommendations to help prevent these types of accidents, with training always listed at the top, as it should be.

Years ago, crane design worked on the principle that the weight of the crane would be heavier than, or equal to, the charted lifting capacity of the machine. This made it less likely that the crane would tip over. However, structural failure was still possible. When hydraulic boom cranes were first introduced, they effectively reversed this rationale, making the maximum charted capacity several times the weight of the crane itself. Because of these remarkable technological advancements in engineering, these more modern machines required more finesse and technical knowledge from their crane operators than the conventional cranes of the past. Though these conventional cranes are still widely used in the industry, they require their own type of expertise to operate. In most cases, little to no comprehensive training for crane operators was available to ensure they had the knowledge, skillset, or capability to safely operate a crane within the specifications provided by the manufacturer. As a result, as many as two workers a week were being killed, contributing to the fatality statistics and reports of the time.

During this time period, mobile cranes were evolving and had already undergone some extraordinary changes. Shifting from steam- to gasoline-powered engines, then to diesel as the current primary fuel source were big changes. The crane industry has witnessed exceptional improvements to hoisting equipment as well. The use of light-weight tensile steel was introduced, making the crane much lighter, yet capable of lifting exceedingly heavier loads while erecting more boom and placing loads at a longer radius than previously thought possible. But no technological advancement had such a far-reaching influence on crane design as the integration of hydraulic power. Engineering principles increased the crane's size, charted lifting capacities, and versatility, allowing them to do much more than ever before. Phenomenal as these improvements in design were, something crucially important had not yet been addressed—the lack of formal, comprehensive training of the crane operator, owner, site supervisor, lift director, rigger, and signal person. Despite crane manufacturers' attempts to help, high-quality training programs were nearly nonexistent.

Outdated or lack of quality information is a large contributor to accidents, lost time, injuries, and fatalities in construction and general industries alike. Staying current and finding quality updated training materials and publications can still be a challenge in the mobile crane hoisting and rigging industries. This is the reason for this book, *Mobile Crane Training and Certification*. I wanted to provide a quality, comprehensive publication geared to help all those who own, operate, or work around mobile cranes. The need for an updated book was important for the mobile crane industry's safety, efficiency, and legal compliance. A work like *Mobile Crane Training and Certification* is important to the industry because it provides a concentrated, up-to-date resource on safety, technology, and best practices. This compares with relying on older materials, books, or crane-specific manufacturers' manuals. The crane industry faces a constant shortage of qualified and certified crane operators, supervisors, lift directors, riggers, and signal persons. This comprehensive book can function as a foundational resource for multiple craft training and certification programs. It consolidates the vast amount of information required for a career in various crafts and trades within the crane industry. This authoritative guide standardizes training practices across our industry, ensuring that crane operators and other craftsmen involved receive consistent, high-quality instruction, regardless of their specific training offered through their trade apprenticeship and training programs or crane manufacturers. Manufacturer manuals are critical for specific makes and models, however, they lack the broad, comparative context necessary to develop a true industry expert. This book will teach principles that apply to most makes and models, helping owners, supervisors, crane operators, and lift directors to become proficient across a wide variety of mobile cranes and supporting equipment. It explains the foundational engineering principles behind crane operations, and the methodical procedures for situations requiring a deep understanding of crane dynamics, lift planning, and industry regulations.

An affiliated website, https://samplecranetesting.com, rounds out this unique resource by providing:

1. Hundreds of additional test questions just like the ones found on the NCCCO exams
2. Even more load chart questions for fixed cab, swing cab, and lattice boom cranes
3. Updates on mobile crane laws and regulations

Having dedicated five decades of my life to the mobile crane industry, I bring a depth of expertise and a lifetime of practical experience to this comprehensive guide. For three decades, I served as a mobile crane instructor and was a staff member at the NCCCO, where I led classes nationwide and represented the organization in Washington State. This project was more than an educational undertaking. It was a testament to the enduring bonds of mentorship and a final collaboration with two of the industry's most respected figures—Paul Parish, my mentor of 30 years, and Ron Cowper, my mentor of 25 years. Each one contributed to and vetted the material in this book. Ron, one of North America's most respected crane consultants and instructors, brought his immense knowledge from his time as chief instructor at the highly regarded Operating Engineers Training Institute of Ontario (OETIO) and co-editor of the work *Mobile Craning Today*. These veterans of the industry poured their collective wisdom into these pages. This work is dedicated to Ron Cowper, honoring his legacy. The pinnacle of a collaborative group effort by Ron, Paul Parish, and my beloved wife, Kathleen, a crucial contributor to everything in my life. Sadly, all tragically passed away as the book neared completion.

The endorsements from industry leaders further underscore this book's importance. Forewords from Thom Sicklesteel, the CEO of NCCCO, Bill Lampson, president of the world-renowned Lampson Crane International, Bruce Stemp, safety director at Lampson Crane International, and Curt Koegen, of the International Union of Operating Engineers confirm the manual's standing as an essential textbook for the mobile crane industry. This book is not merely another training manual. It is a meticulously crafted, operationally tested, and deeply personal chronicle of knowledge passed down through generations of experts. For readers seeking the definitive guide to mobile crane training and certification, this book represents the collective wisdom of an industry and the profound dedication of those who helped shape it.

Danny Thiemens
Fulcrum Crane Training International, LLC
October, 2025

Looking back over time, we can learn a lot about how cranes evolved from those in ancient times to the giant workhorses we use today. The cranes used in ancient Egypt to build the amazing pyramids in 2750 BC worked on the same principles as the giant cranes of today. Look closely at pictures and paintings of the ancient cranes. The pulleys, hoist lines, winches, booms, and counterweights are remarkably similar to those of any modern crane. The main difference is that the cranes of today are powered by diesel fuel and advanced hydraulic systems. Ancient cranes relied on human power.

1 A History of Cranes

Cranes have been part of human history for thousands of years. In ancient Mesopotamia, as early as 3000 BC, a crane-like tool, the *shadoof*, a was used for irrigation. By 2000 BC, shadoofs like the one in Figure 1.1 were found in ancient Egypt. Shadoofs were a combination of a pole with a bucket and a counterweight, essentially nothing more than a pivoting lever with a bucket attached. The operator would lower the bucket into a pond or lake, allowing the bucket to fill with water. Once filled, the operator would raise the lever and spin it to deliver water to nearby crops or irrigation ditches.

Ancient cranes evolved over time to perform the tasks that humans could not carry out because of odd shapes and heavy weights that were impossible to lift by hand. Although early cranes made it much easier to lift and place heavy loads, construction practices of the time stuck to smaller materials that would not overload the crane.

The advancement of the crane allowed construction projects to shift from unskilled to skilled labor during this time. Cranes with crews of specialized workers working with a crane could manage projects much more efficiently than huge teams of workers relying on brute force. Medieval cranes much like the one in Figure 1.2 would continue to improve over time to significantly increase the cranes' capacities and capabilities.

FIGURE 1.1 Ancient crane-like shadoof used to deliver water.

FIGURE 1.2 Image of a medieval crane.

Ancient Greek Cranes

Let's step back from medieval times to the time of ancient Greece, when the crane was invented. Ancient Greece is well known for its extraordinary stone temples. Giant structures such as the Parthenon of Athens would not have been constructed if not for the invention of the crane. The crane is one of the Greeks' greatest innovations.

Some of the first crane activities were performed by ancient Greeks six centuries before Christ. The device got its name because it resembled the long neck of the bird called the *crane*.

Early building methods used by the Greeks required large ramps or scaffolds of earth or mudbrick. According to archaeologists, workers dragged large blocks placed on sleds up the ramps. The invention of the crane allowed heavy blocks to be hoisted and set into place using a small crew of men. In addition to the Parthenon, the crane made it possible to build structures like the Temple of Artemis at Ephesus and the Temple of Poseidon at Sounion. Structures such as these have amazed people for thousands of years.

The Greeks were the first to lift heavy objects utilizing tongs. These lifting attachments, called *Lewis tongs*, were hooked near the center of gravity to grip and hoist heavy loads. Marks on the blocks provide evidence that tongs were used to lift stones as workers placed them to build the Greek temples. According to archaeologists, stones in temples have holes that can still be seen located in the blocks just above the center of gravity where the tongs were placed. Until Greece, there were no civilizations known to use cranes for building purposes. Cranes remained the primary tool used in building construction without major changes for nearly 25 centuries.

Ancient Roman Cranes

The Romans took the crane the Greeks invented and improved the functionality and mechanical advantage so that heavier loads could be lifted. The Romans worked very hard to improve the crane, which ultimately paid big dividends in the end. The cranes that the Romans implemented had much more capacity than the Greek versions. Some archaeologists believe the Romans performed the first-ever multiple crane lifts while constructing the Temple of Jupiter at Baalbek. The golden age of the crane in ancient times was during the construction of the Roman Empire. During this period, building construction was on high, utilizing the improved Greek crane to build bigger, better, and taller buildings than ever before.

A Greek crane called a *trispastos* consisted of a single beam or jib, along with a hoist rope and three pulleys. One man could lift as much as 330 pounds. The Greco-Roman *pentaspastos* was a heavier crane that used five pulleys or sheaves that allowed it to pick up as much as 993 pounds. The largest of all the Roman cranes, the *polyspastos*, utilized three hoist ropes, along with five pulleys or sheaves for advanced mechanical advantage. With four men operating this crane, they could lift a maximum of 6,613 pounds. The polyspastos was innovative and could be configured to use two, three, or four masts and a treadwheel rather than a winch. In this configuration, the maximum load could be doubled to 13,226 pounds. The Romans continued to improve the mechanical advantage of their cranes to achieve a lifting capacity of 12,000 pounds, or 60 tons, over the time of their use. Figure 1.3 shows the names and components of a Roman crane as it performs ancient construction techniques.

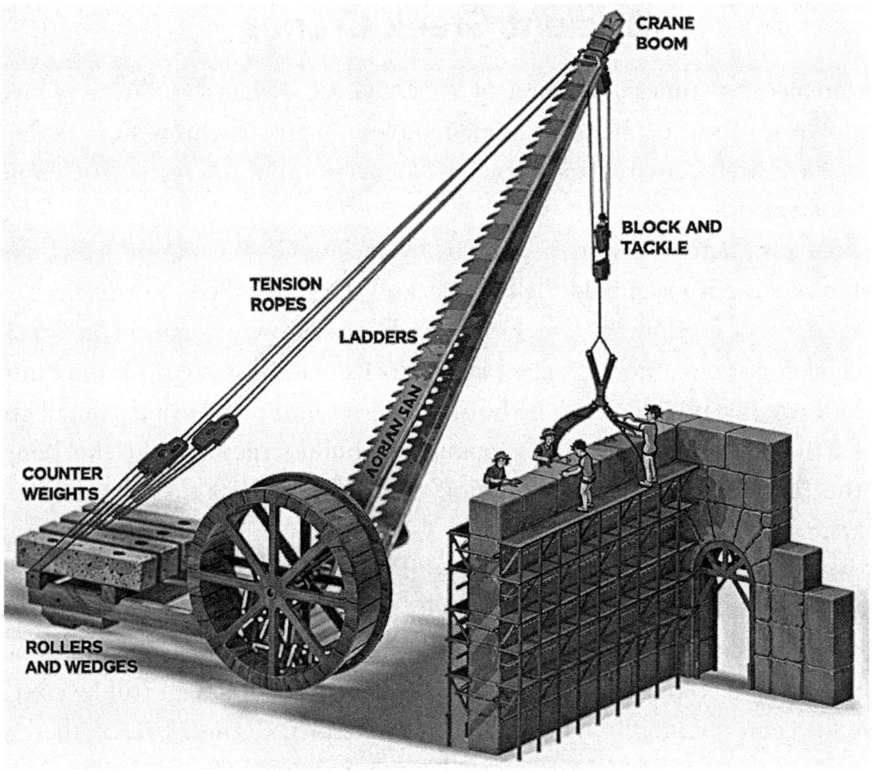

FIGURE 1.3 Roman crane placing blocks.

The Middle Ages

Treadwheel cranes returned to use in the Middle Ages after very little use following the fall of the Roman Empire. Treadwheel cranes were put into production in mines, harbors, and construction building sites all around Europe. The first treadwheel cranes were used in France around 1225. Early use of these cranes can also be traced back to 1244 in Utrecht, Netherlands. The cranes were a crucial tool in the construction of many Gothic cathedrals all across Europe, including the awesome Kolner Dom in Germany. These cranes were operated by windlasses or winches with radiating spokes and cranks. Treadwheel cranes were well known in the construction industry at the time and could be found on jobsites everywhere across Europe.

Treadwheel cranes looked much like a giant hamster wheel (see Figure 1.4). People would walk inside the wheel to hoist or lower loads. Workers inside the wheel had room to walk, and that is all. Because visibility was so poor inside the wheel, workers outside would shout instructions as needed to the human power source inside. Accidents and fatalities were common because of workers getting caught in the lifting mechanisms or falling out of the wheel and landing on the ground. Sometimes the wheels would explode as a result of structural stress and other forces.

Since these cranes were so dangerous, it often became necessary to find new workers. Blind workers were employed as crane operators during the construction of many projects since they couldn't see the dangers involved. They were also hired to work inside the treadwheels because they wouldn't be scared by the height of the buildings and the cranes they were working in. Recruiting blind, homeless people and beggars may have been a way of employing cheap labor because they were paid much less than an able-bodied crane operator.

FIGURE 1.4 Treadwheels were very narrow inside, leaving just enough room for workers to do their jobs.

As noted earlier, being a crane operator in the Middle Ages was dangerous. Cranes were built by trial and error, and they often came apart and killed people. Stopping the wheel was especially difficult because it relied on the strength of workers to stop it. Those cranes were nothing like the cranes today with precision hydraulic levers, brakes, and controls. According to most archaeologists, ancient civilizations had little or no advanced technology that would help them move heavy stones or monoliths, and yet they were able to build spectacular structures and monuments. Our planet is home to numerous amazing ancient structures such as the great Egyptian pyramids, Stonehenge, Sacsayhuaman, and Teotihuacàn, to name only a few. Monoliths like Thunder Stone, Brownshill Dolmen, the Colossi of Memnon, the Stelae of Axum, and the Puma Punku are excellent examples of some of the largest.

We have discussed the ancient cranes of medieval times, the structures built in Greece and Rome, and the capacity of the cranes used to build those structures during those times. Many archaeologists suggest that the massive stones used to build these structures were quarried in rock mines. In some cases, the mines were located hundreds of miles away. Once quarried, the huge stones were placed on sleds and dragged by ancient workers with ropes to the building sites.

It is also theorized that in the case of the great pyramids, the limestones were transported over 500 miles by boat. Some of the stones weighed up to 80 tons, such as those used for the King's Chamber. Some believe the Egyptians pulled the other stones on sleds across the desert to the pyramid construction sites.

The giant structures located on Easter Island are another fascinating case. It is suggested the statues were transported from the South American continent to their placement sites on an island west of what is now Chile. On average, they stand 13 feet high and weigh 28,000 pounds. However, the largest statue is 32 feet high and weighs 164,000 pounds (see Figure 1.5). Island legend claims the statues waddled 11 miles from the quarry where they were mined to their resting place. Others say they were levitated to their spots.

FIGURE 1.5 The largest statue on Easter Island is 32 feet high and weighs 164,000 pounds.

Any way you measure it or calculate it, the stones used to build massive ancient structures, monuments, and monoliths all over our planet were huge and very heavy. Many crane consultants, crane operators, lift directors, and engineers scratch their heads and ponder how, even with using the cranes and methods of today, they would be hard pressed to handle the heavy lifts of ancient times. Could we achieve some of the same lifts using today's cranes and technologies?

As we discuss how cranes evolved from ancient times to the workhorses of today, it is fascinating to explore and look over all the monumental achievements of ancient times. Of course, ancient alien astronaut theorists suggest extraterrestrials came to Earth and offered advanced knowledge to help build ancient sites. It is a challenge to find an explanation of how civilizations of the past managed some of the largest and most mysterious construction achievements on Earth. As you learn more about cranes, you too can develop your own opinions of how these amazing achievements were accomplished.

Evolution of Cranes

The evolution of steam-powered shovels and cranes during the end of the eighteenth century provided these machines with an unprecedented new level of power. Cranes were busy at work in city ports, railway yards, and construction sites all over the world. During the Industrial Revolution that followed new materials were developed to manufacturer the machines, along with gasoline and diesel fuel to offer alternative and better fuel sources to power the machines. New power sources other than steam had a tremendous effect on the evolution of cranes. The cranes became much more mobile and no longer had to be tethered to their power source, the steam engine. During this time with these new developments, cranes flourished with growth in size, design, and capacity capabilities. Cranes were becoming more versatile and powerful than ever before.

Designs with the crane mounted on a track type configurations were first available to the industry in 1911. These machines were originally manufactured for shovels and draglines. Shovels such as these were utilized in the construction of the Panama Canal and burned coal to create steam for their power source. The Bucyrus company, located in Milwaukee, Wisconsin, built the first crane to be mounted on crawler tracks. This groundbreaking, gasoline powered, machine was configured as a "Type 14 Dragline." However, it could also be rigged up with a crane hook. Fitting the undercarriage with crawler tracks gave the machine greater stability and allowed it to move around easier on different terrains

Soon after came designs by Pawling and Harnischfeger, commonly known simply as P&H. In 1914, P&H brought the advancement of the diesel fuel engine to power its 210 crawlers. In 1932, excavator and crane manufacturer Lorain improved on the idea by installing a newer caterpillar diesel engine as the power source in their design. Manitowoc Cranes first began manufacturing cranes in 1825. These cranes were specifically engineered to handle and hoist much heavier loads. The original crane was rated at 65 tons and was noted for being the first crane to utilize a torque converter, providing very precise operation while placing loads These cranes were specially ordered to clear wreckage and remove sunken battleships after the Japanese attacked Pearl Harbor in the summer of 1941 off the coast of Oahu, in the Hawaiian Islands.

Manitowoc cranes were highly respected for their refinement and dependability. They were utilized during construction projects such as the Senate Office Building and the National Gallery of Art in Washington, DC, as well the ship building industry. Manitowoc built a reputation for being tough and reliable, which made them the crane of choice for many industrial and construction projects of the time. Manitowoc cranes became a major supplier of cranes to the U.S. Navy during World War II. They were put to work building submarines and floating dry docks and utilized in harbor salvage operations. Manitowoc started building cranes under an agreement to help Moore Speed Crane under that company's licensing patents. The first cranes models were a reworking of the original Moore Speed Crane. However, in 1932 Manitowoc moved towards their own revolutionary design. The 1932 models can be considered to be the first purpose-built cranes beyond the initial license agreement with Moore Speed Crane. The Manitowoc 3900 much like the 1932 design was extremely popular and sold over 1,000 units during its years of production. The 3,900 had widespread use across North America. Many units can still be found in the crane industry today, and are still in active service. Both World Wars had an extensive impact on the speed of mobile crane progression at that time and continuing for many more years to come.

In the early 1900s the crane operator was lucky to have a plank of wood to sit on. Some of the early mobile cranes required the operator to stand at the controls. Later, when crane operator's cabs became standard equipment, the operator was in a seated position, protected from the environment. Although a far cry from todays climate-controlled cabs, this was a significant improvement in operator comfort. By the mid 1930s, Michigan Power Shovel Company introduced air-controlled clutches improving the way the controls operated. The crane operator's comfort was usually a secondary concern. Cab designs of this era were often small and offered limited visibility. In some cases, the crane operator had to lean out windows to observe what was happening during a lift. Crane cabs of today are equipped with advanced features such as intelligent service panels, remote controls, sophisticated control panels, adjustable seating, easy touch controls, enhanced visibility, and camera systems. Other advancements included climate controls, anti-collision systems, and load moment indicators. Today's mobile crane cabs provide improved operator ergonomics, advanced safety features, and, now more important than ever before, technology enhancing operator comfort, which has been proven to increase safety and, at the same time, productivity.

Since the end of World War II, the craning industry has seen some remarkable changes in design and complexity of its hoisting equipment. The evolution from steam to gasoline and diesel engines, the development of lightweight higher tensile steels, as well as advancements in design and engineering, have all contributed to increased capacity and versatility of modern craning equipment. No single factor however has had such a profound influence on the operational ease of cranes as the incorporation of hydraulic power in the early 1960s and the modernization of steel production methods that was instrumental to manufacturing and improving strength-to-weight ratios compared to previous older material used.

Leading companies across the United States, such as Austin Western, Grove, Bucyrus-Erie, and Drott to name a few, were the trailblazers of the early technological innovations in hydraulic crane theory. The cranes of this era were designed and manufactured to be tough and strong enough to withstand the strenuous work in the construction industry. The weight of the machine was no longer a concern. It wasn't until the 1960s that the first crawler mounted cranes were produced to enhance maneuverability and the amount of weight the crane could pick and carry. Additionally, manufacturers of cranes began to look more closely at the utilization of lighter steel to increase reach and achieve higher capacities in the new cranes they were building.

Hoist ropes used for hoisting heavy loads have had some groundbreaking innovations of their own. Since the 1900s, wire rope has greatly evolved due to advancements in manufacturing techniques and materials, as well as rope design. Steel became the main material for hoist rope because of its strength, durability, flexibility, and resistance to corrosion in comparison to the fiber hemp ropes previously used. Automated machinery used in manufacturing increased efficiency and precision, as well as greater accuracy and speed of manufacture. Design factors include special performance characteristics that minimize rotation under load, including compacted ropes that offer increased tensile strength. Galvanized coatings were adopted, offering corrosion protection while enhanced lubricants and specialty greases added enhanced protection for the hoist ropes. Many advancements have been made since the 1900s, making today's wire ropes stronger, more durable, and safer than ever before.

The 1950s brought the industry the newly developed, rough-terrain cranes of today. Grove manufactured its first model in 1957 and was the first machine to have the cab mounted to the side of its telescoping boom. These cranes were primarily designed to handle the most demanding jobsite conditions. Austin Western Company designed its rough terrain crane with the operator's seat situated ahead of the crane's front axle, the operator's station exposed, with no cab. It featured 4x4 drive and steering for better maneuverability in congested jobsites. Contributions in the 1950s to the industry didn't stop there. Purpose-built crane carrier chassis had renovated lattice truck crane manufacturing, and with this development, larger telescopic boom truck cranes soon followed. As time went by, the number of telescopic boom cranes manufacturers grew to over 20, including Pettibone in 1961, Galion in 1963, Lorain and Shield Bantam in 1967, P&H in 1968, and Link-Belt in 1969, to name a few.

The world's first telescopic boom crane was manufactured by Grove in 1964. Following it came the first 30-ton machine from Pettibone, followed again by Grove in 1966 with their 45-ton machine. Pettibone, not to be outdone, came back strong with a 50-ton crane in 1967, and P&H with a 65-ton crane in 1969. But the true innovative break came from Grove in 1970 with a revolutionary massive 80-ton crane featuring a unique trapezoidal boom. The American companies Austin-Western, Grove,

Bucyrus-Eris, and Drott are all responsible for initiating the many early technological breakthroughs and innovations in the hydraulic crane industry.

Telescopic booms and the incorporation of hydraulics in the 1960s allowed heavier loads to be placed at longer a radius than ever before, following the first 25-ton telescopic truck crane developed by Grove in 1964. The 1960s and '70s also brought powered assisted controls to cranes. The accuracy of hydraulic controlled systems offered smoother control motion, as well as a reduced physical effort needed to operate these cranes. During the 1800s, crane and shovel controls consisted of long mechanical levers and foot pedals that demanded significant physical exertion in order to operate. Since the 1960s, cranes have transitioned significantly due to the advancement in boom design, hydraulic systems, and powered-assisted control levers making the crane operation much smoother. The integration of computers allowed increased operating efficiency and more complex structural designs creating a lighter, yet stronger, crane structure. With the use of computer technology during the 1980s came another paradigm shift in crane functionality and safety. Newly implemented load moment indicators offered assistance to the crane operator, anti-collision systems prevented multiple cranes from colliding in congested construction sites, and automated controls enabled precise movement when placing heavy loads, making today's cranes more productive and safer, thanks to these advancements in technology.

Into the modern day, manufacturers continue to design machines with more and more load capacity, pushing the limits further than ever before. Indeed, the evolution of today's cranes has come a long way, since the ancient Shadoof crane or the tread wheel cranes placing blocks in Rome. As we look at the highly technical, monster-size hydraulic cranes of today, could you imagine where their evolution might take them in another 100 years?

Chapter 1 Review Quiz

1. The first crane-type of tool was used for irrigation and was called the
   - **a.** shadoof
   - **b.** Greek crane
   - **c.** treadwheel crane
   - **d.** Roman crane

2. Who gets credit for inventing the very first crane used for construction purposes?
   - **a.** The Romans
   - **b.** The Greeks
   - **c.** The Germans
   - **d.** The Egyptians

3. The tongs used by medieval cranes to lift heavy blocks are called
   - **a.** treadwheel tongs
   - **b.** block tongs
   - **c.** hay tongs
   - **d.** Lewis tongs

4. Who gets credit for making great improvements to the treadwheel crane, thereby increasing its functionality and mechanical advantage?
   - **a.** The Romans
   - **b.** The Greeks
   - **c.** The Spaniards
   - **d.** The Aztecs

5. In the Middle Ages, operating treadwheel cranes was so dangerous, owners resorted to hiring
   - **a.** blind people because they could not see the dangers
   - **b.** able-bodied people because they could work longer hours
   - **c.** Olympic runners because they could run faster
   - **d.** heavy-set workers because they could stop the wheel from turning more quickly

Summary—and Then Some

Since World War II, the craning industry has seen some remarkable changes in the design and complexity of its hoisting equipment. The evolution from steam to gasoline, and then to diesel engines, as the primary power supply, and then the incorporation of hydraulic power, has led to the heavy-lift cranes of today. However, just like their ancestor cranes of long ago, they lift heavy loads based on the principles of leverage. Figures 1.6 through 1.18 present a variety of cranes from 1874 to present day.

FIGURE 1.6 The first automotive mobile crane with a steam-driven tractor engine first appeared in 1874.

FIGURE 1.7 In the relatively few years since the end of World War II, the craning industry had some remarkable changes in the design and complexity of its hoisting equipment.

FIGURE 1.8 Drott rough-terrain crane.

FIGURE 1.9 Link-Belt hydraulic boom truck crane.

FIGURE 1.10 Broderson carry deck crane.

FIGURE 1.11 National hydraulic boom fixed cab crane.

FIGURE 1.12 Link-Belt rough-terrain hydraulic boom swing cab crane.

FIGURE 1.13 Grove rough-terrain hydraulic boom swing cab crane.

FIGURE 1.14 Link-Belt lattice boom crawler crane.

FIGURE 1.15 Manitowoc conventional lattice boom crane with ringer.

FIGURE 1.16 Liebherr lattice boom crawler crane.

FIGURE 1.17 Liebherr LR 13000 lattice boom crawler crane.

FIGURE 1.18 Liebherr hydraulic boom truck crane swing cab.

Craning and rigging accident and fatality statistics compiled by several entities vary greatly, depending on the time span being considered, the amount of available data used in the research, and the number of jurisdictions that actually contribute information to the investigating agencies.

Providing the correct personal protective equipment for crane operators and the entire lifting team is a critical prerequisite. Crane operators, riggers, and signal persons are highly prone to mechanical failure risks and rigging failure. The Occupational Safety and Health Administration (OSHA) stipulates a comprehensive code for workplace personal protective equipment (PPE). Crane operators and all members of the lifting team must know what gear to wear and how to minimize their jobsite risks.

In recent years, governmental regulations and industry safety standards have recognized that the crane operator is not the only person to carry heavy responsibilities associated with the safe operation of the crane. Other responsibilities have been identified for the crane owner and continue through all those associated with the lift. Along with responsibilities, additional training is also required to ensure that only *qualified persons* are involved with crane and rigging operations. Certification is now a requirement for many personnel whenever a crane is being used for construction activities.

2 Crane Danger, Accident Statistics, Personal Protection, and Responsibilities

Craning and rigging accident and fatality statistics compiled by various entities vary greatly, depending on the time span being considered, the amount of available data used in the research, and the number of jurisdictions that actually contribute statistical information to the investigating agencies. Most of the current statistics compare only the number of reported accidents and the fatalities resulting from them, so it must be noted that not all accidents involving cranes get reported. This would tend to skew the information about individuals injured or killed. Regardless of the number of actual or reported crane or rigging accidents used for the surveys, the number of recorded fatalities remains consistent and must be cause for major concern to everyone involved (Figure 2.1).

FIGURE 2.1 When an accident occurs, it can create devastation everywhere on the jobsite.

Following are the results of some of those pertinent surveys and the agencies conducting them:

Occupational Safety and Health Administration (OSHA)
Bureau of Labor Statistics
1984–1994
In 480 reported incidents, 500 construction workers died.

Center for Construction Research and Training (CPWR)
Bureau of Labor Statistics
Limited statistics for 1992–2006
In 307 reported incidents, 323 construction workers died.

Center for Construction Research and Training (CPWR)
Census of Fatal Occupational Injuries (CFOI)
Revised study for 1992–2006
In 610 reported incidents, 632 construction workers died.

Center for Construction Research and Training (CPWR)
Internet search for January–December 2008
Among the 88 reported incidents, there were 54 deaths and 100 injuries.

According to OSHA statistics:
For 2009, there were 79 fatalities.
For 2010, there were 54 fatalities.

Census of Occupational Injuries

From 2011 to 2017, the Census of Fatal Occupational Injuries reported 297 total crane-related deaths. This is an average of 42 fatalities per year over this seven-year time span. This means 3.5 persons per month, or 0.5 people per week, died in crane-related activities.

In addition, OSHA analysis of crane accidents identified the major causes of crane accidents as:

- Boom or crane contact with energized power lines, accounting for nearly 45% of all cases
- Under-the-hook lifting device failure
- Overturned cranes
- Dropped loads
- Boom collapse
- Crushing by the counterweight
- Poor outrigger deployment

Other Accident Statistics

Crane-related work deaths trended down from 1992 to 2017 (Figure 2-2). There were 33 crane-related fatalities in 2017, the lowest number since we began tracking fatal work injuries in 1992. From 1992 to 2010, crane-related fatalities averaged 78 per year. Between 2011 and 2017, the average fell to 42 deaths a year.

FIGURE 2.2 Due to comprehensive training and crane operator certification as a regulation, accident and fatality statistics are improving.

Here are some additional interesting facts:

- 90% of crane accidents occur because of human error.
- 80% of crane upsets are attributed to crane operators exceeding the crane's charted capacity.
- 54% of incidents are the result of swinging the boom or making a lift without the outrigger being fully extended and set, thus causing turnovers.
- 45% of all mobile crane accidents involve electrocution, which results from the crane contacting a power source during operation.
- 51% of crane accident victims were struck by objects such as an uncontrolled hoisted load or crane component.
- 50% of crane accidents that occurred in the United States that caused injuries in 2009 resulted in fatalities.
- On average, 80 lift and material-handling equipment operators, such as those operating forklifts supporting the crane, are killed each year.

Not only are workers seriously injured or killed, but innocent bystanders are at risk as well.

Did you know that when an accident happens, everyone can be held liable, including, but not limited to, owners, supervisors, crane operators, riggers, and the entire ground crew? The lack of training costs more than tuition for educating operators. With looming dangers and inevitable fines from lack of well-trained crane operators and the entire crane lifting team, it makes sense for companies to invest in education and training leading to qualification as well as certification.

Safety and the prevention of crane and rigging incidents must be everybody's business. Make it yours!

Every year, despite regulations and published information designed to protect them, workers are still injured. These injuries not only inflict short-term pain and suffering but can also produce more serious long-term effects. It is the responsibility of the workers to protect themselves against injury or death by being aware of, and using, the procedures and equipment designed to protect them. You owe it to yourself, your family, and your employer to wear the correct equipment.

Mounting and Dismounting of Equipment

Improper mounting and dismounting of equipment is one of the leading causes of back and leg injuries in the construction industry. To avoid injury to your back, neck, shoulders, arms, knees, and ankles, learn and practice the correct methods of getting on and off your equipment. When climbing on or off of your equipment, always follow the three-point contact rule. Use two feet and one hand, or one foot and two hands (Figures 2.3).

FIGURE 2.3 The three-point contact rule stipulates how professionals mount and dismount equipment.

Eye and Ear Protection

Damage to your sight and hearing can be immediate, or sometimes the effects can go unnoticed for a long time. Protect yourself against sight or hearing loss or injury by knowing when and how to use devices specifically designed to prevent such losses. Wear a safety shield in conjunction with safety glasses or goggles for the best eye protection. When working over the TWA (time-weighted average) noise level of 90 decibels, you must wear ear protection. Never wear headphones as a substitute for hearing protection. One should never exceed eight hours of noise exposure over 90 decibels. OSHA has written citations for this mistake.

Head and Foot Protection

Wearing good foot protection will help maintain balance and help prevent slips, trips, and falls. Employers must provide the correct hard protection boots, and you should wear them. Steel-toed boots are not normally required on most construction sites. The wearing of proper head and foot protection in the form of approved hard hats and work boots is *mandatory* in all phases of construction and related industries. To avoid injury and financial penalties, make sure that all your personal protective equipment is approved, in good condition, and worn at all times. It is important to choose a good hard hat or helmet of the Class B grade. A hard hat should be replaced every two years of regular use or five years from the date it was manufactured. Always wear the necessary protective clothing when there is the possibility of injury to the skin from caustic or hazardous materials. Proper gloves, or mitts, should always be worn as protection against contact with extreme heat or cold, or when handling wire rope or any other potentially sharp or dangerous objects.

The Crane Operator's Responsibilities

The crane operator must be totally qualified to operate the particular crane that will be used (Figure 2.4). The crane operator is responsible for the safe operation of the crane and controls all movements from the hook upward as well as swing and travel motions. The operator must be familiar with the equipment and confirm that the configuration of the crane is appropriate for the loads to be lifted. The

FIGURE 2.4 Any crane operator who will operate this crane must be totally qualified to operate it.

crane operator must be in good physical condition and pass a physical examination every three years in accordance with ASME B30.5. The crane operator must be aware of the site conditions above, at, and below the ground and from which the operator will take direction.

Operators must always take responsibility for those operations that fall under their direct control. According to OSHA regulations and ASME (American Society of Mechanical Engineers) safety standards, the operator should be responsible for the following:

- Conducting pre-operational inspections and ensuring that the crane is in safe working condition.
- Being familiar with, and able to carry out, all regular lubrication and maintenance procedures.
- Knowing the safe operating limits of the crane in all its configurations and ensuring the crane is properly set up.
- Knowing basic rigging procedures and ensuring that they are correctly applied.
- Making sure that the correct load weight has been determined by a supervisor.
- Being familiar with the capacity chart and capable of determining what the crane can do—and cannot do—in all its configurations.
- Being totally qualified to operate that particular crane.
- Performing all necessary pre-lift test procedures such as a load moment indicator (LMI) or anti-two-block devices.
- Being familiar with and following all properly given signals or communications.
- Operating the crane in a smooth and controlled manner.

The crane operator must ensure that no suspended loads are left unattended, including the tackle box, ladder, welders, compressors (including oxygen and acetylene tanks), and spreader bars. Always properly secure the crane when leaving the control station for any period of time.

Always discontinue any operation where there is any doubt about the safety of the lift until the unsafe conditions or procedures have been corrected.

To Whom the Crane Operator Is Responsible

1. The owner of the equipment
2. Other workers and the nearby public
3. Himself or herself

As a member of the construction community, it is your responsibility to be familiar with, and follow, the regulations and standards as they pertain to your particular occupation and to see that other individuals similarly involved are aware of their duties and responsibilities.

Other Important Responsibilities

While the organizational structure of various projects may differ, the following roles are described here for purposes of delineating responsibilities. All responsibilities in the following list shall be assigned in the worksite organization. One person may perform one or more of these roles.

- **Crane owner.** The crane owner has custodial control of a crane by virtue of lease or ownership.
- **Crane owner or users.** Whether or not the crane owner or users are the same entity, their responsibilities are the same. They are required to make sure that the crane being used is in

safe operating condition and is in compliance with the crane manufacturer's requirements and applicable regulations. The crane is operated, maintained, assembled, disassembled, and supported by only "qualified persons."

- **Site supervisor and lift director.** The site supervisor and the lift director may also be the same person, in which case the responsibilities are one and the same. Essentially, the supervisor or director must meet the requirements of a "qualified person" and be responsible for all site preparation and lift procedures. Under the supervision of the lift director, the signaling and rigging personnel are responsible for the operation from the hook downward, including verification of the load weight, attachment of the load to the hook, stability of the load, requirements for tag lines, and load pick-up and set-down procedures. The supervisor or lift director must be aware of any unsafe conditions or any problems with the lift, and must correct any issues immediately with the crane operator and riggers, including members of the lifting team.

  The lift director directly oversees the work being performed by the crane and the associated rigging crew when working independently. The site supervisor exercises supervisory control over the worksite on which the crane is being used and over the work that is being performed on that site.
- **Rigger.** The rigger selects, configures, and assembles the rigging equipment for attachment, support, control, and detachment of the load during lifting activities.

Chapter 2 Review Quiz

1. OSHA analysis of crane-related accidents between 2011 and 2017 identified the major causes of accidents. Nearly 45% of all cases reported were the result of

- **a.** poor outrigger deployment
- **b.** someone crushed by the crane's counterweight
- **c.** boom collapse
- **d.** crane boom contact with energized power lines

2. From 2011 to 2017, in a span of 7 years, the census of Fatal Occupational Injuries reported 297 crane-related deaths. On average, how many people died each month during this time period?

- **a.** 42people died.
- **b.** 52 people died.
- **c.** 62 people died.
- **d.** 72 people died.

3. Can a face shield be used instead of safety glasses?

- **a.** Yes, face shields are efficient protection.
- **b.** No, face shields are not efficient protection.
- **c.** Yes, they can be worn if used in conjunction with safety glasses or goggles.
- **d.** Yes, a face shield is efficient if worn with a Class B hard hat.

4. Who is responsible for seeing that each employee uses appropriate eye or face protection?

- **a.** The safety director
- **b.** The jobsite superintendent
- **c.** The jobsite foreman
- **d.** The employer

5. OSHA sets legal limits on noise exposure in the workplace based on a worker's exposure over what time-weighted average?

- **a.** 2 hours—TWA noise level of 60 dba
- **b.** 4 hours—TWA noise level of 70 dba
- **c.** 6 hours—TWA noise level of 80 dba
- **d.** 8 hours—TWA noise level of 90 dba

6. Can you wear headphones while operating heavy equipment?

- **a.** Only if approved by the safety director.
- **b.** Yes, if approved by the project manager.
- **c.** No, OSHA has issued citations for such practices.
- **d.** Yes, headphones or earbuds offer adequate hearing protection.

7. What class of hard hat is required to be worn on construction projects?
   a. Class A helmets
   b. Class B helmets
   c. Class C helmets
   d. Class D helmets

8. How often should hard hats be replaced?
   a. No more than 1 year of regular use or 2 years from the date of manufacture.
   b. No more than 2 years of regular use or 5 years from the date of manufacture.
   c. No more than 3 years of regular use or 7 years from the date of manufacture.
   d. No more than 5 years of regular use or 10 years from the date of manufacture.

9. Are gloves required to be worn on construction sites?
   a. Only if the safety director requires them to be worn and thermal temperatures are present.
   b. No, it is a personal choice if gloves are worn or not.
   c. Yes, employers must provide the correct hand protection, and gloves should be worn when exposed to hazards.
   d. Only if the company supervisor buys them and harmful substances are present.

10. Are steel-toed boots required to be worn on construction sites?
    a. There is nothing in the rules that requires steel-toed shoes.
    b. Yes, every employee on a construction site must wear them.
    c. Yes, and they must be leather high-top boots.
    d. No, the employer is not required to pay for non-specialty safety toe protection including steel-toed boots.

11. Is the employer required to pay for non-specialty toe protection footwear?
    a. Yes, and they must be steel-toed.
    b. Yes, but only efficient leather work boots are acceptable.
    c. Yes, and the work boots must include metatarsal protection.
    d. No, the employer is not required to pay for non-specialty safety toe protection, including steel-toed boots.

12. What is the main purpose of wearing foot protection?
    a. To maintain balance, preventing trips or slips
    b. To protect the person from falling
    c. To be comfortable and cool the feet
    d. To protect the feet from falling objects

13. The responsibility for the safe operation of the crane belongs to the
    a. crane owner
    b. master mechanic
    c. crane operator
    d. certified rigger

14. According to ASME safety standards, crane operators are required to have a physical examination at least every
  a. year
  b. 2 years
  c. 3 years
  d. 4 years

15. Who has custodial control of the crane by virtue of a lease or ownership?
  a. The project lift director
  b. The certified crane operator
  c. The certified rigger
  d. The crane owner

16. Who arranges the crane's presence on a worksite and controls its use there?
  a. The project engineer
  b. The company superintendent
  c. The crane user
  d. The project manager

17. The person who directly oversees the work being performed by a crane and the associated rigging crew is the
  a. master rigger
  b. certified signal person
  c. certified crane operator
  d. lift director

18. The person who selects, configures, and assembles the rigging equipment for attachment, support, control, and detachment of the load during lifting activities is the
  a. designated lift director
  b. crane operator
  c. rigger
  d. master mechanic

19. What person would exercise supervisory control over the worksite on which a crane is being used and over the work that is being performed on that site?
  a. The project manager
  b. The site engineer
  c. The site lift director
  d. The site supervisor

20. What individual directly controls the crane's functions?
  a. The lift director
  b. The crane user
  c. The crane owner
  d. The crane operator

Summary—and Then Some

Crane and rigging operations expose crane operators and riggers to unique risks and challenges. It is important to timely assess all potential risks that can lead to an accident. Determine the correct PPE for the risk at hand, and wear it whenever there is a chance of exposure.

When all those working around the crane are trained and qualified, including the owner and the supervisors, the crew begins to work as a cohesive unit. Morale is high, production increases, and lives are saved. Safety and training are for everyone. When the entire crew is qualified, everyone achieves more.

Rules and regulations, along with industry safety standards governing mobile crane operations, are long and varied. Guidance depends on whether the work is being performed as part of a construction project or part of a job in general industry. Military projects have their own sets of rules. States, and even in some cases cities, may have different rules and regulations. The American Society of Mechanical Engineers provides 30 unique safety standards, all specialized to the work we perform in order to keep us out of trouble. Industry safety standards help us to improve safety and stay in compliance with governmental regulations. Being familiar with governmental regulations and safety standards is your responsibility. Know which ones pertain to your job.

3 Laws, Regulations, and Standards that Govern Crane Operations

For several decades before 2002, a large discrepancy developed between OSHA 1926.550 and ASME B.30.5, the American Society of Mechanical Engineers' safety standard for locomotive and mobile cranes. The discrepancy was a direct result of many changes in design, technology, and modernized work practices that triggered OSHA to update its safety standards and regulations. And so, in 2002, the Occupational Safety and Health Administration assigned a newly founded group called the Cranes and Derricks Negotiated Rulemaking Advisory Committee (C-DAC) with a very important mission. The committee was assigned to rewrite a very old and outdated standard, known then as CFR 1926.550, for crane and derricks safety. The C-DAC completed its work in July 2004, and the rule was published in the *Federal Register* on August 9, 2010. It became effective on November 8, 2010.

Then on November 17, 2017, after many years of waiting, a nationwide crane operator certification standard was introduced. The scope of the crane standard was limited to powered operated equipment used in construction that can hoist, lower, and horizontally move a suspended load. This would mean that operators of most cranes, when used in construction, would need to be certified by an accredited crane operator testing organization, or qualified through an audited employer program.

The new rule, OSHA 1926.1400, required testing through written and practical examinations and provided certification based on equipment size and type. The entire construction industry was mobilized to meet the new requirements.

OSHA 1926.1400 and 1910.180

Cranes covered by CFR 1926.1400 when working in construction include mobile cranes, crawler cranes, tower cranes, articulating boom cranes, floating cranes, cranes on barges, and locomotive cranes. Also included are industrial cranes like carry decks, pile drivers, mechanic trucks with a hoisting device, monorails, pedestal cranes, portal cranes, overhead and gantry cranes, straddle cranes, and variations of this type of equipment. The rule also includes multipurpose machines when configured to hoist and lower with a winch or hook block and horizontally move a suspended load.

The rule applies to cranes with attachments such as hook blocks, overhaul balls, magnets, grapples, clamshell and dragline buckets, concrete buckets, personnel platforms, and augers and drills, as well as pile-driving equipment.

OSHA 1926.1400 covers cranes in construction only. For cranes working in general industry, the rules can be found in OSHA 1910.180. Crane operators working under this rule are not required to be certified. However, under OSHA 1910.180(a)(35), the crane operator must be *qualified*. Qualified means that the crane operator is able to safely operate the crane and has demonstrated the ability to do so to the employer through training, experience, or instruction.

Owners, supervisors, crane operators, riggers, and all members of the lifting team should be aware of the regulations and all the industry standards that govern work on the jobsite where they work. Also, for the jurisdiction in which the project is located, many industry standards may be relevant all at the same time on one jobsite. It's your responsibility to know how they apply to the work performed on your job.

OSHA is the law across the United States; however, states with their own occupational safety and health plans may have working rules more stringent than OSHA's. When working in these states, those rules will apply. Even certain cities may have their own regulations and safety standards. It's your job to know the laws in the area or jurisdiction you are working in, and how they may change as you move from one location to another during your employment.

Crane and rigging safety standards in previous years oftentimes did not run parallel with associated governmental regulations and standards. Crane manufacturers' recommendations didn't seem similar in some ways. Great progress has been made in recent years to bring everything closer together and easier for owners, supervisors, crane operators, and those working under the hoisting hook to plan and work more safely. Many crane operations bring several industry standards and regulations into play all on one jobsite. It's more important today than ever before for all workers to be familiar with these standards before and during the work they perform.

Other Regulations and Standards

Other important OSHA regulations, OSHA standards, and ASME safety standards to be aware of and familiar with in the crane and rigging industries follow.

OSHA Standards

It is important to be familiar with these OSHA regulations, especially those that pertain to your jobsite:

- **Crawler locomotive and truck cranes.** OSHA Standard 1910.180
- **Slings.** OSHA Standard 1910.184
- **Rigging equipment for material handling.** OSHA Standard 1926.251
- **Boom assembly and disassembly.** OSHA Standard 1926.1404
- **Electric power transmission and distribution.** OSHA Standard 1926.950
- **Wire rope inspection.** OSHA Standard 1926.1413
- **Qualified rigger.** OSHA Standard 1926.1401
- **Signal person qualifications.** OSHA Standard 1926.1428

ASME Industry Standards

Industry standards for the manufacture and operating characteristics of mobile cranes have been established by committee through the auspices of the American Society of Mechanical Engineers as the guide to government authorities having jurisdiction over subjects within the scope of the standards with the expectation that such guidelines will be generally accepted. Standard B30.5 is one of the major sources for the safe operating practices and maintenance procedures for all cranes described within the standards. Several industry safety standards may be relevant to your project. It's important to review those listed below and know which ones are relevant.

- **Overhead gantry cranes.** ASME B30.2
- **Tower cranes.** ASME B30.3
- **Mobile and locomotive cranes.** ASME B30.5
- **Slings.** ASME B30.9
- **Hooks.** ASME B30.10
- **Below-the-hook lifting devices.** ASME B30 20
- **Articulating boom cranes.** ASME B3022
- **Personnel lifting systems.** ASME: B3023
- **Rigging hardware.** ASME B3026
- **Self-erecting tower cranes.** ASME B30.29
- **Ropes.** ASME B3.30

Definitions

Regulations and standards will mention from time to time the individuals listed here and their responsibilities:

- A *competent person* means "one who is capable of identifying existing and predictable hazards in surroundings or working conditions which are unsanitary, hazardous, or dangerous to employees and who has authorization to take prompt corrective measures to eliminate them" (as defined by OSHA).
- A *qualified person* is someone "who by possession of a recognized degree, certificate, or professional standing, or who by extensive knowledge, training, and experience, has successfully demonstrated the ability to solve or resolve problems relating to the subject matter, the work, or the project" (as defined by OSHA).

In a regulation or a standard, the work we are planning will be spelled out if the rule is mandatory or if we have leeway to use our own judgment based on knowledge and experience:

- When the word *should* is used, it means that the rule is a recommendation, the advisability of which depends on the facts in each situation.
- When the word *shall* is used, it indicates that the rule is mandatory and must be followed.

ASME Standard B30.5 (Figure 3.1) is very important to know if you are a mobile crane operator.

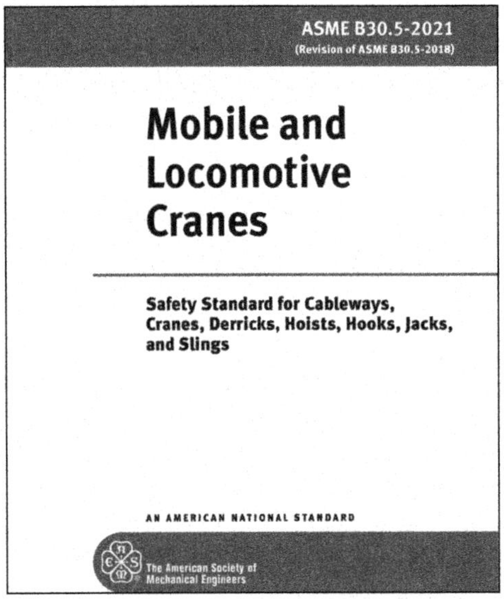

FIGURE 3.1 ASME Standard B30.5 is a major source for the safe operating practices and maintenance procedures for all cranes.

Chapter 3 Review Quiz

1. The OSHA regulation pertaining to the use of mobile cranes used in construction is
   a. 1926.250
   b. 1910.180
   c. 1926.550
   d. 1926.1400

2. What OSHA standard pertains to crawler, locomotive, and truck cranes?
   a. 1926.120
   b. 1926.170
   c. 1926.550
   d. 1926.180

3. What OSHA standard pertains to crane lattice boom assembly and disassembly?
   a. 1926.1404
   b. 1926.1405
   c. 1926.1408
   d. 1926.1412

4. What OSHA standard pertains to qualified riggers?
   a. 1926.1401
   b. 1926.1404
   c. 1926.1406
   d. 1926.1408

5. What OSHA standard pertains to signal person qualifications?
   a. 1926.1416
   b. 1926.1419
   c. 1926.1421
   d. 1926.1428

6. What ASME safety standard pertains to mobile and locomotive cranes?
   a. ASME B30.5
   b. ASME B30.2
   c. ASME B30.10
   d. ASME B30.22

7. What ASME safety standards pertain to crane hooks?
   a. ASME B30.4
   b. ASME B30.6
   c. ASME B30.8
   d. ASME B30.10

8. What ASME safety standard refers to all below-the-hook lifting devices?
    a. ASME B30.18
    b. ASME B30.20
    c. ASME B30.22
    d. ASME B30.24

9. What ASME safety standard pertains to personal lifting systems?
    a. ASME B30.21
    b. ASME B30.22
    c. ASME B30.23
    d. ASME B30.24

Summary—and Then Some

Understanding and being aware of the appropriate safety standards and regulations for the work you are involved in is important when preparing for any national certification written examination. Being savvy and knowledgeable on the job is what makes good supervisors and separates the average craftworkers from the great ones.

Starting a new crane project is fun and exciting. Everyone is excited and eager to get started. Once the crane shows up onsite, the excitement begins; there is a special buzz in the air. However, this is the time to slow activities down and be wise, organized, and safe. Before the first lift comes off the ground, owners, managers, supervisors, crane operators, riggers, and the entire lifting team and ground crew must fully understand the pre-lift plans and all job-specified rules and safety precautions.

4 Making the Lift

Often, those involved in crane operations do not plan to fail; they fail to plan. Preparing formal lift plans, establishing criteria for "critical lifts," and fully understanding the limits of the crane in all its configurations is a very big and complex plan (Figure 4.1). You must make sure that all the support equipment involved is in great shape and only highly trained, qualified, certified crane operators are allowed to operate it. Planning is how we all prevent mobile crane accidents.

The crane operator must determine the weight of the load before making any lift. Before making a lift, the crane operator and lifting team must ensure that the load will not be hoisted and moved over any personnel. All crane lifts require some level of planning regardless of whether the lift is only a few hundred pounds or many tons. Complicated heavy lifts require more extensive planning and closer supervision by experienced, qualified personnel (Figure 4.2). Major crane operations should never get started until one person is appointed to be the lift director. This must be a very knowledgeable person who is totally dedicated to the project's success. The lift director's primary job is keeping every person and every piece of supporting equipment safe during the project.

FIGURE 4.1 Planning for a lift such as this unlimited hydroplane, the lifting team is not so much concerned about its weight but extra care is taken because of the load's value.

FIGURE 4.2 Lift plans become complex when the load is odd-shaped or very heavy or more than one crane is needed to make the lift.

Crane operators must evaluate and make assessments as well as inspections before every work shift. Getting a crane project started correctly is a big responsibility. Meeting deadlines, satisfying material constraints, and locating hazards are some of the main concerns that can delay your project's smooth progress. Crane owners, crane users, project managers, supervisors, and lift directors must work closely with the crane operator and rigging team to form a cohesive unit that works together as a team to fulfill the prerequisites of all the components of the crane operations. Rigging shall be secured and balanced before the load is lifted more than a few inches. Most importantly, never allow any personnel to ride on the load to be lifted (Figure 4.3).

FIGURE 4.3 Never ride on or allow anyone else to ride on the load to be lifted.

In the distant past, riggers working on historical sites such as the Golden Gate Bridge and the skyscrapers in Chicago and New York City often hitched a ride on the headache ball or a load being hoisted as a means of quick transportation. Many of these workers were injured or killed during this time. That is why riding a load or headache ball is strictly forbidden today.

The following procedures and precautions should be observed whenever loads are to be handled:

- Proper personal protective equipment has been provided to all crew members.
- All loads must be properly rigged to prevent the dislodgement of any part.
- Suspended loads should be securely slung and properly balanced before they are set into motion and until landed and brought to a rest.
- A pre-lift checklist has been provided to all members of the lifting team prior to executing the lift.
- The load must be kept under control at all times and whenever necessary, tag lines are used by a qualified person to prevent rotation or uncontrolled motion (Figure 4.4).
- Loads must be safely landed and properly blocked before being unhooked and unslung.
- Avoid dragging slings from beneath loads.

- Never lift loads with one or more legs of a multi-leg sling arrangement until the unused legs are made secure.
- Never ride on, or allow anyone else to ride on, the load being lifted.
- Before making any lift, check to see that slings are properly attached to the load.
- Avoid shock loading caused by sudden jerking when lifting or lowering. Lift the slings gradually until the slack in the rigging is eliminated.
- Never allow the load to be carried over the heads of personnel.
- Never work, or allow anyone else to work, under a suspended load.

FIGURE 4.4 Easy-grip tag lines can be used to prevent the rotation of the load.

Determining a Load's Center of Gravity

Before a load can be lifted or any rigging applied, the center of gravity (COG) must be located. The COG is a theoretical point on an object that engineers use for convenience in calculations as the single point where all of an object's weight is concentrated. The COG can be explained as the tipping point, and so the location of the COG of the load is a very important consideration. If the COG is not exactly centered, it will affect the tension in each sling used to rig the load. If the sling legs are not of equal length, the load will tilt when lifted.

Think of it this way: Imagine a large, heavy cube that we are about to push off a cliff. The COG is the point at which the cube would tip and fall off the edge. The COG is dependent on not just the shape of the object, but also its density, or what material it is made of.

There are many things to consider in finding the exact COG, and that is not always an easy task (Figure 4.5). Complex mathematical methods are available to determine the COG of a load. Crane operators, lift directors, and riggers want to know where the COG is so their lifts are balanced and are more stable in flight.

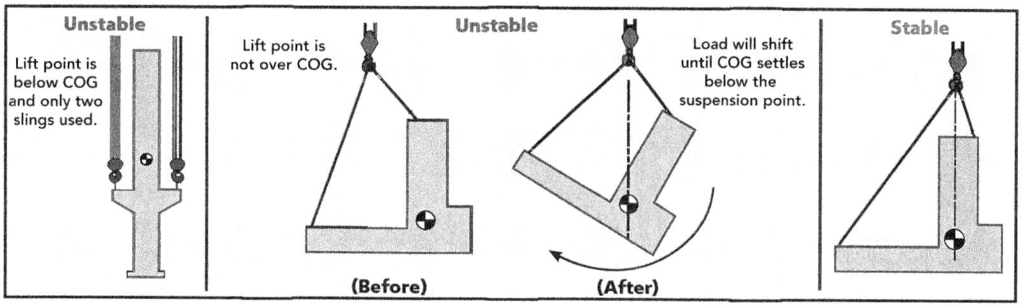

FIGURE 4.5 There are many things to consider when finding the exact center of gravity.

The center of gravity of a load is considered to be the point around which its weight is evenly distributed, or its *balance point*. Whether this point is located within or outside the body of the load itself, it is important that its location be determined and that the crane lifting hook be positioned directly above the COG prior to rigging the load.

The center of gravity of a load depends on its mass. The COG of a load is located where more mass is concentrated. To produce a stable load, or one that will not tilt excessively or slip out of the rigging, the crane lifting hook must remain directly above the load's COG before and during the lifting operation (Figure 4.6). However, the crane operator must be aware of the possibility of crane boom deflection, meaning the load may swing outward, increasing the load radius.

FIGURE 4.6 Keep it simple whenever possible. A very heavy odd-shaped load is handled with four equal-length wire rope slings. A sister hook is used with two slings on each hook to even things out very nicely.

Other forces, such as weather conditions, including snow, ice, and extreme cold and heat, must also be considered. Wind is a big factor that can cause side loading of the load and the crane's boom and must be avoided at all costs.

To safely lift a load, the rigger must locate the COG. This is done through trial and error by lifting the load slowly a few inches from the ground and placing blocking under the object. You can then observe where the load weight shifts by moving the block. Once the COG is determined, the proper rigging can be selected to handle the load (Figure 4.7). A rigger must:

1. Account for any contents of a load that may shift during the lift.
2. Stabilize it before the load leaves the ground.

FIGURE 4.7 If rigged improperly, the suspended load will tilt or shift until its center of gravity comes directly below its suspension point. Shown here, the load is oddly shaped. Through engineering along with some trial and error, the COG was located. The lift was engineered using a spreader bar and the use of both hoist lines to ensure a heavy complex lift was completed safely.

Before hoisting the load, the crane hook must be positioned directly over the COG. When calculating the total weight of the net load, all the slings, rigging hardware, and below-the-hook lifting devices, including the load hook, load blocks, overhaul balls, stowed jibs and rigging, must be added to the total weight of the load.

The total weight must be compared with the charted capacity of the crane to ensure a safe lift. The crane manufacturer's load chart limitations must be considered before hoisting each and every load. Tag lines and swivels may be used to mitigate the effects of inertia on the load. Not paying attention to all the details, along with not rigging properly to the COG, can cause loss of control of the load and, in turn, damage to the load, the crane, and nearby structures as well as possible serious injuries to employees and innocent bystanders on or around the jobsite.

Crane owners, crane users, supervisors and lift directors must ensure the following important details are in place before "making the lift":

- Validate the crane is inspected and in good condition.
- See to it a qualified person is assigned as the lift director.
- Make sure crane operations are coordinated with other jobsite activities.
- Make certain that the area for the crane is adequately prepared, including enough room with firm level ground conditions, and traffic control measures are in place when necessary.

- Make certain any conditions that may adversely affect crane operations are addressed.
- Approve crane operations near powerlines only when the utility has been contacted and all safety precautions are in place and a lift plan has been developed including that a pre-lift meeting has been conducted with all those involved.
- See to it the rigging crew is supervised by a qualified person.
- Confirm crane maintenance has been performed on a frequent and periodic schedule.

Qualifications of Personnel Making the Lift

It is true the crane operator plays a critical role in the operation of the crane and the lift to be made. The operator has significant responsibilities and qualification requirements necessary to making a safe and successful lift, so choosing the right individual for the job is an important part of the planning process. Today, the crane operator's knowledge and skill can be tested before ever reaching the jobsite. OSHA requirements mandate crane operators working in the construction industry be certified by a nationally accrediting agency such as the National Commission for the Certification of Crane Operators (NCCCO). OSHA's 29 CFR 1926.1427 requires crane operators to demonstrate knowledge and skills through written and practical testing. They must be re-tested every five years. This is a wonderful tool for employers to measure the knowledge and basic skills of a crane operator. However, before making the lift, we must know the crane operator not only is "certified" but also has been evaluated for all the cranes that he will be operating during a project to make those lifts. OSHA defines crane operator evaluation as where the employer must ensure that each operator is qualified by a demonstration of:

- The skills and knowledge, as well as the ability to recognize and avert risk necessary to operate the crane safely, including those specific devices, operational aids, software and the size and configuration of the machine, including lifting capacity, boom length, attachments, luffing jib, and counterweight setup.
- The ability to perform the hoisting activities required for assigned work, including relevant blind lifts, multi-crane lifts and personnel lifts.

Clearly, the crane operator must be evaluated before making any lift. OSHA defines three critical areas to assure a crane operator is qualified which include:

- A crane operator training process.
- A licensing or certification process.
- An employer evaluation process.

The crane operator certified and qualified is a great start, however, not enough to be ready for making the lift. The controlling entity of the jobsite also has an important role. An employer, prime contractor, construction manager, and any general contractors play a significant part in planning any lift. The employer's supervisor will exercise control over the worksite on which the crane is being used and over the work being performed on that site. He will appoint a lift director who will directly oversee the work being performed there. Crane owners will have the custodial control of the crane by means of a lease or ownership. If a crane rental is involved, the crane user will arrange the crane's presence on the jobsite and will control its use there. Only highly trained and qualified riggers will be hired. Riggers will select, configure, and assemble any rigging equipment used for the lift. Riggers will look over any attachments needed, including all rigging that will be hitched to the load. Riggers will support the lift, control, and detachment of the load during all hoisting operations. It truly takes a talented, certified and qualified crane operator to safely operate today's cranes. It takes a team including the employer, supervisors, lift director, riggers, signal person, and other members of the lifting team to safely "make the lift."

Company owners and managers are very knowledgeable individuals who will set jobsite rules and ensure they are carried out. The supervisor along with the lift director will work together to implement those rules and all lifting plans throughout the jobsite. Responsibilities they will have include:

- Using crane operators who meet qualification and certification requirements and have been evaluated on the cranes they will operate.
- Designating personnel for maintenance, repair, transport, assembly, and disassembly of the crane.
- Implementing inspection, testing, and maintenance programs and ensuring they are followed.
- Notifying the crane owner if rope on the crane is replaced or shortened.
- Complying with governmental regulations and industry safety standards including any manufacturer's specifications.
- Employing supervisors who are qualified for crane operations.
- Making certain the crane is in good operating condition and seeing to it the crane owner or crane user is provided the required documentation, including that frequent inspections have been performed.
- Making sure the crane has the necessary lifting capacity to perform the work.

Chapter 4 Review Quiz

1. Rigging personnel are allowed to ride on a load to be lifted
   a. only for complex lifts
   b. only for load test purposes
   c. only lift directors are allowed
   d. never; no personnel are ever allowed

2. Crane operators are allowed to hoist and carry loads over the heads of personnel
   a. only when making lifts in the blind
   b. only when safety rigging has been provided
   c. only if tag lines are installed to control rotation
   d. never; loads should never be hoisted and moved over personnel

3. Mobile crane operators should carefully plan
   a. lifts over 60% of charted capacity
   b. lifts over 70% of charted capacity
   c. lifts over 80% of charted capacity
   d. each lift before hoisting the load

4. The rigging shall be secured and balanced before the load is lifted more than
   a. a few inches
   b. 6 inches
   c. 10 inches
   d. 1 foot

5. Before making a lift, the crane operator must first determine the
   a. placement of the rigging
   b. weight of the load
   c. placement of the tag lines
   d. rigging attachment points

6. The center of gravity of a load depends on its
   a. weight
   b. mass
   c. density
   d. shape

7. When planning to lift a heavy load, the crane's hook must be located
   a. directly over the center of gravity
   b. slightly to the left of the center of gravity to allow for side loading of the boom
   c. directly over the face of the load to allow for boom deflection
   d. slightly to the rear of the load to allow for an increase in the radius when lifting

8. When lifting a heavy load and the crane hook is positioned directly over the center of gravity, what must the crane operator be aware of after lifting the object?
- **a.** Swinging out of the load, increasing the load radius
- **b.** Swinging inward of the load, decreasing the load radius
- **c.** Side loading of the crane boom, causing dynamic loading
- **d.** Boom deflection, causing shock loading to the crane

9. The person who directly oversees the work being performed by a crane and the associated rigging crew is the
- **a.** master rigger
- **b.** certified signal person
- **c.** certified crane operator
- **d.** lift director

10. What person would exercise supervisory control over the worksite on which a crane is being used and over the work that is being performed on that site?
- **a.** The project manager
- **b.** The site engineer
- **c.** The site lift director
- **d.** The site supervisor

11. Who has the custodial control of the crane by virtue of a lease or ownership?
- **a.** The project lift director
- **b.** The certified crane operator
- **c.** The certified rigger
- **d.** The crane owner

12. The responsibility for the safe operation of the crane belongs to the
- **a.** crane owner
- **b.** master mechanic
- **c.** certified crane operator
- **d.** certified rigger

13. According to OSHA, crane operators working in construction must be certified by a nationally accredited organization and be re-certified every
- **a.** two years
- **b.** three years
- **c.** four years
- **d.** five years

Summary—and Then Some

Crane owners or crane users, if renting a crane, provide the machines that are used on jobsites to make lifts. They must ensure all the cranes on the jobsite meet the requirements of industry safety standards and governmental regulations. Crane owners are a great resource for technical information on the machines they own by way of operator and shop manuals, as well as many other resources. They will work with company supervisors and the lift director to provide necessary crane attachments and any other equipment required for making the lift, including enough boom and ensuring the crane is within its charted capacity. Owners will generally provide mobilization in and out of the jobsite as needed, including any assembly and disassembly procedures that may be required. When boom is assembled or disassembled a competent and qualified person is required to direct this work. He will serve as the assembly and disassembly director.

Personnel for maintenance and repair programs will be put in place including inspection and testing procedures. Making the lift may sound like an easy and simple concept of picking a load up and lowering it back down to the ground, however, a very talented group of highly trained and knowledgeable people all come together when making the lift and doing it in such a manner that expensive loads are not harmed and cranes are not structurally damaged or tipped over. It is always a number-one priority that the public is never put in harm's way and jobsite personnel go home safely to their families after making each and every lift every day.

Cranes are invaluable tools in the world of construction, saving time and effort and helping to improve the safety of worksites. As with many specialized industries, cranes come with terms and abbreviations. When you are new to cranes, this unfamiliar language is confusing and can lead to misunderstandings among the crew. It is important as a crane operator to know and understand the important definitions used in craning.

5 Definitions Used in Craning

The following list contains the definitions for many of the terms and phrases that are used by the craning industry. These terms and phrases are very important to know and understand when working around the operation of a mobile crane.

Center of gravity. A point relative to a body around which its weight is evenly distributed. Also known as the *balance point*.

Center of rotation. The vertical "axis" around which the crane's superstructure rotates.

Controlling employer. The employer with general supervisory authority over the work site.

Controlling entity. A legal entity such as a general contractor, construction manager, or prime contractor, who has overall responsibility for a construction project.

D/d ratio. The ratio of the diameter (D) of the object around which the sling is bent, divided by the overall diameter (d) of the sling being used.

Dynamic load. The weight of the stationary gross load.

Effective weight. Represents the force an attachment exerts on the crane, which could be more or less than its actual weight relative to its distance from the tipping axis.

Fall zone. The area directly beneath the load in which it is reasonably foreseeable that partially or completely suspended materials could fall in the event of an accident.

Gross load. The weight of the net load plus the weight of all load-handling devices and attachments.

Load moment. The force that causes, or tends to cause, the crane to rotate around its tipping axis.

Load moment indicator. A device that constantly monitors the effect of load movement on the crane and warns the operator of an overload condition.

Net load. The actual weight of the load itself.

Radius. The horizontal distance from the center of rotation of the crane to the center of gravity of the suspended load.

Rated strength. The maximum load that can be applied to any item described as the safe working load or working load limit.

Safety factor. The amount by which the strength is divided to get the safe working load or working load limit.

Side loading. A load applied to an angle to the vertical plane of the boom.

Stability factor. A charted capacity or condition that protects the crane against loss of forward or backward stability.

Static load. The weight of the stationary load plus the effects of inertia. No sudden starting, stopping, or swinging out of the load is allowed.

Stored attachments. Jibs or extensions that are not in use and are stored away from the carrier or boom of the crane.

Stowed attachments. Jibs or extensions that are not in use for the lift but are pinned to the base section of the boom.

Structural competence. The ability of the machine and its components to withstand the stresses imposed by applied loads.

Tipping axis. The point around which the crane tips. Also known as the *fulcrum*.

Ultimate strength. The strength that if exceeded will cause failure. Also known as the *breaking strength*.

Chapter 5 Review Quiz

1. The radius is defined as the horizontal distance from the
   a. boom heel pins to the center of the load
   b. center of rotation of the crane to the center of gravity of the suspended load
   c. front of the crane to the line directly below the boom point
   d. turning radius of the crane

2. The net load is described as the
   a. actual weight of the load itself
   b. weight of the load plus all chockers
   c. weight of the load minus all load-handing devices
   d. weight of the load plus all rigging

3. What best describes the center of rotation of the crane?
   a. The center pin of the house
   b. The vertical axis around which the superstructure rotates
   c. The roller bearings the superstructure rides on
   d. The center of the boom gearbox

4. Which of the following best describes the term *tipping axis*?
   a. The center of the extended outrigger beam
   b. The point on which the cribbing touches the ground
   c. The center of the blocking under the outrigger float
   d. The point around which the crane tips

5. Which best describes the term *load moment*?
   a. A heavy load that forces the crane to lose balance
   b. The force that causes the crane to rotate around its tipping axis
   c. The force that causes the crane to lose stability and tip over
   d. The force that causes structural failure to the crane

Summary—and Then Some

For safe and efficient crane usage, it is imperative to have a full understanding of the terms and phrases used around the crane, including those that apply to the lift. In the crane industry, you will find yourself carrying out extreme lifting challenges on a day-to-day basis. It's important you understand crane definitions because they are an important part of the tools and knowledge you will need to lift your career to a level of success.

A good crane inspection checklist is highly important to any workplace or jobsite. Hoisting equipment is vital to our work whenever we need to lift heavy weights. That comes with significant risk. The crane operator must take those risks seriously. The crane examination is nothing more than a safety audit for the crane operator. It is important that the crane operator check all the vital components before operating the crane. Inspections performed to the manufacturer's requirements daily, and as required, help eliminate significant hazards that the crane endures by the very nature of the heavy lifts it performs. That is why the typical contents of a crane safety checklist help crane operators stay safe and productive.

6 Crane Inspections

The crane walk-around inspection should cover the concerns of the equipment listed by the manufacturer for each specific crane type. Avoid generic checklists that are often found on websites. Generic checklists can often be very lengthy or not thorough enough. Here is the problem: If the crane operator is allotted 30 minutes every shift to complete a checklist a mile too long, the crane operator is likely to "pencil-whip" and check off items quickly without observing the components in an effort to complete the form within the 30 minutes' allotted time. Another problem is that a universal checklist may list items not even on the crane. Worse yet, it misses vital components on the crane that are important to look at but are not listed at all. If the manufacturer of your crane does not provide a checklist specific to the make and model of the machine you are operating, you may need to develop your own checklist and add it to your log book. You will find that the manufacturer's operator or shop manuals will list the items and components it wants you to look at every day or frequently, as well as every month or periodically. A separate checklist for each type of inspection works best, one for frequent and another for periodic.

Checklists can be developed to be very thorough but also efficient at the same time (Figure 6.1). Consider listing all the components you will need to look at with a starting point such as the crane cab or engine compartment, and continue with the items as they appear to you while you are making one trip around the machine, such as cranes shown in Figures 6.6, 6.7, and 6.9. Items are checked, and components observed, just like the manufacturer asked, quickly and efficiently. Always remember to put ink on the pages. There is nothing wrong with stating good things on your report, such as "The crane appears to be well serviced and lubricated" or "The cab is in a neat and orderly condition." It is important to make your checklist "come alive" with some ink and a few comments here and there. Always sign and date it when you are done.

FIGURE 6.1 Developing your checklist on "write-in-the-rain"–treated paper can help avoid smudges, tears, and ugly paperwork.

A crane safety inspection should be performed by a *qualified person*, a well-trained and certified crane operator who possesses the knowledge and experience needed to properly complete these inspections. It is important to know what inspections are required and how to complete them. All inspections, tests, repairs, maintenance, and modifications to the crane must be recorded in an equipment log.

The three inspections all crane operators and maintenance personnel must be familiar with are as follows:

- **Initial Inspections.** Initial inspections are applied to new or altered cranes and will be performed by a qualified person as designated by the crane owner. Every initial inspection will include a load test to verify the crane's structural integrity ratings.
- **Frequent inspections.** Frequent inspections will be performed daily prior to each use, during the operation, or at the beginning of each shift as the operator deems necessary. A strict program of preventive maintenance, daily service, and thorough pre-operation checks will result in properly operating safer equipment with lower operating costs.
- **Periodic inspections.** Periodic inspections may be regularly scheduled on a monthly, quarterly, or annual basis, but not to exceed annually (Figure 6.2). The intervals are determined by the crane owner based on the crane's usage. These inspections may be carried out by either a qualified operator or inspector, depending on the length of the interval and the intensity of the inspection.

FIGURE 6.2 Periodic checklists you use on a monthly basis will require more time to complete than frequent-inspection checklists because the manufacturer will ask you to look over more items and components.

Remember that mobile crane pre-operational observations are a regulatory requirement. Check the rules for an annual inspection in your area. In some jurisdictions, a qualified person may be sufficient. Some owners may insist the annual audit be conducted by an outside independent party. In some states, it may be a regulation that the crane be inspected by an accredited crane inspector. Regulations can differ from one state to another. Check the regulations where you are for using the crane. Annual inspections may require that a load test be performed on the crane. Load tests must never exceed 100% of the charted capacity of the machine.

Plan your crane analysis whenever possible to take place during daylight hours (Figure 6.3). Early start times and late-running shifts could make this difficult. However, going over the crane in the dark runs the risk of missing something vital. In some cases, light plants or other means may be your only solution. If there is a warning sign on the switch or engine starting controls, do not start or shut off the engine until the sign has been removed by an authorized person.

Always consider an inspection before the shift and again afterward. Running hot cycle times while pouring concrete and doing dragline and clamshell procedures are all very hard on the crane. Pile-driving operations are even worse. How about that heavy lift over 70% of the charted capacity of the crane? For sure, that's time to get down and conduct an inspection immediately afterward. Keeping spare parts on hand for your machine is another good idea worth considering. A spare engine belt or hydraulic hose on hand may significantly reduce downtime.

It is your responsibility as a crane operator and a qualified person to know the condition of your equipment and to never operate any crane or equipment that is unsafe or hazardous. Inspecting cranes or similar equipment requires considerable skill and knowledge. If the information or procedure is

FIGURE 6.3 Members of the lifting crew can be trained to watch for issues on the crane. As shown in this figure workers, can keep a close eye on the wire rope hoist lines.

unclear, ask your supervisor. Well-trained crane operators are always alert to the condition of their crane. The pre-operational checklist becomes a routine and constant check of that condition. Crane operators "get to know" their crane—how it feels, how it smells, how it sounds, and, most importantly, how it operates. Visual checklists provide excellent safety. However, always remember the things you smell, hear, and feel can be a big sign that something is not quite right. Use and trust all your senses, and when they tell you something might be wrong with the crane you are operating, there is a good chance there is. Shut the machine down and do a good once-over.

An important part of a properly designed survey is to ensure the correct load chart is in the crane as well as the required operator's manual. Often overlooked are the original manufacturer's safety placards and warning decals. The operator's manual is often missing, and decals are faded, gone, or impossible to read. Both can be replaced at a reasonable cost by contacting the manufacturer. Inspections are not only great for conducting routine checkups; they help ensure compliance with the Occupational Health and Safety Administration's regulations and various industry safety standards.

A special note regarding wire rope: Frequent inspections are visual inspections that a competent person must begin prior to each shift during which the crane is being used. Shift inspections do not require opening of the wire rope, nor do they require the operator booming down the crane to look at all the hoist rope. The inspection must consist of the crane's running and standing ropes that are likely to be used during the shift for apparent deficiencies. Monthly or periodic inspections may require a deeper and closer look, including all the rope on the crane. Of course, this doesn't just apply to wire rope. It also applies to other parts of the crane (Figure 6.4). Always refer to the manufacturer's recommendations and OSHA regulations for specific requirements for the crane you are operating.

FIGURE 6.4 Periodic inspections may require going deeper into the crane, looking at auxiliary and boom point sheaves that you won't have time for on a frequent inspection.

Wire Rope Annual Comprehensive Inspection

At least every 12 months, wire ropes in use on your mobile crane must be inspected by a qualified person (Figure 6.5). In addition to a frequent type of inspection, an annual inspection must be completed and must thoroughly cover the entire length of the wire ropes. Particular attention must be given to those sections normally hidden during frequent and periodic inspections. Wire ropes are also subject to reverse bends or rope passing over sheaves.

FIGURE 6.5 On a monthly basis, the manufacturer may require an inspection of all the wire rope on the crane.

Exceptions

In the event an annual inspection of wire ropes is not possible due to the setup and configuration of the crane, inspections must be conducted as soon as it becomes feasible to do so, but no longer than an additional 6 months for running ropes. Standing ropes may be inspected at the time of disassembly of the boom and crane. If a deficiency is determined to constitute a safety hazard, operations involving the use of the wire rope in question must be prohibited until the wire rope is replaced. If the problem with the rope is localized, it is corrected by cutting the wire rope into two sections. The undamaged portion may continue to be used; however, joining wire rope by splicing is never allowed. If the wire rope is shortened, the employer must ensure that the hoist drum will always have a minimum of two full wraps of hoist rope whenever the load or boom is in its lowest position. If a problem is identified in the hoist rope, the deficiency must be monitored. The deficiency must be noted in the monthly checklist and entered in the crane log book. Remember, when lubricating wire rope on the crane, never use a type that would hinder a visual inspection.

Keeping the crane clean will pay big dividends with regard to properly observing components lying in oily trays or tight places. When checking hydraulic fluid levels, you must retract all outriggers, boom sections, and other components to ensure hydraulic oil is in the reservoir tank. Hydraulic hoses and fittings that are leaking fluid are especially difficult to troubleshoot. Before inspecting or making repairs, relieve all hydraulic oil pressures from all circuits and components. Always locate leaks with an old piece of cardboard, never your bare hands. Hydraulic hoses that flex under normal operation are under tremendous pressure and should be inspected every day. If hydraulic fluid under high pressure pierces your skin, it can be extremely serious. Hydraulic fluid can quickly destroy tissue, leading to gangrene. The need to amputate occurs in nearly half of all mishaps. Rushing the injured person to a hospital may not be adequate. Hydraulic fluid that enters the bloodstream can quickly result in death. Crane inspectors must always wear the proper personal protective equipment to protect them from the many hazards associated with inspecting the crane.

Wire rope condition, sheaves, hooks, outriggers, wheel bearings, hydraulic hoses, gears, brakes, tires, and wheels—how do you determine where to start and when to stop? Understand, all we need to do is pick up and read the operator's manual. Operator's aids, such as anti-two-block devices, and smaller items, like the fire extinguisher all need to be part of a user-friendly, quick-and-easy organized plan.

Going Paperless?!

Today manufacturers offer digital applications you can purchase for your smartphone. They are designed to help crane operators, owners, and inspectors have a quick-and-easy way to record and view observations. It's a paperless way to perform both pre-shift as well as pre-lift inspections. They can be an up-to-date way to record inspections and be ready for maintenance requirements or repairs. They may help motivate crane operators to perform daily inspections on a regular basis and keep your crane compliant with safety regulations and standards. Technology has come a long way. Digital reporting allows crane operators to complete the same checklist they would on paper. The information can be shared instantly. The traditional paper-and-pencil checklist may soon become a thing of the past. Minimizing the risk of human injury or death and the possible damage to the crane is why it is so vital.

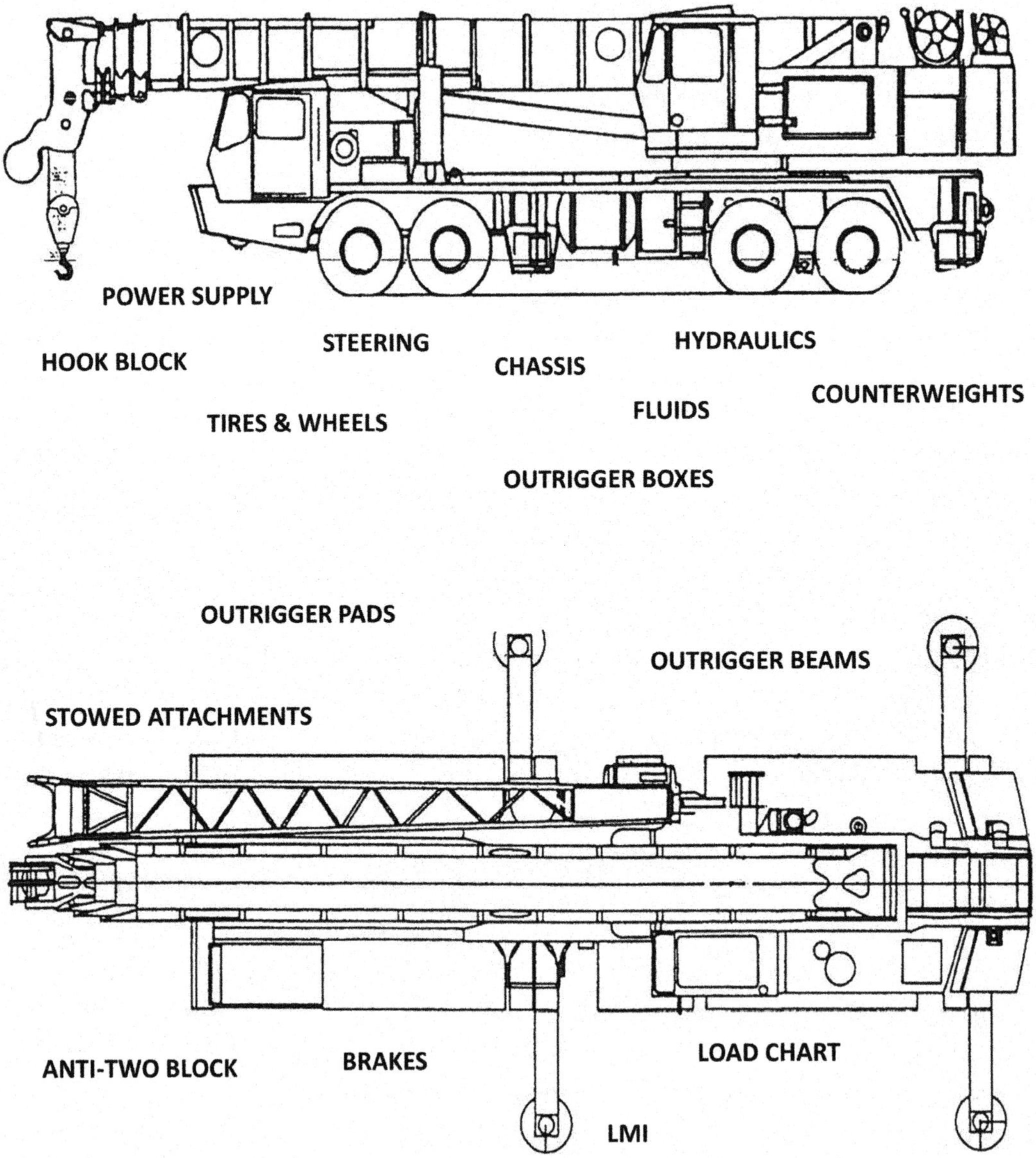

FIGURE 6.6 Typical truck-mounted hydraulic crane.

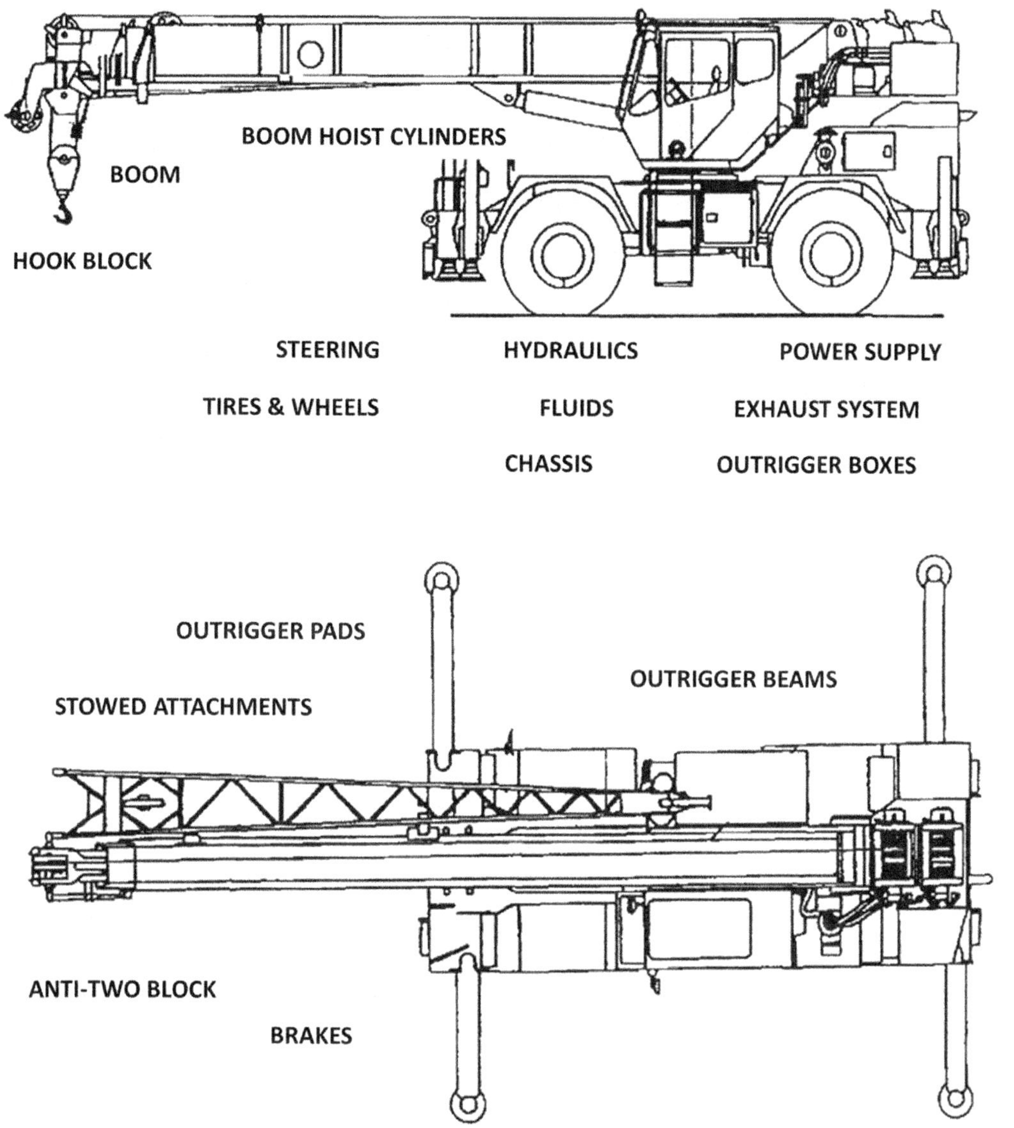

FIGURE 6.7 Typical wheel-mounted rough-terrain crane.

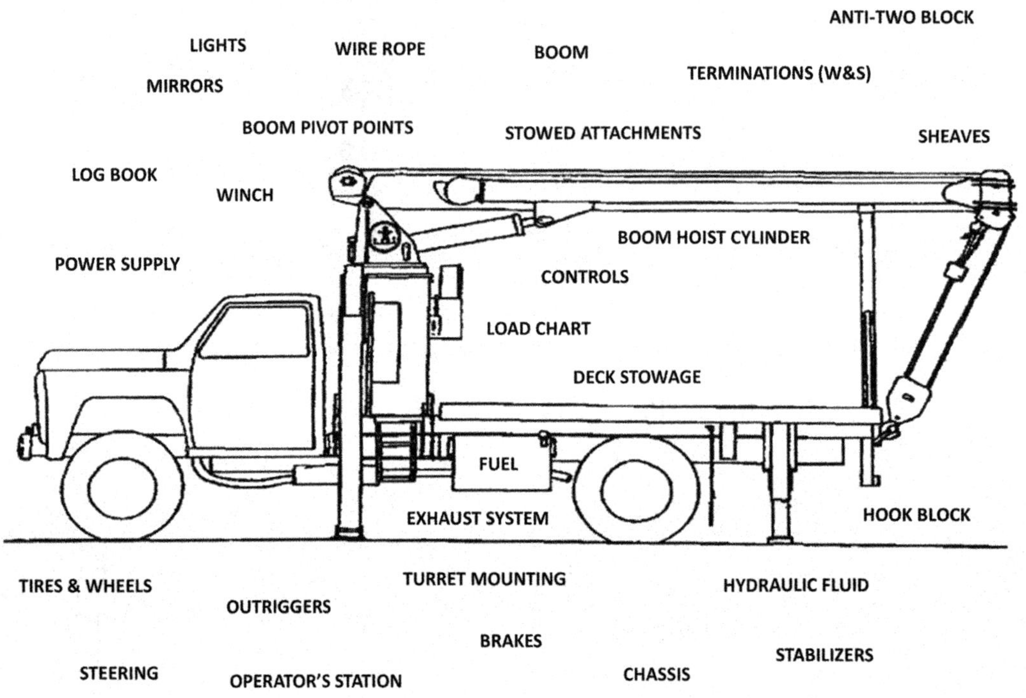

FIGURE 6.8 Typical commercial truck crane with dual operating stations.

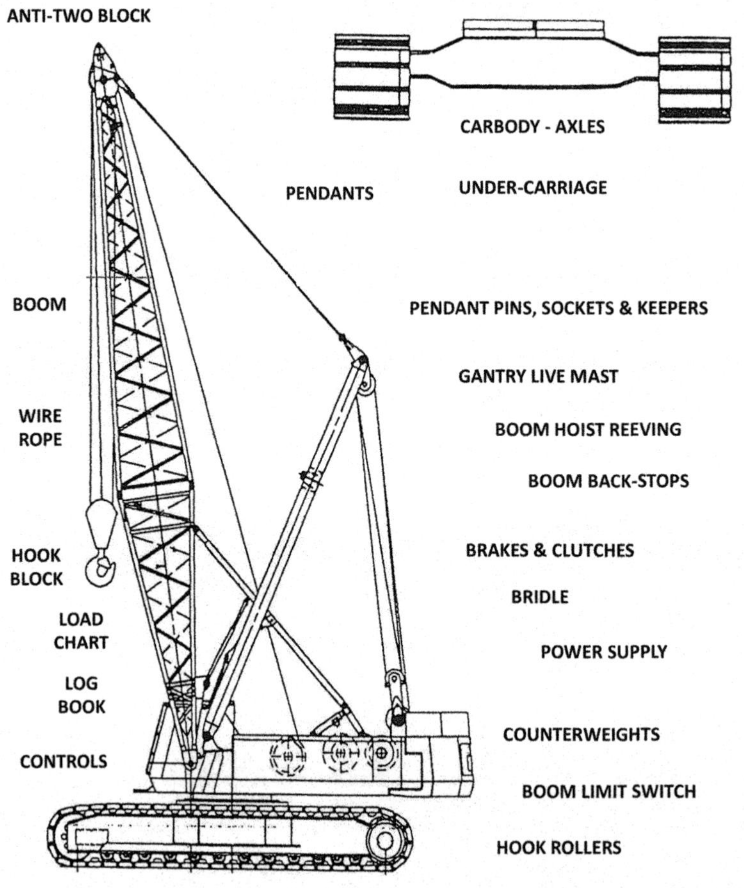

FIGURE 6.9 Typical crawler-mounted lattice boom crane.

Pre-Operational Checklist

Tables 6.1 and 6.2 are typical pre-operational inspection checklists for a typical rough-terrain crane. The lists are not considered to be all-inclusive, nor are they intended to replace any manufacturer's materials or operator's or shop manuals, but they are to be used in combination with this information and your own practical experience. At first, the list items may seem like a lot of things to be checked prior to using your crane every day. However, as you become more proficient, you will find that the time required for checking these items will be greatly reduced, and the thoroughness of your inspections will increase. Before conducting any inspection or maintenance procedures, the operator must ensure that there is no load on the hook, that the boom is in the lowered position or otherwise secured against accidental movement, and the engine is turned off.

TABLE 6.1 Frequent Inspection Report (Grove RT 740; Serial Number 46229)

Performed Daily to Monthly per Manufacturer's Specifications

1.	Light in cab OK.	Accept	Reject
2.	Fire extinguisher in place, fully charged and inspection tag current.	Accept	Reject
3.	Limit switches OK (each motion inched into its limit switch with no load).	Accept	Reject
4.	No leakage in lines, tanks, valves, pumps.	Accept	Reject
5.	All load and boom lifting, towering, extension, refraction, swing, and travel mechanisms properly operate.	Accept	Reject
6.	No cracks or deformation of hooks. Self-closing hook latches covered. Float latches operational.	Accept	Reject
7.	No excessive wear, twisting, or stretching of chains.	Accept	Reject
8.	No more than 6 randomly distributed broken wires in 1 rope lay or 3 broken wires in 1 strand: maximum 5% reduction in overall nominal diameter. Rotation-resistant ropes: no more than 2 random broken wires in 6 rope diameters or 4 random broken wires in 30 rope diameters. Diameters measured monthly.	Accept	Reject
9.	All rope slings, including end connections, are checked before each use for excessive wear, broken wires, stretch, kinking, or twisting.	Accept	Reject
10.	Warning device OK.	Accept	Reject
11.	Hand signals posted onsite.	Accept	Reject
12.	Load rating chart posted in cab.	Accept	Reject
13.	High Power Line warning posted in cab.	Accept	Reject
14.	(Truck cranes) Slow Moving Vehicle emblem posted on rear if highway speed is less than 25 mph.	Accept	Reject
15.	Service brakes, including trailer brake connections.	Accept	Reject
16.	Parking (hand) brake operational.	Accept	Reject
17.	Steering mechanism.	Accept	Reject
18.	Lighting devices and reflectors.	Accept	Reject
19.	Tires properly inflated and in good condition.	Accept	Reject
20.	Horn works.	Accept	Reject

(continued on next page)

TABLE 6.1 Frequent Inspection Report (Grove RT 740; Serial Number 46229) (*continued*)

21.	Windshield wipers work. Cab glass and mirrors free of cracks, breaks, or other deficiencies that hamper view.	Accept	Reject
22.	Testing of anti-two-block system.	Accept	Reject
23.	Guardrails, handrails, steps, platform, walkway, working surfaces.	Accept	Reject
24.	Copy of current log in cab.	Accept	Reject
25.	Guarding on all exposed moving parts.	Accept	Reject
26.	Crane is properly lubricated. The fuel, lubricating oil, coolant, and hydraulic oil reservoirs are filled to proper levels.	Accept	Reject
27.	Improperly marked controls and damaged instruments.	Accept	Reject
28.	Control and drive mechanisms for apparent excessive wear of components and contamination by lubricants, water, or other foreign matter.	Accept	Reject
29.	Air, hydraulic, and other pressurized lines for deterioration or leakage, particularly those that flex in normal operation.	Accept	Reject
30.	Hooks and latches for deformation, cracks, excessive wear, or damage such as from chemicals or heat.	Accept	Reject
31.	Ground conditions around the equipment for proper support, including ground settling under and around outriggers/stabilizers and supporting foundations, accumulation of water on the ground, or similar conditions.	Accept	Reject
32.	The equipment for level position within the tolerances specified by the equipment manufacturer's recommendations, both before each shift and after each shift.	Accept	Reject
33.	The condition of rails, rail stops, rail clamps, and supporting surfaces when the equipment travels on rails.	Accept	Reject
34.	For safety devices and operational aids for proper operation.	Accept	Reject
35.	For derricks, guys for proper tension [Section 1436(p)]	Accept	Reject
36.	For deficiencies identified during the most recent annual inspection that the inspector determined must be monitored in the monthly inspections.	Accept	Reject
37.	Indicators: boom length/angle, jib angle, crane, level, load weighing device (LMI)	Accept	Reject

Report any hazardous conditions immediately to you supervisor, and DO NOT operate the crane.

Comments:____

____

Print Name of Inspector

Signature of Inspector

Date

TABLE 6.2 Periodic Inspection Report (Grove RT 740; Serial Number 46229)

Performed Monthly to Yearly per Manufacturer's Specifications

1.	Check control mechanisms for excessive wear. Visual and operational.	Accept	Reject
2.	Check all safety devices for malfunction. Visual and operational.	Accept	Reject
3.	Check for deterioration or leakage in air or hydraulic systems such as tubes, fittings, hoses, and seals. Visual and operational.	Accept	Reject
4.	Check the hydraulic system for proper oil level. Visual.	Accept	Reject
5.	Check for smooth and correct operation for each function, including remote controls. Operational.	Accept	Reject
6.	Do a visual inspection of crane hooks for wear and measure for size.	Accept	Reject
7.	Visually inspect all wire rope, clips, preventors, and terminations including wedge and socket assemblies.	Accept	Reject
8.	Visually check for loose bolts, nuts, snap rings, and all other devices	Accept	Reject
9.	Visually check for cracked or worn sheaves.	Accept	Reject
10.	Visually check that all safety and warning decals are in place and are legible. Check to ensure all control knobs are in place.	Accept	Reject
11.	Visually examine all exterior welds for cracks, separations, and breaks.	Accept	Reject
12.	Visually and operationally check for worn, cracked, or distorted parts such as pins, bearings, shafts, boom lock, and rollers.	Accept	Reject
13.	Check electrical apparatus for signs of pitting or any deterioration of controller contractors, limit switches, and push-button stations.	Accept	Reject
14.	Check leakage in lines, tanks, valves, and pumps.	Accept	Reject
15.	Check hooks, chains, ropes. Hooks with throat opening of 15% or more, or hook twists of 10 degrees or more, must be removed from service.	Accept	Reject
16.	Check for deformed, cracked, or corroded members.	Accept	Reject
17.	Check for loose bolts and rivets.	Accept	Reject
18.	Check for cracked or worn sheaves or drums, pins, bearings, shafts, gears, rollers, locking or clamping devices, hook roller brackets, removable outrigger attachments, lugs, and welds.	Accept	Reject
19.	Check contacts on controller, limit switches, and push buttons.	Accept	Reject
20.	Check travel, braking, and locking devices.	Accept	Reject
21.	Check tested anti-block system (where applicable).	Accept	Reject
22.	Check the boom and jib for straight without evidence of physical damage, such as cracking, bending, or any other deformation of welds.	Accept	Reject
23.	Check that the counterweight is secure.	Accept	Reject
24.	Check the identification number to ensure it is permanently and legibly marked on jibs, blocks, equalizer beams, and all other accessories.	Accept	Reject

(continued on next page)

Report any hazardous conditions immediately to you supervisor, and DO NOT operate the crane.

Comments:____

____

____

____

____

Print Name of Inspector

________________________________ _______________________

Signature of Inspector Date

Chapter 6 Review Quiz

1. Before setting a periodic inspection schedule for any crane, you should
   a. consult the mechanic
   b. refer to the manufacturer's manual
   c. check the crane's log book
   d. ask the owner

2. During routine inspection of the crane, running rope should be
   a. visually inspected
   b. load-tested
   c. cleaned
   d. lubricated

3. Crane inspections that take place daily to monthly are classified as
   a. irregular
   b. initial
   c. periodic
   d. frequent

4. On a monthly basis, you should inspect several items on your crane, one of which would include
   a. shake out hooks and 4-way sling
   b. synthetic slings and shackles
   c. sheaves and drums
   d. wire rope slings

5. An inspection of all running rope in service on your crane should occur
   a. daily
   b. weekly
   c. monthly
   d. by a certified crane inspector

6. All inspections, tests, repairs, maintenance, and modifications to the crane must be recorded in the
   a. operator's manual
   b. daily time sheet
   c. equipment log book
   d. annual report

7. If there is a warning sign on the switch or engine starting controls, the operator shall not shut off or start the engine until the sign has been removed by the
   a. authorized person
   b. oiler
   c. master mechanic
   d. manufacturer

8. Load tests that are conducted on cranes must not exceed
   a. 85%
   b. 100%
   c. 75%
   d. 65%

9. It is acceptable to load a crane beyond the rated load
   a. if the lifting attachment is rated for the load
   b. never
   c. when approved by the site supervisor
   d. if deployment of extensions is not required for lift

10. Before adjustments and repairs are started on a crane's hydraulic components,
   a. review the maintenance records
   b. consult the manufacturer
   c. review the shop manual
   d. relieve the hydraulic oil pressure from all circuits and components

Summary—and Then Some

Because mobile cranes hoist up and place very heavy loads constantly, parts of the crane wear out, and components are sometimes broken. A good crane inspection can verify that equipment complies with federal and state regulations as well as industry safety standards. Understanding these laws can help companies avoid costly fines and disruption of important work. Government agencies such as OSHA are placing tough regulations on owners and users of mobile cranes. It's important that you take the time to find out what the rules require. Cranes are essential machines in both industrial and construction industries because they can lift and place massive loads. Due to their heavy use, these machines experience a lot of wear and tear, leading to equipment failure that could endanger everyone on the worksite. A thorough inspection of a mobile crane helps detect any problems in these machines to mitigate potential worksite risks and accidents, which will save lives.

Before setting up any mobile crane, it's important that we take the time to select the correct machine for the job. A hydraulic telescopic boom crane and lattice boom crawler crane each have a specific purpose. Selecting the correct crane for ground conditions, tonnage, and boom length are some of the first things to consider. Once the correct crane has been determined, a good through inspection should come next.

Planning is the next important step. Lifting heavy loads is hard work. Preparing for the lift can take significantly more time than the lift itself. A lift plan must be prepared for every project. But first we have to get the crane onto the jobsite. Travel permits and a well-planned route are part of the planning process in order to get the crane safely to the job. Once the crane and crew are onsite, the mobile crane setup procedures can take hours, or in some cases days or even weeks, depending on the size and type of crane. One crane can have as many as a dozen people all working together to make sure it is properly assembled and the lift can proceed safely. Setup procedures may be a bigger part of the overall lifting plan than the actual lift and placement of the load itself. The time the crane operator is in the seat and actually pulling the control levers can be as little as 25% of the job.

7 Setting Up Cranes

A good lift plan will include a map of the entire jobsite, including all its surroundings. To assemble a crane, you need a lot of space. Included in the plan should be access for trucks and pickups, as well as any support equipment that's involved. Once the crane is assembled, every step from start to finish must be planned out. A meticulous lift plan will prevent disasters from taking place. Great planning will help avoid any damages or accidents associated with the crane or lifting team. Well-thought-out plans will reduce accidents and save lives.

Setup procedures must include a crane placement assessment. Some companies will use ground-penetrating radar to locate any underground abnormalities such as sewers, electrical vaults, water lines, and conduits, to name a few. A member of the lifting team must walk the jobsite making note of any concerns both underground and above, such as overhead power lines, traffic lights, trees, and other obstructions. Always make contact with an onsite foreman or supervisor who is familiar with recent and current construction activities. A jobsite foreman will most likely know and be able to point out any areas recently excavated or ground poorly compacted, including the location of underground water lines, conduits, and other buried services. A person who has "boots on the ground" and has been actively involved with any work performed on the site is the best person to ask. The structural stability of the jobsite ensuring that the ground will support the crane under full load is paramount. There is no substitute for crane safety, so check the details of your lift plan, and recheck them before you make the lift. Some organizations recheck the lift plan before every lift. Progressive companies pre-plan their lifts using innovative 3D lift plan software that is very effective (Figure 7.1). Some lifts are simple and routine, and others are complex and heavy; but no matter how routine the lift, there still must be a lift plan, even if it's a simple plan in the crane operator's head. You can organize a pre-lift meeting in the field by using a simple company toolbox safety form anytime a lift needs extra attention. The key to success is planning and communicating before every lift, heavy or light, complicated or not.

FIGURE 7.1 An image created by 3D lift plan software to plan crane lifts.

Setting Up Wheel-Mounted Cranes

When setting up mobile cranes, you must ensure adequate swing clearance for boom, load, counterweight, and fully extended outrigger extensions and floats. OSHA regulations clearly state that accessible areas within the swing radius of the rear of the rotating superstructure of the crane, either permanently or temporarily mounted, shall be barricaded in such a manner as to prevent an employee from being struck or crushed by the crane (Figure 7.2).

Cranes must be set up a minimum of 2 feet from adjacent structures, such as buildings, to prevent workers behind the crane from being crushed between the counterweight and the wall of a structure. Always require all ground personnel to get "eye-to-eye contact" with the crane operator before entry into the roped-off area of the crane. Crane operators have many blind spots and may not see workers around the crane. The crane operators must indicate they see you and acknowledge to you it is safe to enter the lift zone. When barricading your crane, get the minimum plus whatever else you can reasonably get. Your best practice is to allow those—and only those—involved with the lift within the work zone. When working near energized electrical powerlines, nonconductive barricades must be used to form a perimeter to restrict access to the crane work area. Other important zones to be aware of are:

- **Fall Zone.** The area where suspended materials could fall in the event of an accident. This includes the area directly beneath the load, but can also extend beyond that. OSHA recommends keeping a minimum distance of 10 feet from suspended loads, and increasing that distance if the load is higher than 15 feet.

FIGURE 7.2 OSHA says the minimum a crane is to be barricaded is the swing of the superstructure or rear of the counterweight.

- **Drop Zone.** The area in the vicinity of a suspended load where a worker could potentially be struck if the load were to drop or invert. While the crane is in motion, this includes the area below an empty hook or any lifting attachments.
- **Exclusion Zone.** A defined area where workers are not allowed to go when particular work is being done to avoid the risk of being injured by the crane or falling objects.

Ensure the Crane Is Level

Regardless of the method used, wheel-mounted cranes must be level to within 1% of grade or ½ of 1 degree. Both OSHA regulations and ASME safety standards state the crane must be within 1% of grade. Firm, level ground is the foundation for the success of any lift. When a crane is out of level, it can severely impact the crane's stability and charted load capacity. Most cranes come from the factory with a simple bubble level located in the cab or near the outrigger controls. Levels of this type will get you close. Mobile crane professionals will insist on a good-quality, full-length carpenter's level that can be placed across the truck frame as close to the turntable as possible. Always swing the crane and check for level in multiple quadrants, including front, side, and rear (Figures 7.3, 7.4, and 7.5).

FIGURE 7.3 Check the crane for level over the side quadrant.

FIGURE 7.4 Check the crane for level over the rear quadrant.

FIGURE 7.5 Check the crane for level over multiple quadrants.

The importance of having the crane level at all times cannot be overstressed. Checks should be made from time to time during lifting operations to make sure that the crane remains level. Structural damage, increased radius, and loss of stability will result if the crane is being operated out of level and load chart capacities cannot be relied on. Manufacturers' analyses say poor setup is the cause of 60% of all hydraulic crane tip-overs in North America.

Regardless of the ground conditions, it is important to have the crane's centerline at right angles to the horizontal. Use the in-cab levels as a start followed by a carpenter's level at or near the crane boom base in all operating quadrants. The crane tires must be relieved of all weight by the outrigger jacks and set up on a firm, level supporting surface, and raised just enough to see daylight under the tires. The ground conditions must have good stability and bearing pressure capacity to support all loads placed on it by the crane. Always be very cautious and use great care when setting up the crane around newly constructed buildings with possible uncompacted backfill. Trench lines for sewer, water, and electrical conduits are another area to be concerned about. When a crane is set up, its bearing surfaces can exert high pressures, depending on ground conditions and the quadrants of operation it is working in. Often when crane operators are about to lift a very heavy load, they will make the pick over the corner directly over an outrigger float or, in the case of a crawler crane, directly over the corner track pad. Lifting over the corner seems to give us a direct advantage but also produces the maximum ground pressure, earning it the most dangerous position. This is why it is so important to use matting, blocking, or pads under outrigger floats, which provide sufficient strength and thickness to prevent crushing, bending, or sheer failure, and shall be of the appropriate thickness, width, and length to completely support the float.

Table 7.1 may be used as a guide to determine the appropriate ground-bearing values for different types of soils. If there is doubt about the actual type or condition of the soil, a lower value should be assumed. From this, the size of the support pads may be determined.

TABLE 7.1 Approximate Soil-Bearing Values

	Material	Tons per sq ft	Mg/m	kPa
1.	Loose saturated sandy clay soils; medium-soft clay	1	10	96
2.	Firm or stiff clay	1.5	15	140
3.	Loose fine sand, medium compact sand; inorganic silt solid	1	10	96
4.	Compact sand-clay soils	3	30	290
5.	Loose coarse to medium, medium compacted fine sand	2	20	190
6.	Hard, dry consolidated clay	5	50	190
7.	Loose gravel and sandy gravel; very compact sand and inorganic silt soils, compact sand and gravelly sand	5	50	480
8.	Soft rock	8	80	770
9.	Compact gravel and boulder gravel formations; very compact sandy gravel	10	100	960
10.	Hard pan overlying rock	12	120	1150
11.	Medium-hard rock	40	390	3830

If the water table is at or near the bearing surface in granular soils, the load figures should be reduced by 50%. It is not safe to assume that conditions at the surface are a guarantee of conditions below the surface.

Soil pressures in the metric system are expressed in kilopascals, where a pascal is a pressure of 1 newton per square meter.

Conditions can change. Frost does melt, especially under pressure, and a few days of rain can drastically reduce the bearing value of most soils. If the crane remains in one place for an extended period of time, examine the outriggers and reset them if necessary.

The load chart ratings on a crane are based on the crane being level in all directions. A crane that is not level causes slide loading of the crane boom, which will reduce the rated capacity significantly. A crane that is 3 degrees out of level may reduce the rated capacity by as much as 50%, as seen in Table 7.2. This applies to all lifting operations, including on rubber, outriggers, and crawler cranes.

TABLE 7.2 Charted Capacity Lost When Crane Is Out of Level

	1°	2°	3
Short boom minimum radius	10%	20%	30%
Short boom maximum radius	8%	15%	20%
Long boom minimum radius	30%	41%	50%
Long boom maximum radius	5%	10%	15%

Figures 7.6 through 7.8 show examples, respectively, of poor-quality blocking, no blocking, and good blocking.

FIGURE 7.6 Two examples of poor-quality blocking.

FIGURE 7.7 Poor blocking resulting in complete collapse and failure.

It is important during your daily pre-operational inspection to make sure the outrigger pad and hydraulic jack are pinned or latched together, or else they could separate during operations. Blocking must always be leveled under the float on the ground or supporting surface to a right angle of 90 degrees between the cylinder and the float. The recommendation for calculating blocking dimensions is to take the crane capacity in tons and divide by a safety factor of 5. This figure will give the required blocking in square feet. The square root of that number will provide the dimensions of the pads in feet that you need for your crane. An example of a calculation is shown in Figure 7.9. It is advised that the area covered by the outrigger blocking be at least 3 times the area of the outrigger pad or be sufficient to support the float over the ground-soil type the crane is set up on.

FIGURE 7.8 Good blocking, level to provide a right 90 degree angle between the hydraulic cylinder rod and the float and pad.

1. Calculate the FORCE

47,800 Crane weight
+ 30,000 Load weight
+ 850 Accessory weight
+ 175 Rigging weight
78,825 LBS. = FORCE

2. Obtain the soils GBC (PRESSURE)

Lift Supervisor informs Operator the GBC is 40 psi

3. Determine AREA required

78,825 lbs ÷ 40 psi = 1,970.6 inches²

4. Find the square root

√1,970.6 inches² = 44.4 inches by 44.4 inches

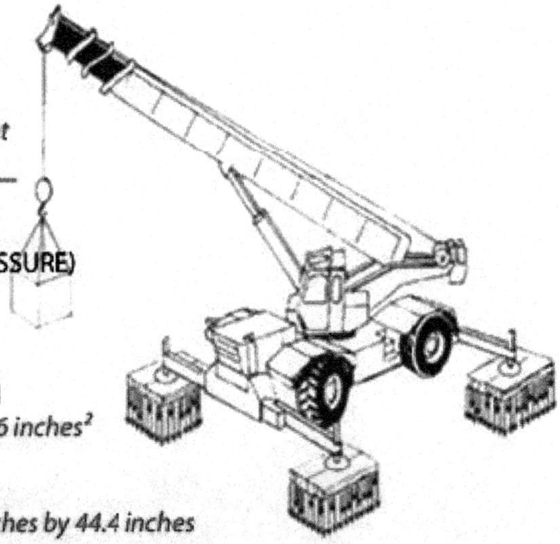

Blocking Dimensions: ***Approx. 4 ft x 4 ft***

FIGURE 7.9 Calculating blocking dimensions.

Ensure that the outrigger beams are properly extended. Always refer to and follow the crane manufacturer's instructions. It is paramount that the crane operator and the supporting crew completely understand the limits of unequal outrigger extension positions. You must use the appropriate load chart for the position at which the outrigger beam is set and pinned when required, as shown in Figure 7.10.

FIGURE 7.10 Most importantly, never, ever place blocking under an extended outrigger beam during setup procedures. These beams have very little structural integrity and cannot handle the pressure of the blocking. These beams are designed to handle hydraulic hoses and components only.

Certain cranes can operate with the outriggers partially extended or just straight down, but only at a fraction of the charted capacity. Some manufacturers provide stability charts for partially deployed outriggers. In these instances, it is acceptable to follow those instructions. If no instructions exist, outriggers must be fully deployed for all lifts.

FIGURE 7.11 Most manufacturers require outrigger beams to be pinned out when performing partial extension work.

The crane operator will move the crane into position as identified on the lift plan. Once the outriggers are properly deployed, the rest of the crane can be assembled. This may take plenty of personal and support equipment that is part of your highly organized lift plan. Once the crane is properly assembled, the lifting can begin. The lifting process can take only a few minutes or as long as many days, weeks, or months, depending on jobsite requirements.

Crawler-Mounted Cranes

Because of their size, configuration, and apparent extra weight, crawler-mounted cranes are generally considered less likely to upset while making a lift or traveling with a load. This is simply not true, because these cranes are more likely to be operated on soft or uneven ground where they have more opportunity to get off level and overturn during operation. The ASME B30.5 mobile crane safety standard provides important information regarding load ratings where stability governs lifting performance. Based on the crane-mounting maximum load rating, wheel-mounted cranes using outriggers fully extended and set with tires off supporting service have a safety rating of 85%. In comparison, crawler cranes without outrigger support have a safety rating of only 75%. The reason is that crawler cranes are much harder to get level. This is why crawler cranes are de-rated more safer than wheel-mounted cranes. De-rating crawler cranes provides an extra safety margin. When crawler cranes are out of level the boom can be side loaded, causing structural failure or a tip-over.

Crawler cranes are designed to have an operator's cab mounted on a crawler undercarriage. The four main modules are the crane's upper works and lower works and its two tracks. Crawler cranes are an ideal choice for large-scale construction projects that require lifting and moving heavy loads. The crawler crane should always be moved on a horizontally level surface with adequate load-bearing capacity. The crawler's travel gears are designed for moving cranes on sites, with a full load on the hoist hook. While wheel-mounted mobile cranes typically struggle with uneven surfaces and loose terrain, crawler cranes can traverse nearly any surface with ease due to their large tracks and even weight distribution across a large platform. The larger platform also provides enhanced stability, eliminating the need for outriggers, saving valuable space, and reducing setup time. In some cases, timber matting under crawler tracks is needed when ground conditions are poor and undertrack pressures are high. The use of timber matting will help distribute high track pressures outward over a larger area (Figure 7.12).

When setting up crawler-mounted cranes, these important actions must be taken:

- Ensure adequate swing clearance.
- Provide enough room to swing the load and boom without obstruction.
- Barricade to prevent entry of the swing radius of the counterweight or superstructure.
- Apply brakes and chalk the tracks to add stability.
- Assemble timber matting to cover at least the area of the tracks (Figure 7.13).

FIGURE 7.12 Timber matting under crawler crane tracks distributes the undertrack ground pressure outward over a larger area.

FIGURE 7.13 When assembling timber matting, it must cover at least the width of the tracks.

Chapter 7 Review Quiz

1. According to OSHA regulations, what is the minimum accessible area of operation a crane must be barricaded?
   a. The swing radius of the boom
   b. The swing radius of the jib
   c. The swing radius of the rear of the superstructure
   d. The swing radius of the front of the operator's station

2. You are setting up an all-terrain crane with all three outriggers extended. In this configuration, when are you allowed to place blocking under the extended outrigger beam?
   a. You are never allowed.
   b. For maintenance and repair.
   c. To change the fulcrum.
   d. To improve stability.

3. You are setting up an all-terrain crane on outriggers. How should the wheels on your crane be positioned in this configuration?
   a. Tires in contact with the ground
   b. Tires elevated high off the ground
   c. Tires high enough off the ground to relieve crane weight
   d. Tires firmly on the supporting surface enough to see daylight

4. Blocking under outrigger floats should cover an area approximately
   a. the same as that of the outrigger floats
   b. one times the area of the outrigger floats
   c. two times the area of the outrigger floats
   d. three times the area of the outrigger floats

5. Before placing the crane in a new work area, operators should confirm the ground support capabilities by
   a. relying on their own experience
   b. checking with the site supervisor
   c. consulting the office manager
   d. consulting the project engineer

6. The reason for using blocking under crane outrigger floats is to
   a. raise the crane to shift the center of gravity
   b. decrease the applied ground pressure
   c. increase the crane's structural integrity
   d. increase the applied ground pressure

7. The stability safety margin for crawler-mounted cranes is greater than that of truck-mounted cranes on outriggers because

- **a.** they have a greater distance from the center of rotation than rubber-tired cranes
- **b.** they have a longer tipping axis
- **c.** they are more difficult to get level
- **d.** they have higher undertrack ground pressures

8. You are operating a 30-ton all-terrain crane with the outriggers fully extended and set. What quadrant of operation would produce the *maximum* ground-bearing pressure?

- **a.** Directly over the rear
- **b.** Directly over the corner
- **c.** Directly over the side
- **d.** Directly over the front

9. You are operating an all-terrain crane in a newly constructed housing development. What would cause you the most concern?

- **a.** The location of the houses
- **b.** The location of the sidewalks and driveways
- **c.** The location of the recent backfill and the amount of compaction
- **d.** The location of the field office

10. When operating a crane on outriggers, for safety purposes it is important to check

- **a.** the blocking under the outrigger beams
- **b.** that the outrigger floats are properly secured to the jacks
- **c.** that the crane is on a firm surface and within 3 degrees of being level
- **d.** that plywood is placed under the floats

Summary—and Then Some

Regular maintenance of the crawler crane is always an important consideration of your daily setup procedures. Clean the crawler chains at regular intervals, and remove any materials that may increase wear, such as sand, mud, and rocks. Also check the chain tension regularly. Make sure all the track components are properly lubricated, including track rollers and sprockets. Preventive maintenance is an important part of the overall setup and lifting plan. Crawler cranes can perform lifts on a wide variety of surfaces including soft soil and gravel. Due to their size, these machines primarily work in a confined work area. Despite this limitation, crawler cranes still provide a great option for projects requiring heavy lifting. Their design and weight distribution allow for a wide work radius without the use of outrigger support.

A crane's "load" is defined simply as the total superimposed weight endured by the load block or overhaul ball hooks. Yet, hoisting a load safely doesn't come as easy as defining it. A crane operator must know if the load falls within the crane's charted capacities and the crane's load chart is their crucially important guide. In it holds all the necessary information required to calculate the crane's lifting capacities, ensuring the operation's safety and its success. A load chart is a synopsis of what a crane can or cannot lift. It provides the crane operator with an overview of the crane's specifications, as well as various features unique to the crane, provided by the manufacturer. A highly trained and knowledgeable crane operator will utilize the load chart to determine the safety of a lift by considering the radius of the crane, boom angle, and boom length, as well as the weight of the load to be lifted. Also, the configuration of the crane and how it is set up.

Each crane has a load chart that specifies its capabilities and details its features and its capacities based on those important factors. The load chart is the most important resource a crane operator must understand to plan a lift and place a load safely. This dangerous work is done with riggers and supporting crew on the ground and the general public nearby. This is why understanding a Mobile crane manufacturer engineers a load chart by calculating and combining two primary limits, these are called "structural capacity limits." These calculations are based on engineering principles and safety factors, accounting for variables like boom length, radius, and counterweight configurations. The resulting load chart visually represents the crane's maximum lifting capacity under specific ideal conditions, which crane operators use to ensure safe operation. Manufacturers design and engineer load charts through a rigorous process of engineering and physical testing, adhering to industry safety standards like ASME B30.5. The charts are built on the core principles of structural integrity and stability. Engineers use computer aided design (CAD) and simulation software to model the crane's way of acting under various load conditions. They perform complex calculations based on rigid body kinetics and electrostatic analysis to determine the stresses on structural and mechanical components.

These calculations determine two main capacity limits for every possible configuration. The first is "structural strength" at which a crane's components such as its boom, jibs, and other parts might break or bend under load. The second is stability or tipping point, at which the loads overturning moment exceeds the crane's stabilizing moment, which is provided by the weight of the crane itself and its counterweights. The design process accounts for numerous variables of the crane's operation that could affect capacity in real construction and industrial site scenarios. This includes counterweight mass and placement, outrigger spread including fully extended, partially extended or working on its rubber tires, as well as the use of extension jibs or other attachments. The relationship between boom length and load radius is analyzed. A longer radius or boom length significantly reduces lifting capacity due to increased leverage. The weight of all load handling devices, such as load blocks, headache balls, slings, rigging hardware and wire rope, is calculated and factored in so the crane operator can determine the net load limit.

After initial design, physical prototypes undergo extensive testing. Actual weights are lifted under controlled conditions to validate the theoretical calculations and see to it the crane performs as expected within its design limits. The validated data from analysis and testing is compiled into a clear, standardized format such as a table or curve diagram. This document visually represents the crane's maximum safe lifting capacities under many operating conditions. Capacities based on structural are listed above a bold line or flagged by asterisks while those based on stability strength are below the line or not flagged by asterisks. One manufacturer may identify these areas differently from another, which we will discuss later in this chapter. The final load chart is fastened to the crane in the view of the crane operator, as well as in the operator's manual. It becomes a mandatory safety guide for the crane operator to see to it the crane is always operated within its certified working load limits (WLL) or charted capacity. The rigorous engineering and testing process ensures the load chart is an accurate and reliable safety guide, maximizing the crane's capabilities while preventing dangerous overloading or tipping incidents.

Always remember, crane load charts are designed and calculated in ideal static conditions while the crane is operated on projects and jobsites with real world operations with dynamic forces. Mobile cranes from Europe are often found here in the US. Mobile cranes from Europe and the United States are similarly engineered under the assumption of static conditions for their primary load capacities. The key difference between US safety standard ASME B30.5 and European standards, such as EN13000, can be found in the specific safety factors applied to those static tipping loads and how they account for dynamic forces and fatigue over the crane's lifespan. European standards, through the EN13000 series, often use more detailed cycle-based calculation methods called "limit state method" to predict fatigue life and structural strength under various dynamic scenarios, including hoisting, slewing or swinging and wind. US standards also ensure structural integrity but have historically been less stringent in their published duty cycle classifications, compared to the European Union.

8 Determining Crane Capacities

Operating a crane is a very complicated job. Reading a load chart must be left to those who are highly trained, qualified, and certified. In addition to the crane operator, other specialists are required to ensure safe hoisting during the lifting process. This team includes the owner, supervisors, the lift director, riggers, and other members of the supporting ground crew. The U.S. Department of Labor's OSHA regulations and ASME industry safety standards require that cranes *not* operated past their capacity as indicated on the machine's load chart. Mobile cranes are load tested and inspected, once a year, by a certified crane inspector. They are tested to 100% of the crane's charted capacity, but never over load chart ratings. Doing so could cause structural or mechanical failure as well as loss of stability, causing a tip-over. Any of these conditions could lead to serious injury or death to all those near the machine, certainly leading to legal consequences.

The employers of crane operators must insist that the operators understand the crane's load chart and its limitations. The load chart must be in the cab or secured to the machine for the crane operator's reference, as well as for reference by the supervisors, lift director, riggers, and other members of the lifting team. Each crane manufacturer and model of crane has a unique load chart. A crane may have one or several load charts, depending on its size and configurations. Many cranes today are configured with a load moment indicator (LMI) or rated capacity indicator (RCI). These devices are like an onboard computer that warns the crane operator if the weight of a load exceeds the charted capacity of the crane. Regardless, even though your crane is equipped with an LMI or RCI, the crane operator is still required to understand and use the crane's capacity chart. Using the load chart properly, along with a good lift plan, ensures that the hoisting operation is safe from the start of the lift all the way to load placement. Also mounted on the crane are load rating charts and instructions that define the crane's lifting limitations. Before operating any crane, the operator must be familiar with the location of this information (Figure 8.1).

BOOM MODE "A"	Maximum Allowable Lifting Capacities Rated Lifting Capacities In Pounds On Fully Extended Outriggers See Set Up Note 2.						
55 Ft. To 60.3 Ft. Main Boom							
	55 Ft.			60.3 Ft.			
Load Radius In Feet	Loaded Boom Angle (Deg.)	360°	Over Front	Loaded Boom Angle (Deg.)	360°	Over Front	Load Radius In Feet
10	77.0	79,700	79,700				10
12	75.0	72,400	72,400	76.5	61,400	61,400	12
15	71.5	63,500	63,500	73.5	57,600	57,600	15
20	66.0	52,300	52,300	68.5	47,100	47,100	20
25	60.0	44,200	44,200	63.0	39,500	39,500	25
30	53.5	37,800	38,000	57.5	33,900	33,900	30
35	47.0	27,900	32,300	51.5	27,700	29,700	35
40	39.0	21,500	24,900	45.0	21,400	24,800	40
45	29.0	17,000	19,700	37.5	16,800	19,600	45
50	14.5	13,500	15,800	28.5	13,400	15,800	50
55				15.0	10,800	12,800	55
Min. Boom Angle/Cap.	0°	9,000	9,000	0°	7,100	7,100	Min. Boom Angle/Cap.

FIGURE 8.1 Here is a good view of a typical load chart of a crane manufactured by Link-Belt cranes.

It's important that the crane operator, lift director, and supervisor choose the correct crane for the job. An important step to consider is the size, shape, and weight of the load to be hoisted. Selection of the wrong crane will certainly waste time and money. The height of the load is always an important consideration: You must select a crane with enough boom length to make the lift, but even more importantly, enough to place it, too. You should analyze the load and what it's made of, including if it contains any internal liquids like gasoline or diesel fuel. Dissimilar loads may require different load handling devices, equipment, or accessories to assist the hoisting operations and are part of your planning process. Ground and weather conditions will help determine the type of crane used for the job at hand. Always choose a machine that has a higher capacity than the heaviest lift you plan to make to allow a safety margin in all the lifts you have planned. Also choose a crane that can maneuver efficiently and safely based on the ground conditions on your jobsite: Professional crane operators faced with a heavy load will often lift over the corner of a track mounted crane or over a corner outrigger float when operating a rubber-tired crane on outriggers. They understand hoisting over a corner gives them the best leverage. Operators also know that when lifting over a corner, the ground pressure under a track or float will be dangerously high. That is why proper blocking, along with excellent crane setup procedures, must be followed.

FIGURE 8.2 Auxiliary boom head installed on the main boom of the crane—143 pounds.

FIGURE 8.3 Shown here is a tele-boom extension stowed on the main boom of a crane. The extension can be set at 35 feet or extended out as far as 56 feet, depending on the requirements of a lift.

Technical Notes

Grove, the manufacturer of this crane, lists the extension having an effective weight of 293 pounds. Capacity deductions are made from the gross capacity of the crane to account for items like the stowed jib, hook blocks, headache balls, and rigging, to name a few.

FIGURE 8.4 Displayed here is a three-sheave hook block.

Technical Notes

The manufacturer lists the hook block weighing 1,095 pounds. It has a maximum capacity of 15 tons. As you review Figure 8.4, you will also see the hook block is reeved with four part of hoist rope by counting the lengths of wire rope. Also note the plate mounted on the hook block. All hook blocks or headache balls will have both their actual weight and capacity displayed on the unit's plate.

In mobile crane operations, "falls" and "parts of line" refer to the reeving of the wire rope, as displayed on the three-sheeve hook block in Figure 8.4. "Falls" and "parts of line" are used interchangeably—both refer to the number of individual lines of rope supporting the load block and hook. The "whip line" operates with a single part of line, or one "fall," and is a completely separate, secondary hoist system equipped with its own, smaller, headache or overhaul ball. The whip line is designed for lifting lighter loads more quickly and more efficiently than the main hoist, which often has many parts of line reeved for heavy lifting, also shown in Figure 8.5.

FIGURE 8.5 Shown hanging off the crane boom point slightly ahead of the three-sheave hook block with cheek plates is a headache ball. It has a maximum capacity of 7.5 tons and weighs 380 pounds. This heavy spherical weight attached to the end of a crane's single-line hoist rope or whip line above the hook is used to maintain tension on the rope, ensuring it doesn't become slack and keeps the line taunt, preventing uncontrolled hook movement.

Technical Notes

Also note on the three-sheave hook block shown in this figure, the side plates fastened to the hook block on both sides. These are referred to as "cheek weights." On multi-sheave blocks, adding cheek plates increases the overall mass of the block maintaining proper line tension and preventing issues like rope winding. Cheek plates are an integral part of the crane hook block, contributing to its strength and efficiency.

- Before making any lifts, it is the operator's responsibility to determine whether or not the crane is capable of making the lift within the crane's charted capacity in all its configurations. This is done by comparing the sum of all the components that make up the load with the crane's charted capacity for the boom length being used at the intended operating radius.
- When calculating the total weight of the load to be lifted, the operator must remember to include the weight of the load itself, including all load handling devices, slings, hook blocks,

overhaul balls, and rigging hardware, plus equalizer beams or spreader bars. It is important to remember the rated capacity of rigging or any load handling attachment, such as a spreader bar, can be a limiting factor for a lift when its working load limit is less than the cranes. Examples of these weights are found in the load chart criteria in Charts 8.1 and 8.2.

- Extra care must be taken when calculating near-capacity loads.

32 Fixed Boom Extensions with 35 Feet–110 Feet Boom	
Stowed	267 lbs.
Erected	4,250 lbs.
32–56 Feet Tele Boom Extension with 35 Feet–110 Boom	
Stowed	293 lbs.
Erected (retracted)	6,368 lbs.
Erected (extended)	8,460 lbs.

CHART 8.1 Load chart for boom extensions.

(Information will include the weights of jibs or extensions erected or stowed.)

Auxiliary Boom Head	143 lbs.
Hook Blocks and Headache Balls	
50 ton, 4 sheave	1,285 lbs.
45 ton, 3 sheave with cheek plates	1,095 lbs.
45 ton, 3 sheave without cheek plates	895 lbs.
15 ton, 1 sheave	380 lbs.
7.5 ton, 4 headache balls	338 lbs.
10 ton headache ball	560 lbs.

CHART 8.2 Load chart for hook blocks and headache balls.

(Information will include the weights of the lifting components.)

Understanding Wire Rope Weight Philosophies

The differences from one mobile crane manufacturer to another, regarding what they require for the deduction of wire rope hoist line weight, come from variations in a manufacturer's specific design, philosophy, and how they define the crane's rated capacities in their load charts. The typical weight of wire rope used in mobile crane deductions ranges from 0.25 to over 2 pounds per foot, depending on its diameter and construction. The specific weight for a given crane is provided in the manufacturer's documentation and load chart notes. Weights vary slightly based on the specific wire rope construction, such as rotation resistant, 6 x 19 class, and core type.

The exact weight of the wire rope is an important piece of information found in the specific crane manufacturer's manual load chart notes. Crane operators must always use the value provided by the manufacturer for their specific model to ensure accurate load calculations and safe operations. The manual will explicitly state which items including the wire rope hoist lines, if it is considered a "deduction" from the crane's charted capacity. The specific implementation in load charts can vary by manufacturer and region. The load chart is the governing document, and crane operators are legally and safely bound to follow its specific notes and instructions.

Understanding Load Moment and Other Forces That Influence the Load Chart

In Chapter 5, we learned the definition of "load moment." The load moment is the turning or tipping force that a load exerts on a crane. It is a measure of the crane's stability, and it is the product of the force or weight and distance or radius from the crane's center of rotation.

The basic principle is:

Load Moment = Weight of the load (plus all deductions) × Horizontal Distance (Radius)

Wire rope weight creates a load moment and is an important factor in crane stability and capacity calculations. The weight of everything hanging from the boom tip, load creates a load moment. This includes all rigging, the hook block, and wire rope itself, all contributing to the overall moment the crane must counteract.

Load moment is a measure of the rotational force or torque applied to a mobile crane by a suspended load calculated as the product of the load's weight and its horizontal distance from the center of rotation. The load moment is a critical factor for ensuring crane stability. The principle is based on the physics of a lever and fulcrum system.

- **Stability principle.** For the crane to remain stable, the counteracting moment generating by the crane's own weight, the boom, and any installed counterweights must be greater than the load moment. The tipping axis is the fulcrum point around which these moments are balanced.
- **Effect of distance.** As the boom extends outward, increasing the load radius, the load moment increases even if the actual weight of the load remains the same. The increased leverage puts more stress on the crane and reduces its lifting capacity at that specific radius.
- **Instability.** If the load moment exceeds the crane's stability limit or its maximum rated capacity in that configuration, the crane will become unstable and tip over. In addition to load moment, mobile crane operators must be aware of various environmental, dynamic, and operational forces that significantly impact stability and safety.

In addition to load moment, a crane's capacity and load chart are significantly affected by environmental conditions, dynamic forces, and the physical setup and condition of the crane itself. Environmental factors introduce external forces and stability challenges not accounted for in ideal test conditions. High wind speeds exert significant pressure on both the boom and the suspended load. The "sail effect" creates additional side forces and increases the overturning moment. Most manufacturers specify maximum permissible wind speeds for operation often requiring de-rating or reducing capacity or stopping crane operations entirely if limits are exceeded. This is why the crane operator must know and understand the many conditions they may deal with in the field.

The stability and bearing capacity of the supporting surface are very important. Soft, uneven, or improperly prepared ground can cause the crane to become unlevel or sink, which dramatically reduces stability and capacity. Cranes are rated for operation on a firm, level surface with outriggers fully extended and set. Extreme high or low temperatures can affect the mechanical properties of the crane's steel and the performance of hydraulic fluids and lubricants, which may require a reduction in lifting capacity. Dynamic forces are generated by the motion of the crane and load during operation and can temporarily exceed static load chart ratings. Shock loading occurs with rapid acceleration, sudden braking, or "snatching" a load off the ground. These sudden impact loads place extreme, unmeasured stress on the boom and rigging, which is a leading cause of boom failure. Moving the boom too quickly introduces inertia and pendulum action, or load seeing, that creates significant side loading and increases the operating radius beyond load chart values. The load chart is based on the assumption that the load is freely suspended directly beneath the boom tip. Any force that pulls the load sideways, including wind, unlevel ground, or dragging a load, introduces dangerous side loading forces that the boom structure is not designed to withstand.

The physical setup and maintenance of the equipment also dictate the safe limits specified in the load chart. The load chart capacity is determined by how the crane is configured. Operating "on rubber" or without outriggers extended or with outriggers only partially extended drastically reduces the capacity compared to a fully setup configuration. Using the incorrect amount or configuration of counterweights as specified by the manufacturer will alter the crane's stability and tipping fulcrum, making the load chart invalid. The use of jibs, extensions, or other auxiliary attachments adds weight to the boom system and changes the leverage dynamics, requiring the use of a specific load chart for that configuration. The physical condition of the crane itself and any bent boom sections, worn wire rope, or proper tire pressure affects its structural integrity and stability. Visual indicators that a crane operator is encountering hazardous dynamic forces or side loading involve observing the movement of the load, the behavior of the crane's structure, and external factors.

Ultimately, a crane operator's proficiency falls in the understanding that the load chart represents the most ideal and stable scenario. The true knowledge and skill out on the jobsite are recognizing and mitigating the countless other on-the-job factors, from the invisible forces of wind and sudden movement, to the substantial risk of poor ground conditions, that constantly threaten the safe balance between the crane's charted capacity and the workplace where it operates. Staying alert to these dynamic forces is not only about governmental regulations and compliance, it is the basic professional practice of preventing catastrophic failure.

How to Determine Line Pull

It is important for the crane operator, lift director, and ground crew to understand line pull and the proper number of parts of line needed to have enough capacity to lift the weight of the load you are planning to hoist (Figure 8.6). The more parts of line you have, the more you increase mechanical advantage and capacity. However, the more parts of line reeved up, the slower the line speed, and you

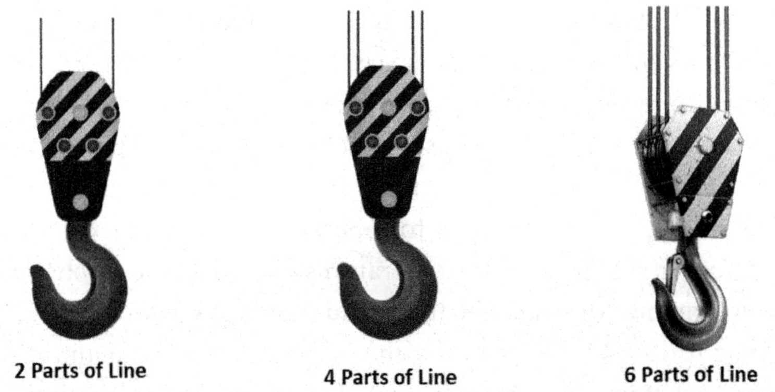

FIGURE 8.6 Examples of parts of line and what they would look like for 2, 4, and 6 parts of line.

have a decreased ability to raise and lower the overhaul ball or hook blocks quickly. Fewer parts of line will increase line speed as needed for production work. It is important that the crane operator and the supporting team understand that reeving up multiple parts of line may require the wire rope weight to be deducted from the crane gross capacity. Crane manufacturers of larger-capacity cranes may make this requirement based on the diameter of the wire rope on the machine. Manufacturers of smaller-capacity cranes may not make this requirement. Always refer to the manufacturer's requirements for any wire rope deductions. When choosing the number of parts of line for the crane, try to keep the number of lines even if possible. Odd numbers of parts of line may cause the hook block to tilt slightly. Every crane has a specified wire rope diameter and capacity, so always use the original capacity and the wire rope specified by the manufacturer of your crane.

Prior to making any lift, the crane operator will need to determine the number of parts of line that will be required to reeve the crane's hoist block to safely lift and place the load. This decision will be based on certain available information that the operator will need to evaluate. In general, the fewer the parts of line used, the faster the hoisting speed, but with limited capacity. Too many parts of line may restrict the hook's length of travel and add unnecessary weight to the load. Obtain the manufacturer's suggested parts of line from the load chart.

Some manufacturers can make this process easier than others. As displayed in Chart 8.3, this manufacturer has provided the allowable line pull for 1 part through 4 parts of line. The math is already done for us, and the type of wire rope is identified on the crane when using IWRC, or rotation-resistant wire rope. Also important to note, this manufacturer illustrates when the auxiliary block needs to be installed on the crane boom. When more than two parts of line are required, the auxiliary boom head must be installed in order to have the extra sheaves to reeve more parts of line, as shown in Chart 8.3. Overhaul balls or headache balls come in many capacities and weights, as shown in Figure 8.7.

ALLOWABLE LINE PULL

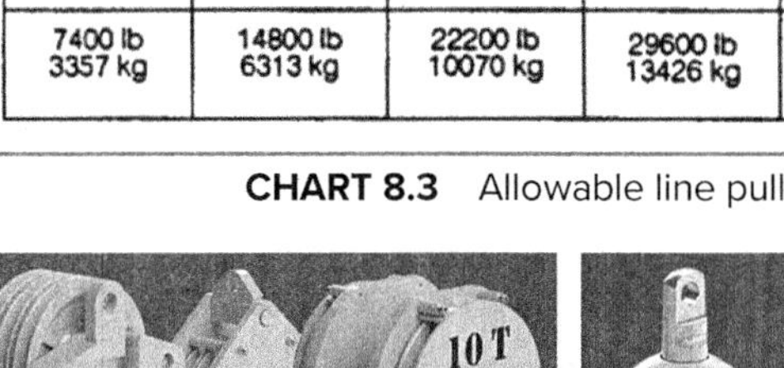

1 PART LINE	2 PART LINE	3 PART LINE	4 PART LINE	WARNING Anti-two-block system must be in good operating condition before operating crane. Refer to owner's manual. Keep at least three wraps on load line on drum at all times.
8500 lb 3856 kg	17000 lb 7711 kg	25500 lb 11567 kg	34000 lb 15422 kg	9/16" (14.29 mm) 6.25 MRC (3.5:1 SF). 29750 lb (13494 kg) Minimum breaking strength.
7400 lb 3357 kg	14800 lb 6313 kg	22200 lb 10070 kg	29600 lb 13426 kg	9/16" (14.29 mm) Rot resistant (5.0:1 SF). 37000 lb (16783 kg) Minimum breaking strength.

CHART 8.3 Allowable line pull chart.

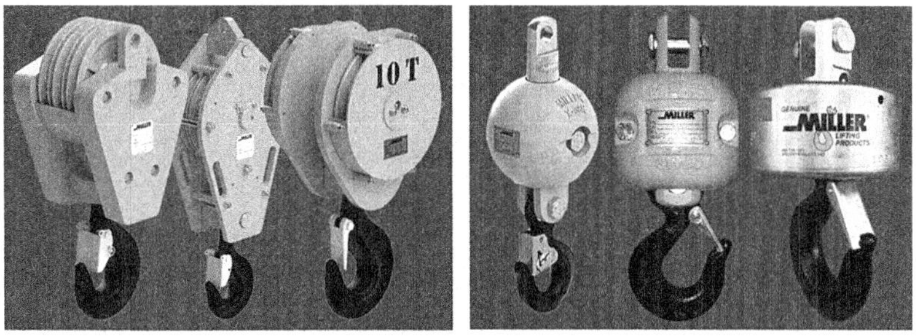

FIGURE 8.7 Typical overhaul weights.

Now that you have looked over the allowable line pull information in the chart, carefully read the following specifications for this crane regarding deductions as listed in the load chart criteria:

Manufacturer's Deduction Requirements

Now that you have had the opportunity to review how to determine line pull on a mobile crane, it's time to discover the five different scenarios a crane operator can experience in the field when dealing with line pull, hoist lines, and hook blocks depending on the size of the crane and its manufacturer. In the following few pages, prepared for you are five questions specifically crafted to have you experience each of these real-world situations. To do so, you will need to imagine yourself as a crane operator behind the controls of three different cranes: fixed cab, swing cab, and lattice boom cranes. Each crane has a set of questions that provides you with a different hoisting and rigging operation you will need to deal with. Before we get started, it's important you read and understand the manufacturer's deduction requirements for your first crane, a fixed cab, in order to answer questions one and two.

Deductions must be made from rated loads for a stowed jib. Optional attachments, hooks and load blocks, weights of slings, and all other load handling devices should be considered part of the load.

QUESTION 1:

You are planning a lift of 100 each 6 × 6 × 20 wood boards from the bed of the boom truck you are operating. The gross load of the lift adds up to 17,000 pounds. Your auxiliary block is installed, and your crane is reeved up with 3 parts IWRC 9/16 wire rope. In this configuration, how much weight should be deducted from the gross capacity for the weight of the wire rope that weighs 0.6 pound per foot?

ANSWER 1:

This manufacturer makes no mention of making a deduction for the weight of wire rope. No deduction is required per manufacturer's specifications.

QUESTION 2:

A crane operator is planning a lift that weighs 6,300 pounds. The overhaul ball on the crane was purchased aftermarket and has a capacity of 5,000 pounds and weighs 120 pounds. In this configuration, can the crane operator safely make this lift?

ANSWER 2:

Load hooks and overhaul balls have a load rating that cannot be exceeded. The rating may be more or less than the original equipment, so that must be considered when making the lift. The aftermarket overhaul ball has a capacity of 5,000 pounds, which is 1,420 pounds less than what is needed to safely make this lift. Therefore, the crane operator cannot safely make this lift in this configuration.

As shown in Charts 8.2 and 8.3, the crane manufacturer will always provide a chart listing the weights of all load handling devices for the crane, including effective weights for any stowed jibs to be used on the machine. Be careful of any aftermarket lifting devices you may want to use. The weights for these components will not be listed in the crane's load chart criteria and will be found on the components themselves. Always be aware of the capacity of aftermarket items. They may be more or less than the crane's original equipment for similar units used on the crane. The information is often stamped or casted on the items and is easily found.

Remember: All load blocks and overhaul balls have a capacity that cannot be exceeded. Aftermarket equipment may have a load rating more or less than the original equipment provided by the manufacturer. The weight of the equipment should also be carefully noted for making proper load chart deductions.

The manufacturer of every crane has engineered and designed each to use a defined type and size of hoist rope. The manufacturer of the hoist rope will provide its WLL, or "working load limit," as well as other important information for the crane operator and other members of the lifting team to ensure the rope they choose is compatible with the crane manufacturer's specifications. Catalogs are also provided to help with the selection process. There are many manufacturers of wire rope hoist line so working with your wire rope dealer can be beneficial to help pick the correct type and use for the crane you plan to be using. Choosing the wrong type of rope, installing it on the crane, and putting it to work will certainly destroy the rope and possibly damage other crane equipment in short order, so getting it right is important. Failure of a hoist rope in some cases can cause loss of a load, structural damage to the crane, or loss of stability as shown in Figure 8.8.

FIGURE 8.8 The loss of a mobile crane's stability can lead to a structural collapse, or the crane tipping over. Both can cause severe damage to the crane, the connected load, and surrounding property, all of which potentially resulting in serious injury to, or death of, workers and civilians.

When a mobile crane has a large amount of boom extended, and the hoist block is reeved up with many extra parts of line, that is when you can accumulate a significant amount of wire rope weight, adding up to hundreds of pounds.

Coming up in our exercises, you will learn how to make wire rope measurements and calculate load weights using both a range diagram and a load chart. You're about to see why wire rope on a crane can develop significant weights that must be dealt with while other times the rope doesn't have a lot of influence, depending on the size of the crane, and the size and diameter of the hoist rope and its weight per foot. You must read the manufacturer's specifications in the load chart criteria to learn the wire rope weight deduction requirements for each crane you operate.

In preparation for your next set of questions, it is important to read the manufacturer load chart specifications and deductions for the crane you will be operating in order to make the lifts described in questions 3, 4A, and 4B. For these next questions, you will be operating a large 100-ton, hydraulic, telescopic boom, swing cab truck crane manufactured by Grove. Pay particular attention to Figure 8.9 in preparation for these questions.

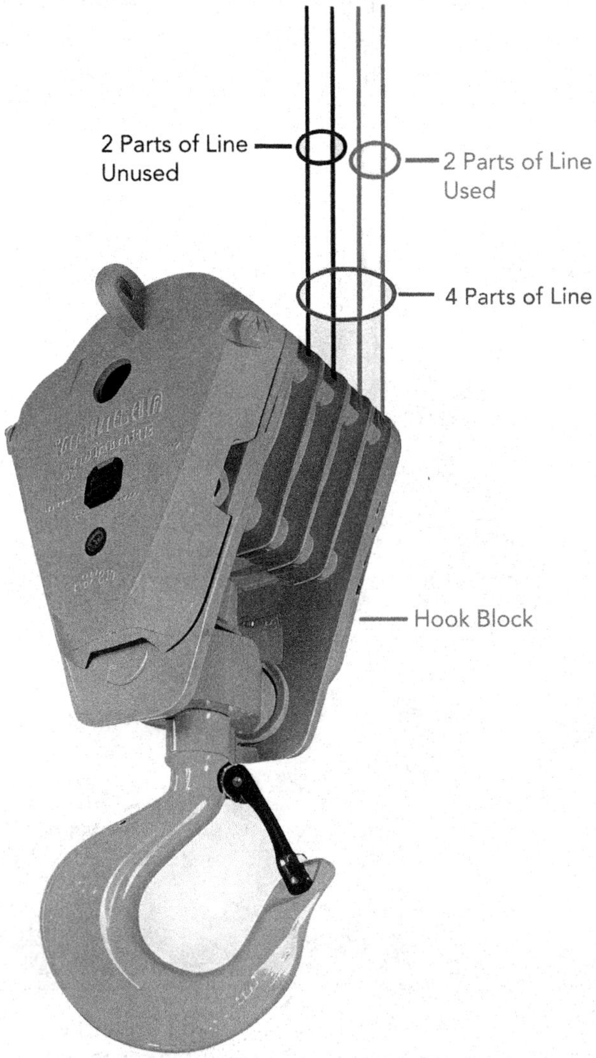

FIGURE 8.9 A hook block, reeved to four parts of line.

Technical Notes

Some manufacturers will require that a portion of—or sometimes all of—the hoist lines, whether being used to make the lift or left hanging unused, have their weight be considered and included as part of the load, as shown in Figure 8.9. The weight is calculated by multiplying the number of parts of line.

In the next set of questions, you will be shown information found in Chart 8.4. In this chart, the crane manufacturer provides specifications for a permissible line-pull of 12,290 pounds for a single full length of hoist rope. With this chart, the operator can do the mathematics to determine line-pull as more falls, or parts of line are reeved and added to the hook block. For example, if the rigging crew were to reeve 8 parts of line to the hook block, the hoist drum would provide 98,320 pounds of line-pull.

CHART 8.4 Line pull and reeving information.

Hoists	Cable Specs	Permissible Line Pull
Main and Aux. Model 30	3/4" (19mm) 18×19 Class or 35×7 Rotation Resistant; Minimum Breaking Str. 64,600 lbs.	12,920 lbs.

Note: Cranes with lower capacities, such as fixed cab boom trucks or carry decks, come supplied with smaller-diameter wire rope hoist lines from the factory than larger-capacity cranes, such as a bigger telescopic boom, swing cab truck or rough terrain cranes. The smaller-capacity cranes, for example, may come with $\frac{9}{16}$-inch diameter wire rope hoist line installed on their hoist drums that weigh as little as 0.59 pounds per foot. If that crane were set up with 200 feet of hoist rope hanging over the boom point that represents only 48 pounds of hoist rope which isn't significant enough for crane manufacturers to require the deduction of the wire rope weight from the gross capacity of the crane. On the other hand, a larger capacity crane such as a rough terrain swing cab crane that has $\frac{5}{8}$-inch diameter wire rope hoist line installed, on its hoist drum, weighing 0.72 pounds per foot would be looked at differently because of the amount of weight that can be accumulating under the boom point. Only 200 feet of hoist line are used in this example and don't convey the importance of making wire rope weight deductions. When the crane has a large amount of boom extended and the hook block is reeved up to so many parts of hoist line, the calculations made to determine hoist line weight will really get your attention as it can add up to hundreds of pounds or more. Manufacturers require for hoist line in use, when wire rope deductions are required, the rope length shall be measured from the boom tip to the ground. Coming up in our exercises, you will learn how to make these measurements and calculations using an actual crane load chart. You're about to see why wire rope on a crane can develop significant weights that must be dealt with while other times the rope doesn't have a lot of influence, depending on the size of the crane and the size and diameter of the hoist rope and its weight per foot. You must read the manufacturer's specifications in the load chart criteria to learn the wire rope weight deduction requirements for each crane you operate.

In the next set of questions, you will be shown information found in Chart 8.4. In this chart, the crane manufacturer provides specifications for a permissible line-pull of 12,290 pounds for a single full length of line. With this chart, the operator can do the mathematics to determine line-pull as more falls or lengths of line are reeved up and added to the hook block; for example, if the rigging crew were to reeve 8 parts of line to the hook block, the hoist drum would provide 98,320 pounds of line pull.

Technical Note

Duty cycle operations refer to work that's especially hard on the crane. This includes pouring concrete while using a bucket attachment, drag line, clamshell, and pile driving. Many manufacturers direct the crane operator to deduct 20% from the crane's charted capacity. This deduction ensures the crane will not be subjected to any excess wear and tear. Duty cycle work is often high-speed operations that degrade the brakes, clutches, and motors of the crane.

Now take a moment to carefully review the manufacturer's deduction requirements in the box below for questions 3, 4A, 4B, and 4C.

Rated loads will include the weight of any hook blocks, slings, and auxiliary devices included, and their weight will be subtracted from the listed ratings to obtain the net weight of the load to be lifted. When more than the minimum required amount of hoist reeving is used, the additional rope will also be considered a part of the load being handled.

QUESTION 3:

The crane operator has some duty cycle operations coming up next week for their crane. In an effort to prevent the wear and tear on the expensive running rope currently on the crane, the workers remove it for safe keeping and replace it with a less expensive rope for the next week's work. Suddenly, a truck arrives with a load of iron beams to be unloaded. The truck has 20 beams, each weighing 1,800 pounds. The operator's crane is currently set up with a single section of line with a ¾-inch rope diameter with a working load limit rating of 9,520 pounds.

Can the crane safely unload the truck in this configuration?

ANSWER 3:

A single section of wire rope hoist line, ¾-inch diameter, has a working load limit (WWL) rating of 9,520 pounds, which is 7,720 pounds more than the 1,800 pound beams. The beams, if correctly rigged, can be hoisted with 5 in a bundle or any other combination less than 5 beams. No extra deduction is necessary if the crane is lifting and not performing any duty cycle operations, as in this example.

QUESTION 4A:

You are making a lift from the main boom over the side of your Grove TM5750B truck crane. Your permissible line pull is 12,920 pound capacity, your gross capacity is 18,050 pounds, and your 45-ton block with cheek plates is reeved to 4 parts of line. How many parts of line do you need to make this lift in this configuration?

ANSWER 4A:

Two parts of line × 12,920 pounds of permissible line pull equals 25,840. The gross capacity is 18,050 pounds, so 2 parts of line are needed to make this lift.

QUESTION 4B:
Your crane is set up on firm, level ground. Your boom is extended to 100 feet, and your radius is at 80 feet. Your 45-ton block with cheek plates is reeved to 4 parts of line at 1 pound per foot, and the minimum required hoist reeving for this lift is 2 parts of line. How many parts of line are additional, and should they be considered part of the load? Use the crane range diagram in Chart 8.5 to assist you in your calculations.

ANSWER 4B:
Two parts of line are additional or considered reeving used. The crane operator must deduct the weight of the 2 parts of wire rope reeving using the working range diagram in Chart 8.5.

Using the manufacturer's crane range diagram, find 100 feet of boom on the right-hand side of the diagram. Next, locate 80 feet of radius at the bottom of the diagram. Then follow the arc from the 100 feet of boom until it intersects the vertical line of the 80-foot radius, and mark the location with an "X." Now follow the intersecting point left across the diagram to 60, the height from the ground in feet.

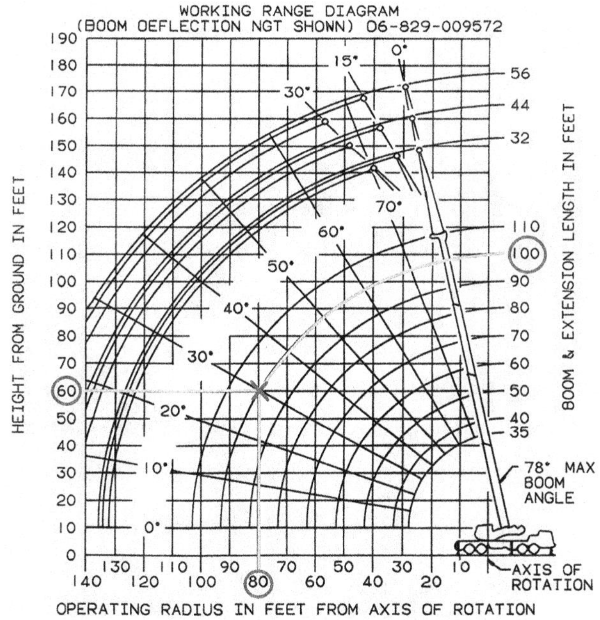

CHART 8.5 Crane working range diagram.

Technical Note

For "line in use," when deductions for wire rope are required, the rope 's length will be measured starting from the boom tip, and ending at the ground while using the crane range diagram.

QUESTION 4C:
How much wire rope weight should be deducted? Use the crane range diagram in Chart 8.5.

ANSWER 4C:
Using the working range diagram in Chart 8.5, the boom point is 60 feet from the ground. Two parts of line reeved up are unused, so 60 feet × 2 parts of line unused equals a total of 120 feet of wire rope. The wire rope weighs 1 pound per foot. Therefore, the deduction is 120 pounds.

To prepare for your next set of questions, it is important to read the manufacturer's load chart deductions and specifications for the crane that you will be "operating." You are now operating a large Manitowoc truck crane.

Now take a moment to carefully review the manufacturer's deduction requirements in the box below for question 5:

> The weight of the jib, load blocks, headache ball, slings, or hoist lines below the boom and jib point sheaves must be considered a part of the main boom load. The boom is not to be lowered beyond the radius where combined weights are greater than rated capacity. When no capacity is shown, operation is not intended or approved.

QUESTION 5:
You are making a lift from the main boom with 42.5 feet erected at a 26-foot radius. Your 225-ton block is reeved up with 10 parts of line weighing 2 pounds per foot. In this configuration, how many parts of line are considered to be part of the load? Use the information in Chart 8.6.

ANSWER 5:
The crane operator must deduct all 10 parts of wire rope reeving per manufacturer's specifications. For 425 feet of boom erected at a radius of 26 feet, your boom elevation is 44.6 feet, and so:

$$10 \text{ parts of line} \times 44.6 \text{ feet} = 446 \text{ feet of rope} \times 2 \text{ lbs per foot} = 892 \text{ lbs}$$

Your wire rope deduction in this configuration is 892 pounds.

Manitowoc Cranes, Inc.
Manitowoc, Wisconsin 54220 U.S.A.

Liftcrane Boom Capacities

Meets ANSI B30.5 Requirements

Boom No. 78T With Heavy Lift Top
0 Lb. to 95,400 Lb. Crane Counterweight
With 25,000 Lb. Front Bumper Counterweight
Rating Over Rear or 360 Degrees On Fully Extended Outriggers

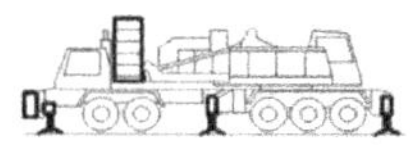

42.5 Ft. Boom

Oper. Rad. Feet	Boom Ang. Deg.	Boom Point Elev.	95,400 Lb. Crane Ctwt. 25,000 Lb. Bumper Ctwt.		78,400 Lb. Crane Ctwt. 25,000 Lb. Bumper Ctwt.		61,400 Lb. Crane Ctwt. 25,000 Lb. Bumper Ctwt.		44,400 Lb. Crane Ctwt. 25,000 Lb. Bumper Ctwt.		Oper. Rad. Feet
			Boom Capacity 360 Deg. Pounds	Boom Capacity Rear Pounds	Boom Capacity 360 Deg. Pounds	Boom Capacity Rear Pounds	Boom Capacity 360 Deg. Pounds	Boom Capacity Rear Pounds	Boom Capacity 360 Deg. Pounds	Boom Capacity Rear Pounds	
12	81.6	50.3	364,000*	364,000*	364,000*	364,000*	364,000*	364,000*	358,100*	358,100*	12
13	80.2	50.1	364,000*	364,000*	364,000*	364,000*	348,400*	348,400*	332,500*	332,500*	13
14	78.8	49.8	354,900*	354,900*	340,000*	340,000*	325,100*	325,100*	310,200*	310,200*	14
15	77.4	49.6	332,600*	332,600*	318,600*	318,600*	304,600*	304,600*	290,600*	290,600*	15
16	76.0	49.3	312,800*	312,800*	299,600*	299,600*	286,400*	286,400*	273,200*	273,200*	16
17	74.6	48.9	295,200*	295,200*	282,700*	282,700*	270,200*	270,200*	257,700*	257,700*	17
18	73.1	48.6	279,400*	279,400*	267,500*	267,500*	255,700*	255,700*	243,800*	243,800*	18
19	71.7	48.2	265,100*	265,100*	253,800*	253,800*	242,500*	242,500*	231,000*	231,000*	19
20	70.2	47.8	252,100*	252,100*	241,400*	241,400*	230,600*	230,600*	218,000*	218,000*	20
22	67.2	46.9	229,500*	229,500*	219,700*	219,700*	208,600*	208,600*	195,700*	195,700*	22
24	64.1	45.8	210,400*	210,400*	200,900*	200,900*	189,100*	189,100*	177,300*	177,300*	24
26	61.0	44.6	194,100*	194,100*	183,500*	183,500*	172,700*	172,700*	161,900*	161,900*	26
28	57.7	43.3	178,800*	178,800*	168,800*	168,800*	158,800*	158,800*	148,800*	148,800*	28
30	54.2	41.8	165,400*	165,400*	156,100*	156,100*	146,800*	146,800*	136,700	137,500*	30
32	50.6	40.0	153,700*	153,700*	145,000*	145,000*	136,300*	136,300*	122,000	127,600*	32
34	46.8	38.1	143,400*	143,400*	135,200*	135,200*	127,100*	127,100*	109,900	118,900*	34
36	42.7	35.8	134,300*	134,300*	126,600*	126,600*	117,100	118,900*	99,700	111,200*	36
38	38.2	33.2	126,100*	126,100*	118,800*	118,800*	107,100	111,600*	91,000	104,300*	38
40	33.1	30.0	118,800*	118,800*	111,900*	111,900*	98,400	105,000*	83,400	98,100*	40
45	12.7	16.0	98,100*	98,100*	93,500	96,900*	80,800	90,800*	68,000	84,700*	45

CHART 8.6 A load chart for a lattice boom truck crane, manufactured by Manitowoc.

Technical Notes

As indicated in question 5, you can see that the boom length as 42.5 feet. Where it appears circled, the chart indicates a 26-foot radius, a boom angle of 61 degrees, and a boom point elevation of 44.6 feet. Additional details provided are capacities for 360-degree rotation, or for lifts that will be taking place over the rear of the crane. Capacities for different rear or front counterweights are also listed.

Structural or Crane Stability Limitations

The capacities shown on many crane load rating charts are based on either structural or stability limitations. Care must be taken to never exceed these charted capacities. Doing so could result in the structural failure or loss of stability of the crane, the loss of the load, and possible injury or death. Manufacturers provide indicators on the load chart warning the crane operator of a limit of structural or stability strength. When you are operating a crane when a stability or a tipping condition starts, it happens quickly, providing the crane operator with little or no time to react and compensate, and little chance of recovery. Different manufacturers' limitations of the structural and stability constraints on the load chart are displayed in several different ways. You should be aware of these different indications you might see laid out on a crane load chart, such as the following:

- The manufacturer uses shaded areas to indicate structural integrity, and areas not shaded on the load chart are based on tipping or stability (Chart 8.7).
- The manufacturer uses a solid bold line to separate capacities on the load chart (Chart 8.8). The capacities shown above the bold line are based on structural integrity. The capacities shown below the line are based on stability or tipping.

- The manufacturer uses an asterisk next to the values on the load chart to indicate that those numbers are based on structural integrity (Chart 8.9). The values without an asterisk next to the number are based on tipping.
- If a crane operator has a given boom length, the crane's structure has limitations on how much weight it can hoist at a certain radius. A longer radius length requires a stronger structure to support the lift. If the length of the radius is too long, the crane risks exceeding its structural limitations, possibly causing an accident.

Shaded Structural

Load Radius In Feet	95 Ft. Loaded Boom Angle (Deg.)	95 Ft. 360°	95 Ft. Over Front	105 Ft. Loaded Boom Angle (Deg)	105 Ft. 360°	105 Ft. Over Front	110 Ft. Loaded Boom Angle (Deg.)	110 Ft. 360°	110 Ft. Over Front	Load Radius In Feet
20	77.5	31,800	31,800							20
25	74.5	28,300	28,300	76.0	25,700	25,700	77.0	22,500	22,500	25
30	71.0	25,300	25,300	73.5	23,100	23,100	74.5	22,200	22,200	30
35	68.0	22,900	22,900	70.5	20,900	20,900	72.0	20,100	20,100	35
40	64.5	20,800	20,800	67.5	19,000	19,000	69.0	18,300	18,300	40
45	61.5	19,000	19,000	65.0	17,400	17,400	66.0	16,700	16,700	45
50	58.0	16,500	17,500	61.5	15,900	15,900	63.5	15,200	15,200	50
55	54.0	13,800	15,900	58.5	13,900	14,700	60.5	13,900	13,900	55
60	50.0	11,800	13,500	55.0	11,900	13,600	57.0	11,900	12,500	60
65	45.5	10,000	11,700	51.5	10,100	11,800	54.0	10,200	11,200	65
70	41.0	8,600	10,000	48.0	8,700	10,100	50.5	8,700	10,100	70

Unshaded Tipping

CHART 8.7 Shaded areas indicating structural integrity and unshaded areas indicating tipping or stability limitations. Structural limitation refers to the maximum load weight the structural components, such as the boom and jib, can bear without breaking.

Structural

20	44,950 (44.5)	43,000 (52)	39,400 (61)	33,600 (66.5)	30,500 (70.5)	28,000 (73.5)	25,500 (75.5)	23,300 (77.5)	
25	35,200 (28.5)	35,050 (41.5)	32,400 (54)	27,750 (61)	25,200 (66)	23,800 (69.5)	22,000 (72)	20,400 (74.5)	18,500 (76)
30		28,400 (27)	27,150 (46)	23,350 (55)	21,100 (61)	20,400 (65.5)	19,300 (68.5)	17,550 (71.5)	15,750 (73.5)
35			21,650 (36.5)	20,000 (49)	18,050 (56)	17,400 (61.5)	16,400 (65)	15,200 (68)	13,650 (70.5)
40	See Note 16		17,200 (24)	16,850 (41.5)	15,650 (50.5)	15,050 (57)	14,150 (61.5)	13,300 (65)	12,000 (68)
45				13,350 (33)	13,600 (45)	13,200 (52.5)	12,350 (57.5)	11,550 (61.5)	10,600 (65)
50				10,650 (21.5)	11,450 (38.5)	11,600 (47.5)	10,850 (53.5)	10,100 (58.5)	9,500 (62)
55					9,400 (31)	10,000 (42)	9,630 (49.5)	8,950 (54.5)	8,470 (58.5)
60					7,760 (20)	8,300 (36)	8,570 (45)	7,950 (51)	7,500 (55.5)

Stability

CHART 8.8 Ratings above the line are limited by the crane's structural strength. If these capacities are exceeded, the crane and its components may break. Those that are below the line are limited by the crane's stability; meaning the crane could tip over if capacities are exceeded.

Oper. Rad. Feet	Boom Ang. Deg.	Boom Point Elev. Feet	95,400 Lb. Crane Ctwt. 25,000 Lb. Bumper Ctwt.		78,400 Lb. (25,000 Lb. B
			Boom Capacity 360 Deg. Pounds	Boom Capacity Rear Pounds	Boom Capacity 360 Deg. Pounds
26	82.3	157.4	161,100*	161,100*	161,100*
28	81.5	157.1	155,100*	155,100*	155,100*
30	80.7	156.7	148,000*	148,000*	148,000*
32	80.0	156.4	138,900*	138,900*	138,900*
34	79.2	156.0	130,900*	130,900*	130,900*
36	78.4	155.5	123,900*	123,900*	123,900*
38	77.6	155.1	117,600*	117,600*	117,600*
40	76.8	154.6	112,000*	112,000*	111,700*
45	74.9	153.2	100,200*	100,200*	97,300*
50	72.9	151.7	90,700*	90,700*	83,500
55	70.8	150.0	81,500*	81,500*	72,200
60	68.8	148.1	72,000	73,300*	63,200
65	66.7	145.9	63,900	66,500*	55,900
70	64.6	143.6	57,100	60,600*	49,800
75	62.5	141.0	51,400	55,500*	44,700
80	60.3	138.1	46,600	51,000*	40,300
85	58.0	135.0	42,300	47,100*	36,500
90	55.7	131.7	38,700	43,600*	33,200
95	53.3	128.0	35,400	40,500*	30,300
100	50.9	123.9	32,500	37,700*	27,700

Structural

Stability

CHART 8.9 Values on a load chart are indicated with and without asterisks; those with asterisks are based on structural limitations that cause breakage, while those without are based on stability limitations causing tipping that leads an overturn if exceeded.

Structural Limits

These limits are based on the actual tested strength of the materials used to build the crane. The concern is in a component failure, such as the boom collapsing, or a wire rope hoist line breaking. Structural limitations are found in the upper sections of the load chart where shorter radii and wider boom angles are found. Structural limitations can be indicated in different ways, depending on the manufacturer. They can be indicated by appearing above a bold line, in a shaded area of the load chart, or marked using asterisks. Structural ratings are designed with larger safety margins when compared to stability ratings, in accordance with the ASME engineering standards.

Stability Limits

These limits are based on the crane's ability to maintain balance and avoid tipping. The primary concern is the crane's center of gravity, and preventing its load from moving beyond the tipping axis, or fulcrum, causing the machine to become unstable and overturn.

Stability ratings are found in the lower sections of the load chart where longer radii, lower boom angles can be found, such as below a bold line, not in a shaded area, or not marked by an asterisk, depending on the crane manufacturer. Stability ratings have a smaller margin of safety, making it critical to follow these limits. The rating is a percentage of the actual tipping load.

The crane's safe lifting capacity for any given configuration is the LOWER of the two limits for that specific lift or operation. The asterisks, shaded areas or bold lines on the load chart shows where the controlling factor shifts from structural strength to stability or vice versa. Crane operators must always check the load chart and totally understand which limitation is the governing factor for their planned lift. Regardless of the specific visual method used all modern load charts make this distinction to see to it crane operators know during operations if they are in a structural or stability area of chart during a lift.

A crane's structural strength is the machine's ability to handle loads without any of its components breaking or failing. Structural failure can include these problems:

- Boom bending or breakage
- Hoist line failure
- Hydraulic system or component failure

For areas of operations that are limited by stability, there must always be greater leverage on the crane side of the fulcrum or tipping axis than on the other side of the fulcrum where the load is located. Balance, leverage, and gravity are the key crane stability factors. Tipping incidents are the most common type of mobile crane accidents. Stability greatly depends on adequate ground conditions and proper crane setup. Best practices to improve crane stability can include these factors and must be considered by the crane operator when operating the machine:

- Avoid lifting loads of unknown weight.
- Properly deploy all outriggers on mobile cranes. Always use proper blocking under outrigger floats.
- Never exceed the charted capacities of the crane load chart.
- Understand the crane manufacturer's specifications and limitations.
- Use timber matting under the tracks of crawler cranes and blocking under the tips to add stability and spread ground pressure out over a larger area.
- Always use qualified and certified crane operators.
- Watch weather reports with high-wind warnings.
- Operate the crane slowly and smoothly to avoid inertia forces caused by fast acceleration and quick deceleration.
- The cranes stability is influenced by it's radius. A long radius with a heavy load can cause the crane to tip over if the weight is not evenly distributed while the crane is erected on firm, level ground and within 1% of true level.

According to ASME B30.5 Safety Standards for mobile crane as well as crane manufacturers specifications the maximum load rating percentage for wheel mounted cranes using outriggers fully extended and set with tires off supporting surface is 85%. Hydraulic systems on this type of crane allows the crane operator to easily raise the machine until the tires leave the ground. Using the hydraulic system on the cranes outrigger extensions and floats, the crane operator can raise or lower the machine to within 1% level, the same requirement for all mobile cranes including crawler cranes. The maxi-mum load rating for crawler cranes without outrigger support is 75%. Crawler cranes are mounted on rigged frames and tracks making the machine much harder to meet the within 1% requirement of being level. This is why manufacturer's load rating for wheel mounted cranes is lowered 15% for an 85% maximum load rating while crawler cranes are lowered 25%t for a 75% maximum load rating. It's all because the crawler cranes are much harder to get level in comparison to wheel-mounted cranes, which use outriggers to make themselves level with the assistance provided by an efficient hydraulic outrigger system helping the operator when setting the machine up. In many cases, the use of other heavy equipment may be necessary, such as bulldozers or motor graders, to build a crane pad that a crawler would sit on in order to make sure the crane stays within the 1% level grade requirement.

It is important to understand that the mobile crane can become unstable in certain conditions or configurations, causing the crane to be turned over backward. Most crane operators are aware the loss of stability can turn the crane over forward toward its boom point, but what about backward towards its counterweight? A mobile crane can be overturned backward due to several factors listed here:

- Excessive boom extension
- High boom angles
- Unstable ground conditions
- Bad weather and high winds
- Improper use of outriggers

Unbalanced rope reeving where the load is not evenly distributed on the boom can contribute to instability and increase the risk of a backward overturn. Backward stability for cranes has always been a serious risk and often never considered. There are more crane accidents due to loss of backward stability than there are for all the other causes combined. Fully understanding the crane's capacity chart will lower these risks.

FIGURE 8.10 The stability factor is a charted capacity or condition that protects the crane against loss of forward or backward stability. The crane shown here has lost stability.

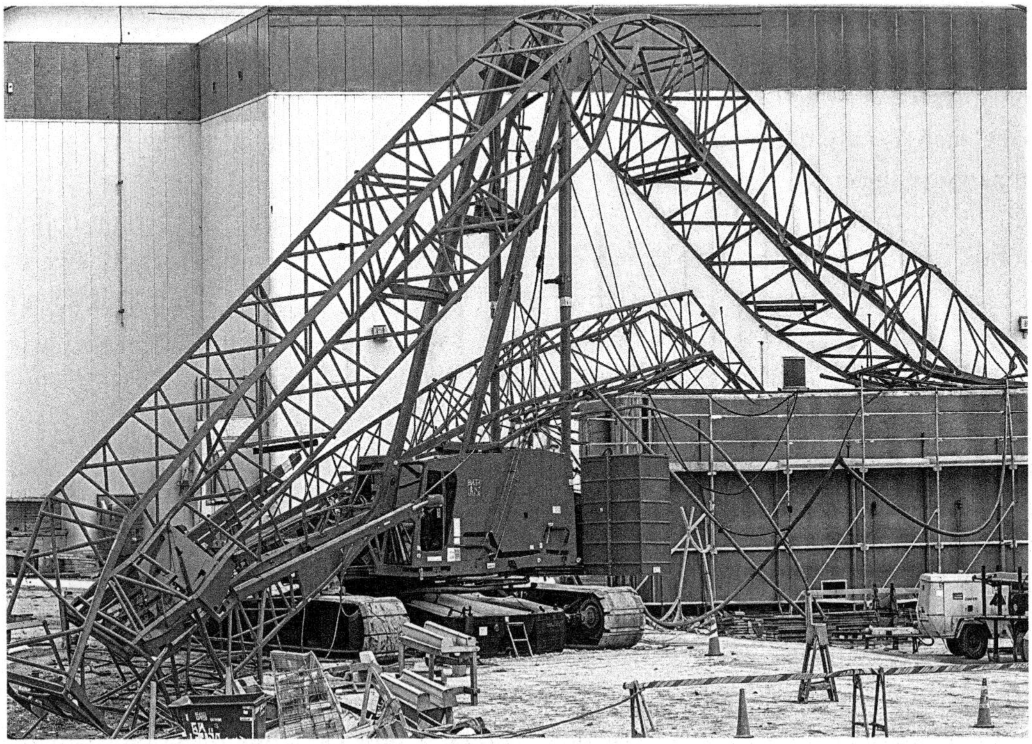

FIGURE 8.11 Structural competence is the ability of the machine and its components to withstand the stresses imposed by applied loads.

Compressive, tensile, bending, and buckling are four types of structural failure. Professional crane operators will always ask themselves these questions before operating the crane to avoid structural problems:

- Is there a jib stowed on the crane, and do I need it there for this operation?
- Are the tires properly inflated and in good condition when performing pick and carry operations?
- What is the configuration of the wire rope reeving on my crane?

Other important questions include:

- Is it possible to control any shock loading caused by rapid acceleration or sudden stopping of the load?
- Is it possible to control any side loading caused by load dragging or pulling of a load sideways?
- Has the crane been properly assembled and rigged?
- Is the crane properly set up on firm, solid ground within 1% level?
- During inspection, were any cracks in welds or any other signs of structural failure located?

One more consideration: Be very cautious when performing duty cycle operations. The forces generated by this type of work can cause side loading as well as higher temperatures in vital components such as hydraulic pumps, motors, clutches, hoses, and brakes. Most manufacturers require the load chart to be reduced by 20% when conducting duty cycle operations.

Determining Capacities Among Values

There are many types and sizes of mobile cranes in the industry today. In the cab or on the crane, the professional crane operator expects to find a load chart that includes several important specifications and the machine's gross capacities. Gross capacity is not the actual load that can be lifted. The weights of any jibs stowed on the boom or erected on the boom unused, including hook blocks, overhaul balls, rigging, and all load handling devices, must be subtracted from the gross capacity. These components are called *capacity deductions*. These items are listed in the load chart along with what they each weigh, including:

- Stowed jibs and attachments
- Overhaul balls and hook blocks
- Rigging, including any load handling devices
- Wire rope hoist line if directed by the manufacturer

The load chart includes other additional important information valuable to the crane operator. This includes boom length and radius measured in feet, as well as boom angle measured in degrees. Looking at Chart 8.10 as an example, the top row of values in cells refers to the working length of boom available for the crane. The values you see in the cells in the far-left vertical column represent the distance from the center of rotation of the crane to the center of gravity of the suspended load. This is known as the *operating radius*.

60 Feet | 65 Feet | 70 Feet

Radius in Feet	Main Boom Length in Feet								
	35	40	50	*60	70	80	90	100	110
9	©100,000 (67.5)								
9	90,000 (67.5)								
10	80,000 (66)	68,000 (69.5)	58,150 (74)						
12	67,400 (62)	63,800 (66)	55,450 (71.5)	44,600 (75)					
15	58,050 (56)	54,700 (61)	48,050 (67.5)	42,250 (71.5)	35,600 (74.5)	33,000 (77)			

CHART 8.10 Longer boom length between values example.

The numbers you see displayed to the right of the radius column are known as *gross capacities* of the crane. They are listed where boom length intersects with the radius distance. Below each gross capacity on the load chart, you will see numbers placed inside parentheses. These values represent the crane's gross capacity at a specific crane boom angle to the ground or the horizontal. Different manufacturers may use different symbols for boom angle values. To determine a gross capacity for a lift, you must first decide how much boom will be needed for it. For a quick example, let's use 50 feet of boom extended. Next, we need to know the distance from the center of rotation of the crane to the center of gravity of the suspended load, this will give us the radius of the machine. In this exercise, let's say we have a 12-foot radius row straight across the load chart until it intersects with the vertical column for 50 feet of boom. We find a value of 44,600 pounds. This represents the gross capacity of the crane with 60 feet of boom extended lifting the load at a 12 foot radius. Notice inside the parentheses the boom is positioned at 75 degrees.

Sometimes finding the gross capacity on a load chart isn't that easy. You will notice that manufacturers of cranes provide boom and radius values placed in cells in nice neat rows and columns at specific distances. The question is: What should a crane operator do if the length of boom they are dealing with is between the boom lengths listed on the load chart? Looking at Chart 8.10, for example, let's say the crane you are operating has its boom extended to 65 feet. As you can see, you are between boom length distances listed on the load chart. In this scenario, what is the correct thing to do? The answer is when the actual boom length you are operating is between values on the load chart, use the rating for the next-longer boom lengths. You are between ratings of 60 and 70 feet, so you would choose the next-longer rating of 70 feet. Please take a moment to notice that the gross capacity for 60 feet of boom at a 15-foot radius is 42,250 pounds and then again at 70 feet of boom and a 15-foot radius is 35,600 pounds. What is important to realize is that the values to the right are lesser capacities. Manufacturers direct us to do this with all mobile cranes because by doing so, we are building a safety margin into our calculations from the very start.

Next let's imagine the crane you are operating is properly set up and ready to go to work. In preparation for the next lift, you measure the radius from the crane to be 22 feet from the center of rotation of the machine as shown in Chart 8.11. When you go to the crane's load chart, you discover the distance of 22 feet is between listings of 20 and 25 feet. Manufacturers will direct you when the actual radius is between those listed on the load chart. You must use the rating for the next-longer radius listed. In this case, you should choose 25 feet. Let's take a look at the numbers. As you can see, the gross capacity for this crane is less at the 25-foot radius listing of 35,200 pounds than the gross capacity listed at 20 feet 44,950 pounds. As required per manufacturer's specifications, we must choose the longer radius listing when between values on the load chart. This will start us out with a smaller number as we start our calculations.

The left side of a crane's load chart represents shorter radii. The gross capacities are greater, allowing more weight to be lifted due to the load being closer to the crane's center of rotation. As the radius increases, and you move to the right on the load chart, the gross capacities become smaller, representing a reduced charted capacity of the crane.

Now you can see that by going to the longer boom length, it sets you up to select a smaller value on the load chart than the numbers to the right. This helps design a safety factor into the lift immediately, from the very start of your calculations.

Next, let's imagine the crane you are operating is properly set up and ready to go to work. In preparation for the lift, you measure the radius on your crane to be 22 feet, as shown in Chart 8.11. You go to the load chart and discover the distance is between listings of 20 and 25 feet. When the actual radius being used is between those on the load chart, use the rating for the next-longer radius listed. In this case, you should choose 25 feet. Once again, by choosing the larger value, you are building a safety margin into the lift that you are planning.

- **Radius.** If the actual working radius is between those listed, use the capacity for the next-longer radius on the load chart.
- **Boom Length.** If the actual boom length is between those listed on the load chart, use the capacity associated with either the next longer or shorter boom length, whichever is less.
- **Lower Boom Angle.** Mobile cranes are equipped with boom angle indicators. If the boom angle reading is between values on the load chart, use the next-lower value. The principle of using the lower boom angle capacity is a safety measure. It ensures that the crane's charted lifting capacity is not exceeded.

ON OUTRIGGERS FULLY EXTENDED - 360°

Radius in Feet	#0001 Main Boom Length in Feet						
	35	40	50	*60	70	80	90
9	90,000 (67.5)						
10	80,000 (66)	68,000 (69.5)	58,150 (74)				
12	67,400 (62)	63,800 (66)	55,450 (71.5)	44,600 (75)			
15	58,050 (56)	54,700 (61)	48,050 (67.5)	42.250 (71.5)	35,600 (74.5)	33,000 (77)	
20	44,950 (44.5)	43,000 (52)	39,400 (61)	33,600 (66.5)	30,500 (70.5)	28,000 (73.5)	25,500 (75.5)
25	35,200 (28.5)	35,050 (41.5)	32,400 (54)	27,750 (61)	25,200 (66)	23,800 (69.5)	22,000 (72)
30		28,400 (27)	27,150 (46)	23,350 (55)	21,100 (61)	20,400 (65.5)	19,300 (68.5)
35			21,650 (36.5)	20,000 (49)	18,050 (56)	17,400 (61.5)	16,400 (65)
40	See Note 16		17,200 (24)	16,850 (41.5)	15,650 (50.5)	15,050 (57)	14,150 (61.5)
45				13,350 (33)	13,600 (45)	13,200 (52.5)	12,350 (57.5)
50				10,650 (21.5)	11,450 (38.5)	11,600 (47.5)	10,850 (53.5)
55					9,400 (31)	10,000 (42)	9,630 (49.5)

20 feet
22 feet
25 feet
Left=13,360
Right=13,600

CHART 8.11 Longer radius between values example.

As you make crane boom calculations during your career, and you find that you are between the boom lengths and values on the load chart, you will almost always find the lesser or smaller value for gross capacity by moving to the right on the load chart. However, very rarely you might find a smaller capacity moving to the left on the chart as shown in Chart 8.11. Please look at the values listed on the chart for both 60 and 70 feet of boom extended at a 45-foot radius. It is important to note the gross capacity for 60 feet of boom is 13,350 pounds, while the capacity for 70 feet of boom is 13,600 pounds. This is a rare occasion of finding the lesser value listed on the chart to the left rather than the right. This is due to complexities of the manufacturer's engineering. Regardless, if you ever find a smaller value listed on the left than on the right, we know that we must use the lesser value.

In some cases, a mobile crane's load chart may show a lesser load capacity on the left, closer to the crane, compared to the right being further away from the crane. This is due to the crane's charted capacities decreasing as the radius increases. A longer radius requires the length of the boom to be extended and causes the crane to work harder to lift the same load weight and ultimately reduces the load capacity as the machine reaches its tipping point. Realistically the left of a load chart represents lifts closer to the crane where it has its greatest charted capacity. Lifts represented on the right side have longer radii where the crane has its lowest lift capacities.

Previously in this chapter, we discovered how manufacturers of cranes indicate on the load chart the area based on structural integrity or stability by using bold lines, shaded area, or asterisks. On a mobile crane load chart, the lower gross capacity (or maximum weight) a crane can lift is found on the left compared to the right. This is due to the cranes stability when working at closer radius from the crane's center of rotation to the center of gravity of the suspended load or its radius. As the crane moves the load close to itself or the carrier, the weight distribution becomes more balanced, allowing the machine to lift heavier loads without compromising its stability. The higher capacities to the right listed on the load chart indicate the crane's structural integrity allowing it to lift heavier loads without much anxiety.

From time to time, a professional crane operator will use the boom angle indicator on the crane to plan work or calculate capacities. This is done by reading a pendulum attached to the crane boom or by reading the load moment indicator, an onboard computer display on many cranes. Boom angles can fall between the listings on the load chart as well. When the actual boom angle being used is between those shown on the load chart, use the rating for the next-lower boom angle. Using Chart 8.12 as an example, the boom angle indicator on your crane reads 68 degrees. You are between listings of 70 degrees and 66 degrees on your load chart. In this case, you would choose a boom angle of 66 degrees.

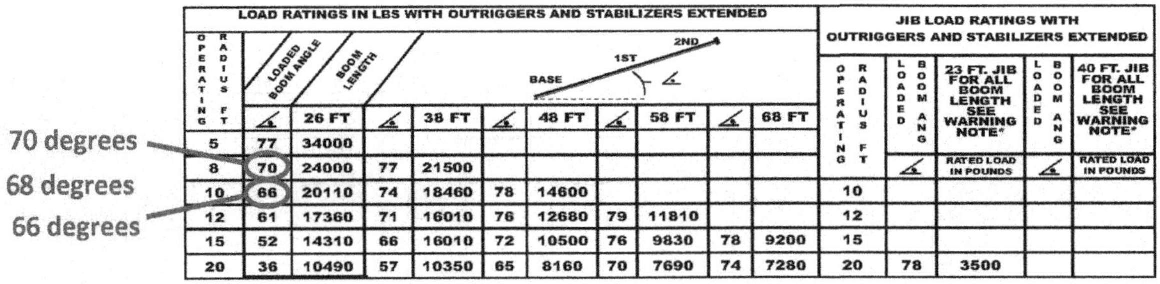

LOAD RATINGS IN LBS WITH OUTRIGGERS AND STABILIZERS EXTENDED										
OPERATING RADIUS FT	LOADED BOOM ANGLE ∠	BOOM LENGTH 26 FT	∠	38 FT	∠	48 FT	∠	58 FT	∠	68 FT
5	77	34000								
8	70	24000	77	21500						
10	66	20110	74	18460	78	14600				
12	61	17360	71	16010	76	12680	79	11810		
15	52	14310	66	16010	72	10500	76	9830	78	9200
20	36	10490	57	10350	65	8160	70	7690	74	7280

JIB LOAD RATINGS WITH OUTRIGGERS AND STABILIZERS EXTENDED				
OPERATING RADIUS FT	LOADED BOOM ANG ∠	23 FT. JIB FOR ALL BOOM LENGTH SEE WARNING NOTE* RATED LOAD IN POUNDS	LOADED BOOM ANG ∠	40 FT. JIB FOR ALL BOOM LENGTH SEE WARNING NOTE* RATED LOAD IN POUNDS
10				
12				
15				
20	78	3500		

CHART 8.12 Example of boom angles between values.

FIGURE 8.12 A crane that failed to lift a swimming pool from a backyard and ended up splitting the roof of the house in half.

Conditions for Calculating Crane Capacities

Here are eight things to know when calculating crane capacities:

1. Rated lifting capacities must not be exceeded. They are based on freely suspended loads when the machine is traveling and when it is standing on a firm supporting surface. Ratings must include outriggers being extended to their maximum position, and tires raised free of crane weight before extending the boom or lifting loads.
2. Gross capacity is obtained from the load chart using boom configurations and the operating radius for the quadrant of operations the crane is working. Load charts provided specifically for each mobile crane are the supreme source and most valid way to determine the safe lifting capacity established on the operating conditions, configuration, and setup of the machine. Crane operators should not use the working load limit (WLL) of hoist lines to determine gross loads. WLL represents the breaking strength of the hoist rope itself and does not consider into the equation factors like the crane's comprehensive stability and boom structural integrity.

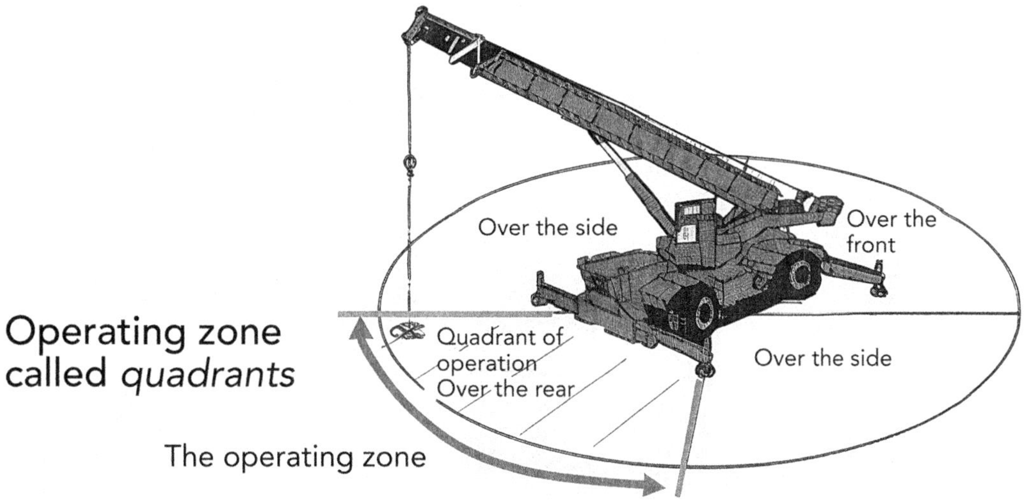

FIGURE 8.13 The load carrying capacity of the crane varies according to the crane's quadrants.

Remember, the gross capacity of a crane represents the maximum weight it can lift, and that's under ideal conditions with the crane being set up on firm ground that's level, accurate to within 1% with no wind or rain—essentially a clear, beautiful day. Now ask yourself: In the real world, how often do cranes operate in ideal conditions? This is why the charted capacities of a crane should never be exceeded. You must also consider the condition and age of the crane. Has it been regularly serviced, maintained correctly every year? Has it been pushed to its limits too many times?

So, should you ever exceed your crane's weight capacities? The answer is no!

3. All load handling devices and boom attachments are considered part of the load, and suitable allowances must be made for their combined weights.
4. Boom extensions and jibs are *not* considered to be deductions from the crane's capacity if a lift is being made over them.

FIGURE 8.14 Poor setup was the cause of this accident. The outrigger pad sank into the ground, causing this crane worth $3.2 million to turn over.

FIGURE 8.15 The number-one cause of crawler cranes to overturn is operating on an unstable supporting surface that isn't level, which is often the result of inadequate ground preparation and planning.

5. When calculating capacities and when the actual boom length being used is between those listed on the crane's load chart, you must use the capacity listed for the next-longer boom length shown.
6. When the actual working radius being used is between those listed on the load chart, you must use the capacity listed for the next-longer radius shown.
7. When the actual boom angle being used is between those listed on the load chart, you must use the capacity listed for the next-lower boom angle shown.
8. The maximum net capacity that can be listed is the result of the gross capacity minus all load handling devices and attachments.

Definitions the crane operator must know and understand include:

- **Gross load.** The sum of all the components of the load.
- **Net load.** The weight of the load itself.
- **Net capacity.** The maximum that the load itself could weigh.
- **Gross capacity.** The capacity shown on the load chart.
- **Radius.** The distance from the center of rotation of the crane to the center of gravity of the suspended load.
- **Static load.** The weight of the stationary gross load.

Steps for Determining the Net Capacity of Cranes

The following are steps for making a drawing to determine the net capacity of cranes (Figure 8.16):

1. **Boom configuration.** Draw a horizontal line at a slight angle to represent the boom length of your crane. Label the boom length in feet.
2. **Radius.** Add a horizontal line to the base of your boom line to represent the maximum operating radius where the load is to be lifted from or placed. Label it in feet.
3. **Boom angle.** Draw an arc from the boom line to the ground to indicate the boom angle in degrees.
4. **Outriggers and operating quadrants.** Draw a triangle to indicate outriggers fully extended and set. Below the triangle, identify the quadrants of operation to be used during the lift, such as "front," "rear," "side," or "360 degrees."
5. **Deductions.** Indicate the weights of all items involved in the actual lift. Draw in all items attached to the boom, such as extension jibs, either erected or stowed. Include the auxiliary boom heads, overhaul balls, hook blocks, wire rope, rigging, and any load handling devices, such as spreader bars or equalizer beams.
6. **Procedure.** Once the drawing is complete, select the gross capacity from the load chart that matches the information on your drawing, such as boom length, radius, or boom angle. Total and label all of your deductions, and keep your math neat and clean.

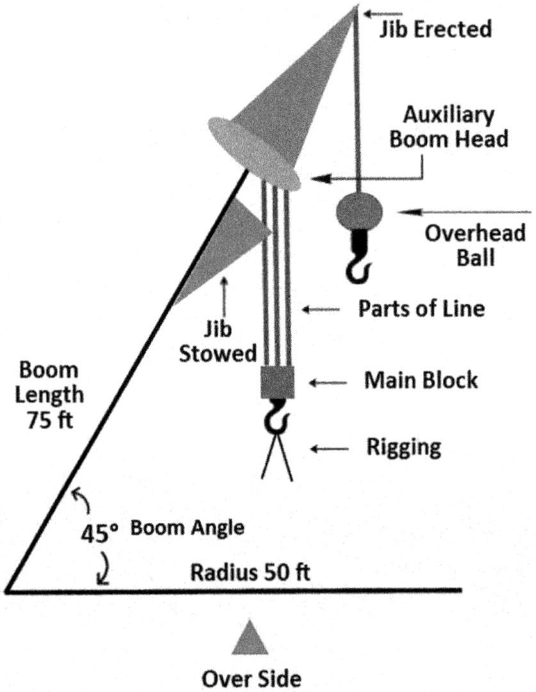

YOUR NET CAPACITY WILL BE THE RESULT OF THE GROSS CAPACITY LESS DEDUCTIONS

FIGURE 8.16 How to draw a model crane.

Learning how to be efficient when determining crane capacities is an important step on the road to becoming a professional crane operator. Assembled for you are a series of mobile crane load chart scenario questions grouped into three separate crane classifications:

- Telescopic boom crane fixed cab (TSS)
- Telescopic boom crane swing cab (TLL)
- Lattice boom crane (LAT)

These questions will help you build your knowledge and skills when interpreting mobile crane load charts. You will find six questions crafted for each crane type. First, you will need to download the load charts for each crane, providing you necessary information and important data needed to correctly answer each load chart question. You can locate the three load charts needed by visiting the National Commission for the Certification of Crane Operator's website (https//www.nccco.org). Search for mobile crane load charts to find all three documents. You can download all three free of charge. With load charts in hand, along with following the steps for determining the net capacity of mobile cranes, you are prepared to solve any of the questions crafted for you in your book. We recommend starting with the questions and load chart for telescopic boom crane (TSS). Before getting started, take the time to thoroughly review your fixed cab load chart information as well as the questions you want to solve. Here are some tips to help you get started:

1. Carefully read the question, and have the correct load chart close at hand.
2. Begin reading the information provided in each cell of matrix located above the questions. Always start at the top left hand side of the matrix and work downward, carefully reviewing each cell for information on how the crane is configured. Imagine that each matrix cell is numbered step one, step two, three, four and so on. Draw your stick model crane using

one cell at a time. Much like building a model airplane or model car, you are building a stick model crane using the information found in the matrix cells.

3. Work down the left side of the matrix one cell at a time, continuing to the top right cell and down through the remaining cells, drawing your stick model crane as shown in Figure 8.17.

1	Outriggers	Fully Extended & Set	Extension/Jib	40 ft Erected	9
2	Counterweight	N/A	Extension/Jib Offset	N/A	10
3	Tires	N/A	Block	N/A	11
4	Crawlers	N/A	Overhaul Ball	5 Ton	12
5	Main Boom	48 ft	Rigging	50 lbs	13
6	Main Boom Angle	75	Wire Rope (Main Hoist)	9/16" Rotation Resistant	14
7	Radius	N/A	Wire Rope (Auxiliary)	N/A	15
8	Other	Hose Reel Installed	Other	N/A	16

FIGURE 8.17 Draw your stick model crane using the information in the data table matrix cells one step at a time, as if the cells were numbered 1 through 16, and the steps are leading you to build a sample crane.

4. With your stick model crane complete, your next step will be to go to the load chart and select the gross capacity that matches the information on your drawing, such as boom or jib length, radius or boom angle, as shown in Figure 8.18.

Gross Capacity	=	1,690
Deductions:		
Hose Reel	=	190
Overhaul Ball	=	120
Rigging	=	50
Net Capacity	=	1,330

FIGURE 8.18 Deductions for this crane are shown in the load chart data with each component's weight.

5. The first time you went through the matrix, you carefully looked at each cell. Based on data in each cell you were able to draw out the crane as it was configured in the question. Now it's time to go through the matrix one more time in the same fashion as before, looking closely at each cell. This time, you are looking for any items that would be considered a deduction, such as stowed jibs, overhaul balls, hook blocks, or rigging. As you look at each cell, develop a labeled list of deductions below your gross capacity. The weight of these items can be found in the load chart data as shown in Figure 8.20. Other items may have their weight listed in the matrix, such as rigging 50 lbs, as shown in Figure 8.17.

LOAD RATINGS IN LBS WITH OUTRIGGERS AND STABILIZERS EXTENDED											JIB LOAD RATINGS WITH OUTRIGGERS AND STABILIZERS EXTENDED				
OPERATING RADIUS FT	LOADED BOOM ANGLE ∠	BOOM LENGTH 26 FT	∠	38 FT	∠	48 FT	∠	58 FT	∠	68 FT	OPERATING RADIUS FT	LOADED BOOM ANG ∠	23 FT. JIB FOR ALL BOOM LENGTH SEE WARNING NOTE* RATED LOAD IN POUNDS	LOADED BOOM ANG ∠	40 FT. JIB FOR ALL BOOM LENGTH SEE WARNING NOTE* RATED LOAD IN POUNDS
5	77	34000													
8	70	24000	77	21500											
10	66	20110	74	18460	78	14600					10				
12	61	17360	71	16010	76	12680	79	11810			12				
15	52	14310	66	13320	72	10500	76	9830	78	9200	15				
20	36	10490	57	10350	65	8160	70	7690	74	7280	20	78	3500		
25			47	8310	58	6570	65	6240	70	5910	25	75	3060	78	1940
30			36	6610	51	5420	59	5200	65	4940	30	72	2700	75	1690
35			17	4560	42	4470	53	4390	60	4200	35	69	2400	72	1490
40					32	3580	47	3720	55	3610	40	65	2150	70	1320
45					16	2380	39	3120	50	3110	45	62	1950	67	1180
50							29	2500	43	2670	50	58	1770	64	1060
55							15	1610	36	2240	55	55	1550	61	960
60									28	1780	60	50	1350	58	870
65									14	1400	65	46	1160	55	790
70											70	41	1010	51	730
75											75	36	830	48	670
80											80			44	610
85											85			40	570
90											90			35	520
95											95				
100											100				
	480 LBS.		330 LBS.		260 LBS.		220 LBS.		190 LBS.		DEDUCTIONS FOR STOWED JIB				

FIGURE 8.19 Shown in the fixed cab load chart is the gross capacity of the crane on the 40 ft jib at a boom angle of 75 degrees. The gross capacity is 1,690 lbs in this configuration.

Make your calculations neat and orderly. Gross capacity goes on top with deductions listed in Figure 8.18.

Deductions				
Auxiliary block	50 lb	22.68 kg	Swing-around jib (stowed)	See load rating chart
Overhaul ball	120 lb	54.43 kg		
Single-sheave load block	260 lb	117 kg	Warning: Lifting off the main book point while the swing-around jib is erected is not intended or approved.	
Double-sheave load block	360 lb	158.76 kg		
Hose reel	190 lb	86.18 kg		

FIGURE 8.20 List your labeled deductions below the gross capacity of the crane found in the load chart data.

Always label your deductions in the order they are found in the matrix cells. Never skip a cell, and list deductions in exact order as found in the matrix. Your last step is to total all deductions and subtract that number from the gross capacity, giving you the net capacity as displayed in the sample.

This is a simple, foolproof way of always coming up with the correct calculations for your crane. Regardless of the crane's manufacturer, color, age, configuration, or capacity, this system will work for you. This innovative, but simple, method will work regardless of whether computations are made out in the field, calculated on a napkin from your lunchbox, or in a testing environment while taking a written examination for crane operator certification. This simple process does work, so practice it over and over until you are totally proficient. Solutions for the load chart scenario questions can be found in the answers section in the back of the book. If you would like more of these types of questions to work with, please visit https://samplecranetesting.com and click on the homework study center link. Video instruction is also available upon request.

Load Moment Indicator Systems

Load moment indicators are installed on today's cranes as part of the original equipment more often than ever before. Older machines may have them installed by the owner as a means of safety enhancements. These systems complement the crane operator's skill by providing the information needed to make the complicated lifts of today and displaying it to the crane operator in the cab on a screen or standing at the controls. Regardless of whether your crane is equipped with a load moment indicator or not, you are required as a crane operator to understand and use your load capacity chart. These systems are great tools for the crane operator and offer a continuous reading of the crane's rated capacity as well as other valuable information. If hazardous conditions are approached, the LMI system will warn the operator by sounding an audible alarm or flashing a warning light. The system is calibrated to the load chart of the crane and gives actual readings of boom length, radius, boom angle, and weight of the load to be lifted, including weight of the rigging. The LMI is designed to give the crane operator the information necessary to fully utilize the crane's capacity while keeping within the working limits of the crane in all its configurations.

There are several manufacturers of LMI systems (Figure 8.21). It is important you are familiar with the functions of your unit and know how to properly use it. Information includes data such as main boom length, main boom angle, jib angle, jib length, weight of the load on the hook, line speed, wind speed, total weight being lifted by the crane, and more. The LMI is an electronic device that will need servicing from time to time. It is an operational aid that warns a crane operator of approaching overload conditions that could cause damage to equipment and ground personnel. LMI devices are not a substitute for a well-trained and qualified crane operator's judgment and experience. If the LMI on your crane stops working, tell your supervisor immediately. If you both agree that you can still operate the crane, you can continue to use the crane by making calculations using the load chart.

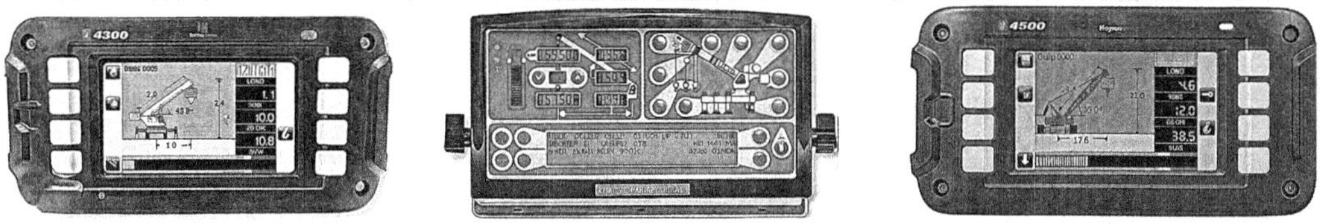

FIGURE 8.21 Views from several different load moment indicator systems.

Other Moments and Forces Equations for Your Reference

- Load moment, also known as *tipping moment*, is the moment attempting to tip the crane over. The primary equation for the load moment (M load) around the tipping axis is:

  M load = total suspended load weight × load radius

  Total suspended load weight includes the payload itself plus the weight of the hook block, overhaul balls, stowed jibs or extensions, auxiliary boom heads, slings, rigging hardware, and any other load handling devices or attachments.

  Load radius is the horizontal distance from the center of rotation to the center of gravity of the suspended load.

- Restoring moment, also known as *counteracting moment*, is the moment provided by the crane's own weight and counterweight that keeps the crane stable. The exact calculation is complex and involves totaling the moments of every major component, including the chassis, counter weights, boom, and other parts acting on the opposite side of the tipping axis:

  M restoring = Σ (component weight × horizontal distance from tipping axis to component's center of gravity)

  Note: Moment example calculation for a crane's load moment. In this exercise, the crane is lifting a load while stationary, so only static forces are considered.

  The crane is set up on firm level ground with all four of its outriggers fully extended and set. The crane is within 1% level. You are the crane operator preparing to make this lift, a large concrete pipe that weighs 10,000 lbs and is sitting at a radius of 40 feet from your crane.

  Calculation = load moment (M load) = W load × R load

M load = 10,000 lbs × 40 ft
M load = 400,000 foot-pounds (ft-lbs)

Interpretation and Answer of Exercise

- The tipping or load moment trying to overturn the crane is 400,000 ft-lbs.
- For the crane to be stable, its restoring moment created by the crane's own weight and counter weights must be significantly greater than 400,000 lbs.
- If the crane's maximum rated capacity at a 40-foot radius was, for example, 8,000 lbs, attempting this lift would be unsafe, because the calculated load moment of 400,000 ft-lbs would exceed the safe operating limits, leading to potential instability.
- The restoring moment in this example calculation is 700,000 ft-lbs, ensuring a stable lift when compared to the 400,000 ft-lbs tipping or load moment.

How to Calculate Restoring Moment

Step 1: Define Assumed Crane Components

Step 2: Calculate Component Moments

To calculate the restoring moment, we assume a simplified crane configuration with the load positioned over the extended right outrigger beam with the fulcrum established under the float touching the ground. We consider the counterweight and main chassis as the primary components providing the stabilizing force on the opposite side.

- Counterweight (W cw): 30,000 lbs
- Counterweight horizontal distance (R cw): 15 feet left of the fulcrum or tipping axis
- Chassis weight (W chassis): 50,000 lbs
- Chassis horizontal distance (R chassis): 5 feet left of the fulcrum or tipping axis

Step 3: Sum Moments and Compare

Totaling the individual moments gives the total restoring moment:

M restoring = M cw + M chassis = 400,000 ft-lbs + 250,000 ft-lbs
M restoring = 700,000 ft-lbs

Interpretation and Answer of Exercise

The total restoring moment is 700,000 ft-lbs. Comparing this to the previous tipping or load moment calculation of 400,000 ft-lbs shows a healthy safety factor of 1.75, meaning the crane is statically stable for this lift.

Net moment is the difference between the overturning moment and the load moment. If the load moment exceeds the overturning moment, the crane tips.

Boom moment is the weight of the crane boom itself plus its jibs or extensions. It creates its own moment that the crane must support, influencing overall capacity.

Other Important Equations to Know

How to Calculate Wind Loading

Strong winds can add significant force to both the cranes' structure and the suspended load, increasing the tipping moment and potentially causing the load to sway uncontrollably. Wind exerts pressure on exposed surfaces. The force can be estimated using this engineering formula:

$$F\ wind = 0.5 \times C \times p \times v^2 \times A$$

C = wind resistance or drag factor

p (rho) = air density

v = wind speed; (note that pressure increases with the square of speed)

A = projected surface area exposed to the wind. This force then creates an additional tipping moment by being multiplied by its lever arm to the tipping edge.

How to Calculate Dynamic Loads

These are additional forces resulting from movement such as sudden braking or acceleration. They are typically accounted for in safety factors, often 1.2 for the load weight built into engineering standards rather than a simple field formula.

For safe lift, the condition provided here must be met, usually a built-in safety margin (stability factor ≥ 1), often around 1.4 for mobile cranes.

$$\text{Restoring moment} > \text{load moment} + \text{wind moment} + \text{dynamic moment}$$

Chapter 8 Review Quiz

1. Load tests that are conducted on mobile cranes must not exceed
   a. 85%
   b. 100%
   c. 75%
   d. 65%

2. The crane's load rating chart must be
   a. available with the master rigger
   b. available with the lift director
   c. accessible to the crane operator
   d. kept in the operator's manual

3. A crane boom should never be used to
   a. make a lift on rubber
   b. make a lift under water
   c. make a lift under ground level
   d. drag a load sideways

4. The capacities shown in the crane's load chart are
   a. the gross capacities
   b. the deducted capacities
   c. the weights of the loads to be lifted
   d. the net capacities

5. Capacities listed above the solid line on the load chart are based on
   a. the SWL of the hoist line being used
   b. the crane's stability factors
   c. the crane's net capacities
   d. the crane's structural limitations

6. Specific details regarding the effective weights of attached crane components such as jibs are found in
   a. the load chart data
   b. the pre-operational checklist
   c. the operator's manual
   d. the rigging handbook

7. When the actual boom length being used falls between those listed on the load chart, use the capacity listed for
   a. the next-longer boom length listed
   b. the next-shorter boom length listed
   c. an estimated length between those listed on the chart
   d. the next-longer or next-shorter boom length, whichever is less

8. Increasing the number of parts of line to the crane's hoist block
   a. increases line pull
   b. increases the lifting speed
   c. increases the mechanical advantage
   d. decreases line pull

9. Deflection of the crane boom when making a lift will increase
   a. the load radius
   b. the stability of the crane
   c. the effective weight of the boom
   d. the boom angle

10. When a jib is erected but not used for the lift, the operator must make a deduction from
   a. the actual weight of the jib
   b. the interpolated values of the jib and the boom
   c. the effective weight of the jib
   d. the difference between the jib's actual and effective weights

11. When the jib is erected from the stowed position and not being used, its effective weight
   a. is deflected
   b. remains the same
   c. decreases
   d. increases

12. The side loading of the crane boom is permitted
   a. when directed by the lift director
   b. as long as there is no jib installed
   c. when all outriggers are fully extended
   d. only with freely suspended loads

13. The stability safety margin for crawler cranes is greater than for truck-mounted cranes with outriggers because
   a. they can reach a longer radius from the center of rotation
   b. they have a longer tipping axis
   c. they are harder to get level
   d. they have higher undertrack ground pressures

14. When making a lift with a crawler crane, ground pressures would be highest when
   a. lifting over the ends of the tracks
   b. lifting over the rear of the crane
   c. lifting over the corner of the tracks
   d. lifting over the side of the crane

15. Mobile crane stability is based on the
- **a.** position of the boom and the load
- **b.** ability of the crane to lift the load
- **c.** principles of leverage
- **d.** proper outrigging deployment

16. As you boom down, the leverage of the load on your mobile crane will
- **a.** equalize ground pressure
- **b.** decrease
- **c.** stay the same
- **d.** increase

17. When operating your crane, rapid acceleration or sudden stopping of the load would
- **a.** produce excessive ground pressure
- **b.** make sure the crane is level
- **c.** produce load forces greater than the load's actual weight
- **d.** produce less ground pressure

18. When is the rated capacity of the rigging or attachment a limiting factor for a lift?
- **a.** It is never a limiting factor.
- **b.** When the rated working load exceeds the crane's capacity.
- **c.** When the rated working load is less than the crane's capacity.
- **d.** When you are required to interpolate between values.

19. When the crane's gross capacity is between two stated values on the load chart, always
- **a.** interpolate values
- **b.** use the next-higher value
- **c.** use the next-lower value
- **d.** use the average between values

20. When the crane is equipped with a manual boom extension that isn't being used, the operator must
- **a.** use the gross capacity listed on the load chart
- **b.** use the net capacity listed on the load chart
- **c.** deduct the weight of the extension from the crane's gross capacity
- **d.** deduct the weight of the extension from the crane's net capacity

21. When an all-terrain crane makes a lift off the ground on rubber, what effect does it have on the crane?
- **a.** The boom angle will increase.
- **b.** The boom angle will decrease.
- **c.** The boom angle will stay the same.
- **d.** The load radius will decrease.

22. You have just changed the jib angle to the ground on your crane from 65 degrees to 35 degrees. You have now

- **a.** increased the capacity of the jib
- **b.** decreased the capacity of the jib
- **c.** increased the length of boom
- **d.** decreased the length of boom

23. You are operating a lattice boom crawler crane. The boom angle on your load chart is measured between the

- **a.** loaded jib and the horizontal
- **b.** unloaded jib and the top of the tracks
- **c.** loaded boom and the horizontal
- **d.** loaded boom and the vertical

24. The large telescopic boom crane you are operating is set up on firm, level ground with outriggers fully extended and set. You are operating with 70 feet of boom at a 53-foot radius. Your load chart indicates a capacity of 11,950 pounds at a 50-foot radius, followed by a capacity of 10,500 pounds at a 55-foot radius. What is your gross capacity at the 53-foot radius?

- **a.** 10,500 pounds
- **b.** 11,950 pounds
- **c.** 13,600 pounds
- **d.** 15,650 pound

25. What is the net capacity of a truck crane with 40 feet of boom extended, unless the manufacturer indicates otherwise?

- **a.** The capacity indicated on the load chart and subtracting the weight of any rigging, jibs, balls, and attachments
- **b.** The capacity indicated on the load chart with the addition of any rigging, jibs, balls, and attachments
- **c.** The capacity indicated on the chart and no adjustments are necessary
- **d.** The capacity indicated on the load chart and subtracting the weight of the overhaul

Now the fun begins. Over the next several pages, we have provided six calculating load chart scenario questions for each of the three mobile cranes currently offered by NCCCO in their specialty written examinations. An NCCCO mobile crane specialty examination is a specific written test that crane operators must pass to demonstrate their knowledge and skills related to the operation of a particular type of mobile crane, such as telescopic boom cranes, both fixed and swing cabs, and including lattice boom cranes. The specialty written examinations complement another written examination known as the *core*. The core examination covers general knowledge crane operation principles. Please visit www.nccco.org to review more about the certification process.

Next you can choose the load chart and questions of your choice or interest, or do them all! The solutions are in the back of your book for your convenience. Visit our new website specifically designed for the owners of this book at https://samplecranetesting.com and then click on the homework study center link.

The website provides many more multiple-choice written examinations to challenge your knowledge. These additional questions support our chapter contents. Also, you will find 20 more calculating load chart scenario questions like those you are about to complete for more practice that you will need towards becoming confident and efficient with mobile load charts. The solutions for all of the 60 additional load chart questions are provided along with lots of other valuable training materials. Invite your coworkers to get their own copy of our book. Challenge them to test their knowledge. Safety directors, lift directors, crane operators, superintendents, your jobsite foreman, and the owner are all invited to benefit from our new book and website. Soon your entire crew will be working at the highest level of safety, production will increase, jobsite morale will be at an all-time high, workplace injuries will decrease, and most importantly, lives will be saved.

So let's get started by downloading your load charts by visiting www.nccco.org/mcocharts0922.

Remember to read the crane manufacturer's notes and specifications in detail. It is fine to use a calculator to make your calculations. Use all the knowledge and steps you have learned, and build confidence as you work.

Telescopic Boom Crane Fixed Cab

A mobile crane is classified as a telescopic boom fixed cab crane (TSS) if it has a non-rotating operator's station or cab, and a telescoping boom. This means the operator's seat and control panel are stationary, not rotating with the crane's upper structure, and the boom extends or retracts hydraulically.

Note: Before you get started, some mobile crane manufacturers deduct the weight of the hoist line from the crane's load capacity while others do not. Some manufacturers choose to list the load capacity, including the hoist line weight to simplify their load charts. Since deductions for hoist lines vary from one crane manufacturer to any other, it is vital to read all the rules and specifications listed in the load chart for the crane you are operating before any calculations are made and prior to putting the crane to work. If the manufacturer doesn't mention hoist line weight, then don't deduct the hoist lines, and try not to overthink the questions or add steps or deductions the manufacturer doesn't instruct you to. For fixed cab cranes, and when multiple sections of line are required, pay particularly close attention to when the axillary boom head needs to be installed and when it doesn't. Remember, the axillary boom head will be a deduction.

FIGURE 8.22 Telescopic boom crane "fixed cab" with boom extended.

Telescopic Boom Crane - Fixed Cab (TSS)

Outriggers	Fully Extended & Set	**Extension /Jib**	23 – 40 ft Jib Stowed
Counterweight	NA	**Extension /Jib Offset**	NA
Tires	NA	**Block**	NA
Crawlers	NA	**Overhaul Ball**	5 TON
Main Boom	Fully Extended	**Rigging**	10 lbs
Main Boom Angle	NA	**Wire Rope (main hoist)**	9/16" Rotation Resistant
Radius	MAXIMUM	**Wire Rope (auxiliary)**	NA
Other	NA	**Other**	NA

Note: For line in use, when wire rope deductions are required, the rope length shall be measured from the boom tip to the ground.

1. Your net capacity at maximum operating radius over the "side" using the necessary hook block could be?

A. 1400 lbs

B. 790 lbs

C. 1080 lbs

D. 2880 lbs

Use for Workspace

Telescopic Boom Crane - Fixed Cab (TSS)

Outriggers	Fully Extended & Set	**Extension /Jib**	40 ft Erected
Counterweight	NA	**Extension /Jib Offset**	NA
Tires	NA	**Block**	NA
Crawlers	NA	**Overhaul Ball**	5 - Ton
Main Boom	48 ft	**Rigging**	50 lbs
Main Boom Angle	75	**Wire Rope (main hoist)**	9/16" Rotation Resistant
Radius	NA	**Wire Rope (auxiliary)**	NA
Other	Hose Reel Installed	**Other**	NA

Note: For line in use, when wire rope deductions are required, the rope length shall be measured from the boom tip to the ground.

2. What is the net capacity when a lift is made on the extension over the "side" of the crane?

A. 1380 lbs

B. 1520 lbs

C. 1330 lbs

D. 1690 lbs

Use for Workspace

Telescopic Boom Crane - Fixed Cab (TSS)

Outriggers	Fully Extended & Set	**Extension /Jib**	23 – 40 ft Jib Stowed
Counterweight	NA	**Extension /Jib Offset**	NA
Tires	NA	**Block**	NA
Crawlers	NA	**Overhaul Ball**	5 - Ton
Main Boom	Maximum	**Rigging**	75 lbs
Main Boom Angle	NA	**Wire Rope (main hoist)**	9/16" Rotation Resistant
Radius	Maximum	**Wire Rope (auxiliary)**	NA
Other	Hose Reel Installed	**Other**	Spreader Bar 110 lbs

Note: For line in use, when wire rope deductions are required, the rope length shall be measured from the boom tip to the ground.

3. What is the net capacity of this crane for a lift over the rear the with the main boom?

A. 715 lbs

B. 1400 lbs

C. 905 lbs

D. 825 lbs

Use for Workspace

Telescopic Boom Crane - Fixed Cab (TSS)

Outriggers	Fully Extended & Set	**Extension /Jib**	23 – 40 ft Jib Stowed
Counterweight	NA	**Extension /Jib Offset**	NA
Tires	NA	**Block**	Operators Choice
Crawlers	NA	**Overhaul Ball**	Operators Choice
Main Boom	26 ft	**Rigging**	150 lbs
Main Boom Angle	NA	**Wire Rope (main hoist)**	9/16" Rotation Resistant
Radius	10 ft	**Wire Rope (auxiliary)**	NA
Other	NA	**Other**	NA

Note: For line in use, when wire rope deductions are required, the rope length shall be measured from the boom tip to the ground.

4. Lifting from the main boom over the "side" of the crane using the necessary hook block or ball, your **maximum** net load could be?

A. 19,000 lbs

B. 19,170 lbs

C. 19,630 lbs

D. 21,000 lbs

Use for Workspace

Telescopic Boom Crane - Fixed Cab (TSS)

Outriggers	Fully Extended & Set	**Extension /Jib**	23 – 40 ft Jib Stowed
Counterweight	NA	**Extension /Jib Offset**	NA
Tires	NA	**Block**	Single Sheave
Crawlers	NA	**Overhaul Ball**	NA
Main Boom	36 ft	**Rigging**	45 lbs
Main Boom Angle	71	**Wire Rope (main hoist)**	9/16" Rotation Resistant
Radius	NA	**Wire Rope (auxiliary)**	NA
Other	NA	**Other**	Lifting Beam 110 lbs

Note: For line in use, when wire rope deductions are required, the rope length shall be measured from the boom tip to the ground.

5. What is the net capacity of this crane when a lift is made off the main boom over the "side"?

A. 15,215 lbs

B. 15,165 lbs

C. 19,630 lbs

D. 15,275 lbs

Use for Workspace

Telescopic Boom Crane - Fixed Cab (TSS)

Outriggers	Fully Extended & Set	**Extension /Jib**	40 ft Extended
Counterweight	NA	**Extension /Jib Offset**	NA
Tires	NA	**Block**	NA
Crawlers	NA	**Overhaul Ball**	5 - Ton
Main Boom	68 ft	**Rigging**	75 lbs
Main Boom Angle	69	**Wire Rope (main hoist)**	9/16" Rotation Resistant
Radius	NA	**Wire Rope (auxiliary)**	NA
Other	Spreader Bar 110 lbs	**Other**	Hose Reel Installed

Note: For line in use, when wire rope deductions are required, the rope length shall be measured from the boom tip to the ground.

6. What is the gross capacity when the lift is made on the extension over the "side" of the crane?

A. 1320 lbs

B. 875 lbs

C. 685 lbs

D. 1180 lbs

Use for Workspace

Telescopic Boom Crane Swing Cab

A telescopic boom swing cab crane (TLL) is identified by two main features. First a cab that rotates with the cranes upperworks allowing 360-degree visibility, and a boom that can extend and retract hydraulically. This combination allows for more precise and versatile lifting and placement of materials.

Note: Before you get started some crane manufacturers consider the hoist line weight as part of the overall load, while others consider it as part of the crane's equipment. The decision to deduct the hoist line weight is a design choice made by the manufacturer, that is why crane operators must understand the load chart, read all notes and specifications and when required make hoist weight deductions correctly using the correct procedures.

FIGURE 8.23 A large telescopic "swing cab" crane mounted on a truck carrier.

Telescopic Boom Crane—Swing Cab (TLL)			
Outriggers	Fully Extended & Set	Extension/Jib	34 ft Fly Erected
Counterweight	Full	Extension/ Jib Offset	15° Offset
Tires	N/A	Block	N/A
Crawlers	N/A	Overhaul Ball	8.5 Ton
Main Boom	110 ft (Mode B)	Rigging	150 lbs
Main Boom Angle	67.5°	Wire Rope (Main Hoist)	18 X 19 Rotation Resistant 1 lb per ft
Radius	65 ft	Wire Rope (Auxiliary)	18 X 19 Rotation Resistant 1 lb per ft
Other	N/A	Other	22 ft. Fly Stowed

1. What ***maximum*** net load can be lifted with the offset fly over the "side" of the crane?

A. 6490 lbs

B. 7000 lbs

C. 7510 lbs

D. 6190 lbs

Note: For line in use, when wire rope deductions are required, the rope length shall be measured from the boom tip to the ground.

Telescopic Boom Crane—Swing Cab (TLL)			
Outriggers	Fully Extended & Set	Extension/Jib	34 ft Fly Erected
Counterweight	Full	Extension/ Jib Offset	15°
Tires	N/A	Block	N/A
Crawlers	N/A	Overhaul Ball	8.5 Ton
Main Boom	Length 110 ft (Mode B)	Rigging	150 lbs
Main Boom Angle	72°	Wire Rope (Main Hoist)	18 X 19 Rotation Resistant 1 lb per ft
Radius	N/A	Wire Rope (Auxiliary)	18 X 19 Rotation Resistant 1 lb per ft
Other	Auxiliary Boom Head Installed	Other	22 ft. Fly Stowed

2. What is the net capacity when a lift is made off the fly over the "side" of the crane?

A. 8860 lbs

B. 8200 lbs

C. 6640 lbs

D. 7240 lbs

Note: For line in use, when wire rope deductions are required, the rope length shall be measured from the boom tip to the ground.

Telescopic Boom Crane—Swing Cab (TLL)			
Outriggers	Fully Extended & Set	Extension/Jib	34 ft Offset Fly Stowed
Counterweight	Full	Extension/ Jib Offset	N/A
Tires	N/A	Block	N/A
Crawlers	N/A	Overhaul Ball	8.5 Ton
Main Boom	Length 110 ft (Mode B)	Rigging	200 lbs
Main Boom Angle	59.5°	Wire Rope (Main Hoist)	18 X 19 Rotation Resistant 1 lb per ft
Radius	N/A	Wire Rope (Auxiliary)	18 X 19 Rotation Resistant 1 lb per ft
Other	N/A	Other	N/A

3. What is the net capacity of this crane for a lift over the rear with the main boom?

A. 10,440 lbs

B. 11,440 lbs

C. 11,000 lbs

D. 9900 lbs

Note: For line in use, when wire rope deductions are required, the rope length shall be measured from the boom tip to the ground.

Telescopic Boom Crane—Swing Cab (TLL)			
Outriggers	Fully Extended & Set	Extension/Jib	56 ft Fly Stowed
Counterweight	Full	Extension/ Jib Offset	N/A
Tires	N/A	Block	40 Ton Reeved to 6 Parts
Crawlers	N/A	Overhaul Ball	8.5 Ton Hanging 10 ft below ABH
Main Boom	Length 75 ft (Mode B)	Rigging	450 lbs
Main Boom Angle	N/A	Wire Rope (Main Hoist)	18 X 19 Rotation Resistant 1 lb per ft
Radius	25 ft	Wire Rope (Auxiliary)	18 X 19 Rotation Resistant 1 lb per ft
Other	Auxiliary Boom Head Installed	Other	N/A

4. What is the net capacity of this crane when a lift is made off the main boom over the side?

A. 37,100 lbs

B. 39,630 lbs

C. 38,650 lbs

D. 39,120 lbs

Note: For line in use, when wire rope deductions are required, the rope length shall be measured from the boom tip to the ground.

Telescopic Boom Crane—Swing Cab (TLL)			
Outriggers	Fully Extended & Set	Extension/Jib	56 ft Offset Fly Erected
Counterweight	Full	Extension/ Jib Offset	30°
Tires	N/A	Block	60 Ton Hook Block Reeved to 4 Parts
Crawlers	N/A	Overhaul Ball	8.5 Overhaul Ball Hook Ball Hanging 15 ft
Main Boom	Length 85 ft (Mode B)	Rigging	210 lbs
Main Boom Angle	N/A	Wire Rope (Main Hoist)	LBCE Type RB 18 X 19
Radius	35 ft	Wire Rope (Auxiliary)	LBCE Type RB 18 X 19
Other	Auxiliary Head Installed	Other	Spreader Bar 450 lbs

5. You are making a lift over the front of the crane using the 60 ton Hook Block. What is your Net Capacity with the crane in this configuration?

A. 15,145 lbs

B. 15,330 lbs

C. 15,540 lbs

D. 15,690 lbs

Note: For line in use, when wire rope deductions are required, the rope length shall be measured from the boom tip to the ground.

Telescopic Boom Crane—Swing Cab (TLL)			
Outriggers	Full Extended and Set	Extension/Jib	56 ft Offset Fly Erected
Counterweight	Full	Extension/ Jib Offset	1°
Tires	N/A	Block	40 Ton Hook Block 4 Parts of Line Hanging 10 ft
Crawlers	N/A	Overhaul Ball	8.5 Ton Overhaul Ball
Main Boom	Length 110 ft (Mode B)	Rigging	150 lbs
Main Boom Angle	N/A	Wire Rope (Main Hoist)	18 X 19 Rotation Resistant 1 lb per ft
Radius	100 ft	Wire Rope (Auxiliary)	18 X 19 Rotation Resistant 1 lb per ft
Other	Auxiliary Head Installed	Other	N/A

6. You are planning a lift over the side from the offset Fly. Your crane is properly set up on firm level ground. What is your net capacity in this configuration?

A. 880 lbs

B. 2120 lbs

C. 2280 lbs

D. 3080 lbs

Note: For line in use, when wire rope deductions are required, the rope length shall be measured from the boom tip to the ground.

Lattice Boom Crane

A mobile crane is classified as a lattice boom crane (LAT) primarily by the structure of its boom, which is characterized by a W or V shaped lattice design. This design is advantageous for supporting heavy loads while minimizing overall weight. Lattice boom cranes are often mounted on crawler tracks for stability and maneuverability, making them suitable for heavy lifting and extended reach on construction sites.

Note: Before you get started: When a crane is lifting a load, the weight of the hoist line can affect the crane's stability, especially at longer boom lengths. Deducting this weight can ensure that the crane maintains a safe operating radius. Deducting the hoist line weight provides a more conservative estimate of the crane's lifting capacity, ensuring that the crane is not overloaded during a lift. For line in use, when wire rope deductions are required, the rope length should be measured from the ground to the tip of the boom. Look over your lattice boom load chart provided by NCCCO, and you will find that this manufacturer provides boom tip elevation to the ground data included on its load chart. Note that the load chart provides gross capacities for the crawler tracks retracted or extended. Read each question carefully to determine the configuration of tracks.

FIGURE 8.24 The crane in this image is a lattice boom crawler crane. It was manufactured by Manitowoc Crane Company.

Lattice Boom Crane			
Outriggers	N/A	Extension/Jib	30 ft Erected 2,700 lbs
Counterweight	Crane counterweight 179,100 lbs carbody counterweight 44,000 lbs	Extension/ Jib Offset	25°
Tires	N/A	Block	35 Ton Reeved 2 Parts 1267 lbs
Crawlers	Crawlers Extended	Overhaul Ball	15 Ton (722 lbs) Hanging 15 feet from jib tip
Main Boom	230 ft	Rigging	160 lbs
Main Boom Angle	N/A	Wire Rope (Main Hoist)	1 in. 6 X 25 EIPS IWRC @ 2 lbs per ft
Radius	110 ft	Wire Rope (Auxiliary)	1 in. 6 X 25 EIPS IWRC @ 2 lbs per ft
Other	N/A	Other	N/A

1. Lifting on the main boom, the maximum radius at which you can place a net load of 10 tons over the side is?

A. 125 ft radius

B. 110 ft radius

C. 130 ft radius

D. 120 ft radius

Lattice Boom Crane			
Outriggers	N/A	Extension/Jib	N/A
Counterweight	Crane counterweight 179,100 lbs carbody counterweight 44,000 lbs	Extension/ Jib Offset	N/A
Tires	N/A	Block	70 Ton Reeved 8 Parts of Line 1711 lbs
Crawlers	Crawlers Extended	Overhaul Ball	N/A
Main Boom	290 ft	Rigging	350 lbs
Main Boom Angle	63.8°	Wire Rope (Main Hoist)	1 in. 6 X 25 EIPS IWRC @ 2 lbs per ft
Radius	N/A	Wire Rope (Auxiliary)	1 in. 6 X 25 EIPS IWRC @ 2 lbs per ft
Other	Spreader Bar 550 lbs	Other	N/A

2. Your crane is set up on firm level ground with your crawlers extended. You are lifting over the rear of the crane. What is your net capacity in this configuration?

A. 8,835 lbs

B. 9,185 lbs

C. 9,735 lbs

D. 14,012 lbs

Lattice Boom Crane			
Outriggers	N/A	Extension/Jib	N/A
Counterweight	Crane counterweight 179,100 lbs carbody counterweight 44,000 lbs	Extension/ Jib Offset	N/A
Tires	N/A	Block	70 Ton Reeved 4 Parts 1711 lbs
Crawlers	Crawlers Retracted	Overhaul Ball	N/A
Main Boom	180 ft	Rigging	195 lbs
Main Boom Angle	N/A	Wire Rope (Main Hoist)	1 in. 6 X 25 EIPS IWRC @ 2 lbs per ft
Radius	80 ft	Wire Rope (Auxiliary)	1 in. 6 X 25 EIPS IWRC @ 2 lbs per ft
Other	N/A	Other	N/A

3. What would your net capacity be when lifting with the main boom over the "side" of the crane?

A. 45,220 lbs

B. 48,500 lbs

C. 48,305 lbs

D. 45,349 lbs

Lattice Boom Crane			
Outriggers	N/A	Extension/Jib	30 ft Jib
Counterweight	179,000 lb Crane counterweight 44,000 lbs carbody Counterweight	Extension/ Jib Offset	5°
Tires	N/A	Block	70 Ton (1711 lbs) Reeved 6 Parts of Line Hanging 30 ft
Crawlers	Extended	Overhaul Ball	15 ton 722 lbs
Main Boom	260 ft	Rigging	225 lbs
Main Boom Angle	63°	Wire Rope (Main Hoist)	1 in. 6 X 25 EIPS IWRC @ 2 lbs per ft
Radius	N/A	Wire Rope (Auxiliary)	1 in. 6 X 25 EIPS IWRC @ 2 lbs per ft
Other	Spreader Bar 550 lbs	Other	N/A

4. You are planning a lift over the rear of your crane utilizing the 30 ft jib. What is your net capacity in this configuration?

A. 16,402 lbs

B. 17,453 lbs

C. 17,846 lbs

D. 18,343 lbs

Lattice Boom Crane			
Outriggers	N/A	Extension/Jib	30 ft Jib
Counterweight	179,1000 lb Crane counterweight 44,000 lbs carbody counterweight	Extension/ Jib Offset	25°
Tires	N/A	Block	70 Ton (722 lbs) Reeved 4 Parts of Line Hanging 20 ft
Crawlers	Extended	Overhaul Ball	15 ton 267 lbs
Main Boom	150 ft	Rigging	175 lbs
Main Boom Angle	N/A	Wire Rope (Main Hoist)	1 in. 6 X 25 EIPS IWRC @ 2 lbs per ft
Radius	83 ft	Wire Rope (Auxiliary)	1 in. 6 X 25 EIPS IWRC @ 2 lbs per ft
Other	N/A	Other	N/A

5. Lifting from the extension jib over the side, what is the net capacity of this crane in this configuration?

A. 37,544 lbs

B. 37,212 lbs

C. 37,052 lbs

D. 39,200 lbs

Lattice Boom Crane			
Outriggers	N/A	Extension/Jib Erected	40 ft Jib
Counterweight	179,1000 lb Crane counterweight 44,000 lbs carbody counterweight	Extension/ Jib Offset	15°
Tires	N/A	Block	70 Ton (1711 lbs) Reeved 8 Parts of Line Hanging 35 ft
Crawlers	Extended	Overhaul Ball	15 ton 722 lbs
Main Boom	120 ft	Rigging	445 lbs
Main Boom Angle	N/A	Wire Rope (Main Hoist)	1 in. 6 X 25 EIPS IWRC @ 2 lbs per ft
Radius	82 ft	Wire Rope (Auxiliary)	1 in. 6 X 25 EIPS IWRC @ 2 lbs per ft
Other	N/A	Other	Equalizer Beam 475 lbs

6. Your crane is set up on firm, level ground with your crawlers extended. What is the net capacity of the crane in this configuration?

A. 36,996 lbs

B. 37,441 lbs

C. 37,151 lbs

D. 35,997 lbs

Summary—and Then Some

For professional, experienced, highly skilled crane operators, managing a crane is an incredibly serious matter. They begin by getting to know the crane until they feel as if it is an extension of themselves. They feel the controls, hear the hydraulic fluid flowing through the system, and smell the fluids of the machine. Most importantly, the operators visualize the way the crane is set up and all the surrounding areas around the lift zone. They paint a picture for themselves not only around the crane but also above and below the machine. These crane operators are checking for any hazards such as trees, light poles, and power lines above and water lines, conduits, and electrical vaults below. And that's just the start of things for crane operators.

Knowing how to read the crane load chart and make all the necessary calculations to ensure a load can be safely hoisted and then be placed is vital. Crane operators are constantly asking themselves questions about the crane configuration, the setup, and the surroundings. They continue to make load chart calculations, as well as analyze the rigging and load handling devices being used to make the lift. Just imagine you are sitting in your crane on a beautiful sunny day waiting for riggers to hitch up the load. However, one of the slings selected for the lift is substandard. You get up on the load, and after it is about ten feet off the ground, the sling fails, and the load drops to the ground. The shock sends dynamic force flowing throughout the crane. No one is injured, nor the crane damaged, and they are extremely lucky.

Professional crane operators always have a close eye on any rigging selected for a lift. Crane operators understand that rigging or a lifting attachment can be the limiting factor for a safe lift. Lift after lift, they refer to the crane's load chart. Repeatedly, the crane operators check the load chart, making sure safety is the top priority of every lift, and must be confident in their knowledge of load charts and ability to make flawlessly accurate calculations.

Operation of mobile cranes near power lines is a hazardous process that must be avoided whenever possible. When that isn't possible by use of alternative equipment, extra safety precautions must be taken—and only when a safe minimum clearance is maintained. According to OSHA, the power line owner and operator or a qualified registered professional engineer determines the minimum clearance distances that must be maintained to prevent electrical contact. If the clearance distances specified by OSHA cannot be maintained, the first option available is to de-energize and visibly ground the power lines. Other protective measures, such as guarding or insulating the lines, help prevent accidental contact with the crane. The most important safeguard starts with a company safety plan that requires all workers to be trained in safety procedures and requirements.

9 Operating Near Power Lines

It is obvious that the combination of mobile cranes and power lines presents a giant hazard. According to the Bureau of Labor Statistics, 818 crane accidents have been responsible for workplace deaths since 1997. OSHA states that hundreds of construction workers die every year in the United States on the job. In 2011, there were reports of over 700 fatalities. The third leading cause of these deaths was electrocution. OSHA revised its requirement for working around overhead power lines in 2010 and took a more proactive approach to preventing power line contact to address this problem. Today's statistics show 45% of all crane-related accidents involve contact with an energized power line. All of them result in one or more fatalities. Even though progress has been made, each year an average of 15 electrocutions are caused by contact between cranes or similar boomed equipment and energized overhead powerlines. These numbers are still alarming, and much more progress is needed.

Prior to beginning crane operations, a hazard assessment inside the work zone must be performed:

- The hazard assessment must identify the work zone and assess it for power lines and then determine how close the cranes can get to the lines. The employer has an option of doing this assessment for the area 360 degrees around the crane or for a more limited, demarcated area.

- If the assessment shows that the crane could get closer than a distance of 20 feet for lines rated up to 350 kV and 50 feet for lines rated over 350 kV, then additional action is required.

Some of the main reasons incidents occur involving mobile cranes when working near power lines are:

- Operating the crane parallel to a power line, and the crane operator accidentally swings in the wrong direction.
- Booming down or swinging into an energized power line that appears farther away because of the difficulty in judging distance by the crane operator.
- Placing or removing concrete or asphalt during highway construction work.
- Installing pipe, traffic barriers, and guardrails on projects with adjacent or crossover power lines (Figure 9.1).

FIGURE 9.1 Note the power lines in the photo. Power line contact often occurs when installing pipe.

Before any major project starts, a professional contractor begins to deal with the issue of power lines on the jobsite. During the bidding process, engineers will identify any powerlines on the jobsite that will need attention. At this time, company engineers should be contacting powerline utility owners to learn all they can about any powerlines in the project area. They will determine if the power lines can be de-energized or re-routed or if other safety precautions will need to be implemented.

Before the actual work begins, a meeting will be scheduled with local property owners, along with contractors performing the work, including company supervisors and crane operators as well as other important members of the crane hoisting and rigging team. Together they will identify any hazards associated with the power lines on the job (Figure 9.2). Safety procedures and a written plan will be developed to prevent accidental contact with all the cranes in operation on the jobsite throughout the duration of the project. Together with power line utility representatives, steps and precautions will be implemented to ensure safety throughout the jobsite.

FIGURE 9.2 Hazards must be identified before work begins.

Some of the safeguards put into effect to reduce the potential for crane power line contact may include:

- Using an insulated link installed on the crane's hoist line above the load block or overhaul ball hook and the load to be lifted, providing protection for the rigging crew when hooking up or detaching rigging equipment.
- Installing a proximity warning device on the crane that sounds a warning to the crane operator when the crane boom is coming too close to an electrically charged power line.
- Placing highly visible flagged warning ropes at safe distances both below and to the sides of electrically charged power lines.
- Erecting goal posts on both sides of the power line with crossbars 4 feet below the power line.
- Using insulated sleeves that can be attached directly to the power line. The sleeve is made of nonconductive material that prevents physical contact between the crane hoist lines and the power line.

- Using nonconductive tag lines when they must be used. Polypropylene offers better nonconductive properties than most other types of rope.
- Choosing a boom gauge guard on some applications. The gauge surrounds the boom of the crane. If power line contact happens, the guard will prevent the boom from becoming energized. The rigging crew is then protected along with the rest of the ground crew.
- Choosing a dedicated spotter who is in continuous contact with the crane operator. A qualified signal person works very well in this application.
- Setting a clearly visible line of safety cones or sanctions placed 20 feet out from the power line on both sides when applicable.
- Using nonconductive rigging such as synthetic slings.
- Ensuring riggers wear insulated gloves rated for the voltage involved.
- Providing insulated mats to stand on that are rated for the voltage when crane operators are operating the crane by way of remote controls.
- Allowing only personnel essential to the operation to be permitted in the work zone.
- Using a line of fluorescent snow fence erected 20 feet from power lines on both sides when applicable.
- Forming or erecting a barricade 10 feet around the crane to prevent personnel from touching the machine.

These safeguards are no substitute for an alert, well-trained, certified crane operator and qualified lifting team and ground crew. Sole reliance cannot be placed on these precautions alone. If one should fail, the results could be deadly.

One of the largest changes OSHA made in its new regulation pertaining to mobile cranes operating near power lines was to increase the minimum approach distance for energized overhead power lines from the longstanding 10 feet out to 20 feet for up to 350-kV power lines. For over 350-kV lines, the distance was extended all the way up to 50 feet. These new distances don't mean mobile cranes cannot get closer than these distance requirements. It does require that if cranes must work closer, then additional safety measures must be put in place. When this is the case, the employer must make some clear choices before any work can begin. Here are the options:

- *Option one* is to work with the power line utility owner to have the power de-energized and grounded or possibly re-route the line. It may be possible to bury the line underground. Clearly this first option is the one OSHA would prefer, avoiding the need for other precautions. But there are times when de-energizing is not practical. When this happens, OSHA has provided two other options the employer can take to safeguard its workforce.
- *Option two* is to work 20 feet away from the power line no matter what. Option two also states that if the voltage of an energized power line is unknown, then a 20-foot clearance distance must be used for up to 350-kV lines.
- *Option three* allows work to take place closer to energized power lines if the correct voltage in the line is known. Then the approved distances provided by OSHA, shown in Table 9.1, can be used if all the required safeguards have been met and the crane operator and ground crew have been properly trained.

TABLE 9.1 Specified Clearance in the Vicinity of Energized Electric Power Lines

Normal Voltage, kV (Phase to Phase)	Specified Clearance, ft (m)
During Operation	
Up to 50	10 (3)
Over 50 to 200	15 (4.6)
Over 200 to 350	20 (6.1)
Over 350 to 500	25 (7.6)
Over 500 to 750	35 (10.7)
Over 750 to 1,000	45 (13.7)
Over 1,000	Determine specified clearance after consultation with utility owner/operator

Remember, before work can begin, training must take place in accordance with OSHA 1926.1408. All employers with a workforce near the vicinity of the power lines, whether they are directly involved or not, must be properly trained to understand how electricity dissipates across the ground and the proper procedures to take if electrical contact is made with a power line and a crane.

Whenever a crane is in transit with no load on the hook, and the boom and boom support are lowered, the clearance distances in Table 9.2 shall be maintained.

TABLE 9.2 During Transit or Travel with No Load and with Boom or Mast Lowered

Normal Voltage, kV (Phase to Phase)	Specified Clearance, ft (m)
During Operation	
Up to 0.75	4 (1.2)
Over 0.75 to 50	6 (1.8)
Over 50 to 345	10 (3.1)
Over 345 to 750	16 (4.9)

When planning transit or travel of the crane and the machine under an energized power line, the effects of speed and terrain on the hydraulic or lattice booms and carriers shall be considered. While traveling, a dedicated spotter must be used if any part of the crane and its equipment will be within 20 feet of an electric power line at any time. Continuous contact with the crane operator must be maintained. This is best accomplished by radio contact using high-quality, hands-free equipment. Headsets for the crane operator and qualified signal person allow for the signal person's hands to be free and to hold a fluorescent flag for additional visibility to help prevent encroachment. The spotter must be positioned to effectively gauge the clearance distance and give timely information to the crane operator. If you plan to move a crane at night or any other time of poor visibility, the power lines may need to be illuminated, or another means of identifying the location of safe path of travel may be used. Cranes are sometimes moved around on construction sites on an hourly basis. Raising and lowering the crane gantry systems are frequent occurrences. Identifying the best and safest location on the construction or industrial site to pass under energized power lines in advance is an excellent idea. Having that location dedicated as the only safe route to use, and marking it with signage helps eliminate confusion and lowers the risk when cranes cross under power lines. The hope is that cranes are always passing under at the best spot rather than anywhere that might be convenient.

Careful consideration must be given when working near transmitters or communication towers. When working near transmitters or communication towers when the equipment gets close enough for an electric charge to be induced to the crane, including materials being handled, such as any rigging and load handling devices, the transmitter must be de-energized, or the crane must be provided with an electrical ground. Clearly, OSHA is getting very serious about anything in our workplace that can induce an electric charge.

Other important safety factors to be considered include:

- Conditions affecting atmospheric conductivity, such as dust or moisture in the air
- The time for the crane operator to bring the crane, load line, and load to a complete stop, with both boom and swing inertia effects
- Weather conditions, such as wind, rain, snow, ice, and extreme heat or cold.
- Degree of sway or sag in the power lines
- Lightning conditions often overlooked in safety plans

A great option is a power line that is equipped with a device that automatically de-energizes the circuit in the event of a power line contact before use of the crane and work begin. The automatic reclosing feature of the circuit-interrupting device makes the line inoperative. The key to safety around power lines is awareness. An experienced crane operator assumes that all lines are energized until it is confirmed they are not. Overhead power lines caused 36% of all crane-related fatalities on jobsites every year, according to the Electrical Safety Foundation International. It is paramount that all companies working cranes around energized power lines have an emergency plan in place before beginning work near these lines. Not all worksites are the same, and so the precautions will be different. Your emergency plan must be specific, not only to the site, but to the crane as well.

Before work begins, a top-notch crew will be selected and thoroughly trained. They should include the owners, supervisors, and a certified crane operator, as well as qualified riggers and a signal person. The entire lifting crew must be trained in electric shock cause and effect. They must fully understand what happens when a crane boom makes contact and becomes energized. Workers must know that the ground around the crane becomes electrified around the point of electrical contact in a rippling pattern know as *ground gradient* (Figure 9.3). These gradient lines are invisible and very deadly. If you throw a rock into a pond of smooth water, the ripples that roll away from the impact of the rock would resemble gradient lines. How far these lines flow away from the crane is an unknown. How much voltage would be in a gradient line is also unknown. Each incident will be different. The voltage in the power line contacted by the crane and the moisture content in the ground along soil types are some of the factors that will determine the power in these gradient lines and how far the flow might travel. We know one thing for sure: They are deadly!

FIGURE 9.3 Power flows away from the crane in a rippling pattern known as a *ground gradient*.

As a worker moves away from the point of contact at the crane, the voltage drops continuously. If a person has one foot on one side of a gradient line and another foot on the other side of a line, the difference in voltage will cause a flow of electricity through the person's body. Electric shock occurs when two parts of the human body, such as an arm and a foot, can concurrently contact two objects of unlike voltages—typically, one object energized and another one grounded. When the voltage difference is high enough to produce electric current in the human body of more than a few thousandths of an ampere, the current will be felt. This can occur at voltages as low as 100 volts. Evidence of electric current in the person can range from mild discomfort to severe pain or death. The most severe effects are loss of the ability to control muscles and involuntary muscular actions. It can also cause fibrillation of the heart, which could stop beating and cause death. Other symptoms are severe burns.

When a crane is accidentally brought into contact with an energized power line, the machine and its hoist line, including any rigging, load handling devices, and the load itself, become electrified to the same voltage as that of the power line. The ground around the crane, including crane tracks, outriggers, floats, and wheels, becomes highly electrified. The ground crew near the crane should avoid touching the machine. A barricade should be formed at least 10 feet around the crane to help workers avoid touching the crane.

Other precautions that must be followed are:

- The folks in the ground crew should put their feet together and shuffle away so their feet are touching one small area of the ground at a time. They should not stop until they are well away from the crane. Hopping away is another option, but the chance of falling over is also a risk factor.
- Normally, the crane operator is not in danger of shock. Like a bird sitting on a power line, the operator is not at risk and must stay in the cab. The crane operator should act quickly to swing the boom and hoist line away from the power line. This is not always possible though, since in some cases the boom or wire rope on the machine may weld up to the power line (Figure 9.4). The crane operator should make every attempt to warn others in the area. If there are workers in the area, they should be told to stay put. If ground personnel are down

and injured, the crane operator or others should call emergency response for medical aid. In addition, call the power utility company and get the power shut off. The crane operator must remain in the cab except where there is an imminent danger of fire, explosion, or another emergency (Figure 9.5). If the crane operator is forced to leave the crane, he or she should jump off the crane, being careful not to touch the crane and ground at the same time. The operator should shuffle away to avoid the danger of the potentially energized zone around the crane and the step potential of energized gradient lines.

FIGURE 9.4 Wire rope on the crane may weld up to the power line, making it impossible for the crane operator to break the crane free. The result may be the entire crane becoming energized, causing fuel tanks to explode.

If members of the lifting team or anyone else becomes electrically injured near the crane or in the work zone, get help immediately. Anyone who is not injured must *not* attempt to rescue or aid them. This can be especially difficult, since the people who were injured or even killed may be members of your work crew, people you have lunch with everyday or may associate with after work. In some cases,

FIGURE 9.5 The crane operator must stay in the cab except where there is an imminent danger of fire or explosion. In this case, the danger of fire caused the operator to abandon the cab.

it could even be a relative you work with. However, you must remember that you cannot help anyone if you become a victim yourself.

As noted earlier, when a crane is in contact with an energized power line, the high voltage difference between the machine and ground will probably badly burn or kill any workers touching the crane and the ground at the same time. When the machine is in contact with an energized power line, the ground will become electrified. All those around the crane, including riggers, signal persons, and ground crew members, standing with their feet apart will be at a very high risk (Figures 9.6 and 9.7). The electric current can cause loss of muscular control. That would cause the workers to fall over, creating a very bad situation. And any personnel who attempt to walk will suffer the same effect. Because of the step potential, a person with both feet on the ground on the opposite sides of a gradient line with a potential voltage differential would be electrified and killed. This is the reason for shuffling with both feet together or hopping. This way, your feet are touching one small area of the ground at a time. Remember, if you choose to hop, you must be careful not to fall, because by doing so you would then be touching the ground at different points several feet apart across gradient lines. The result surely would be bad.

FIGURE 9.6 It can be very difficult for a crane operator to judge the distance from the cab and controls out to an energized power line. Power line grouped closely together makes it very hard for the crane operator to judge one line from another.

FIGURE 9.7 The certified crane operator of this crane is home safe with his family. He was trained, knowledgeable, and smart. He kept his distance, operated his crane by the rules, and stayed alive.

Chapter 9 Review Quiz

1. Tag lines shall be held by an experienced person, and when working in the proximity of energized electric lines
   a. should have a safe working load rating
   b. should have a working load limit rating
   c. should be nonconductive
   d. should have a safety factor rating

2. If the load line of your crane contacts a power line, and there is no sign of any other danger such as fire, you must
   a. carefully climb down from the crane
   b. jump down while keeping your feet together
   c. stay in the crane until the line is de-energized
   d. get as far away from the crane as possible

3. The required clearance for normal voltage when operating a crane near high-voltage power lines over 200 to 350 kV is
   a. 5 feet
   b. 10 feet
   c. 15 feet
   d. 20 feet

4. For cranes in transit with no load or boom lowered, the minimum required clearance up to 0.75 to 50 kV is
   a. 4 feet
   b. 6 feet
   c. 8 feet
   d. 10 feet

5. The third most common cause of death among construction workers is
   a. contact with power lines
   b. crane tip-overs
   c. being crushed by the crane's counterweight
   d. being struck by uncontrolled loads

6. All overhead power lines should be considered energized unless
   a. an appointed person verifies that the lines are de-energized
   b. a designated person verifies that the lines are de-energized
   c. a lift director indicates otherwise
   d. an electrical utility authority verifies the lines are de-energized

7. When determining the safe working proximity to energized electric lines, it is also important to consider the
   a. placement of trees and foliage
   b. location of power poles
   c. sway or whipping caused by wind
   d. location of ground crews

8. The required clearance for normal voltage when operating a crane near high-voltage power lines over 350 to 500 kV is
   a. 20 feet
   b. 25 feet
   c. 30 feet
   d. 35 feet

9. If the use of a tag line is necessary while making a lift with a crane, which of the following types offers better insulating properties than most other ropes when working around power lines?
   a. Manila hemp
   b. Sisal
   c. Nylon
   d. Polypropylene

10. You are setting up an all-terrain crane near an overhead power line that you are told is 850 kV. According to OSHA 1926.1408, what is the minimum clearance distance you must maintain between any part of your crane and the load you are lifting?
    a. 35 feet
    b. 45 feet
    c. 55 feet
    d. 65 feet

Summary—and Then Some

Contact with overhead energized power lines continues to be one of the most common causes of crane-related fatalities. So, as we summarize this chapter, we should ask ourselves why. There has been deficient information and many misconceptions about cranes working around energized power lines for many years. Unfortunately, some of those misconceptions are still heard in industry and construction today. OSHA and ASME have provided much-improved regulations and safety standards to help make working around power lines much safer. However, poor information still exists due to a lack of comprehensive crane training for owners, supervisors, lift directors, crane operators, riggers, and ground support personnel.

If today we could all board a tour bus and take a field trip and visit a few local construction sites where energized power lines exist, it would be fun and very interesting to ask the crane operators and members of the ground crew a few questions regarding work around energized power lines on their job. There is a good chance some of the misinformation in our industries would come to light during our questions.

As we point out an energized power line on the jobsite, we would ask our first question: *Do you workers know how many kilovolts of power are in that power line up there?*

There is a very good chance no one on the job would know, with the exception of the supervisor. And even the supervisor knowing may be wishful thinking. Not knowing has always been a big part of the problem. Most construction workers cannot identify how much power is in one line in comparison with another power line, and so proper distances of approach to dangerous lines are often not taken. Other safeguards are not put in place to help prevent encroachment as well. It all comes down to a lack of knowledge and training.

The next question we would ask our crane operators and ground crew would be: *When you set up your crane, how far away are you required to be from that power line?*

In all likelihood, most of our workers wouldn't have an answer. Those who do would probably answer 10 feet. That's a big misconception that exists in the crane industry and has existed for years. It comes from the minimum clearance distances and tables provided by OSHA and ASME. In fact, the tables do say the minimum clearance distance for lines up to 50 kV is 10 feet. The lack of training comes into play once again because distances of approach in the table don't *end* at 10 feet. Distances in the table continue all the way up to 45 feet for lines over 750 to 1000 kV. In addition, many other safeguards would also be added to prevent encroachment.

Workers have come to think and believe all energized power lines require a distance of approach of 10 feet. They believe this to be true regardless of their power ranking. In recent years OSHA regulations and ASME industry safety standards have provided a lot of good information (Figure 9.8). Tables and illustrations to help workers understand are out there for our review. Good information has been provided. Now it is up to our employers and all those workers working around energized power lines to look up and live. Find out how much power is in the power lines on your job. Call the utility owner and find out, step back, and stay alive. Together with the utility owner, have safeguards put in place, and most importantly train yourself and your workforce about the dangers associated with cranes working around energized power lines. It will be then, and only then, that we see the high fatality reports of the past finally improve.

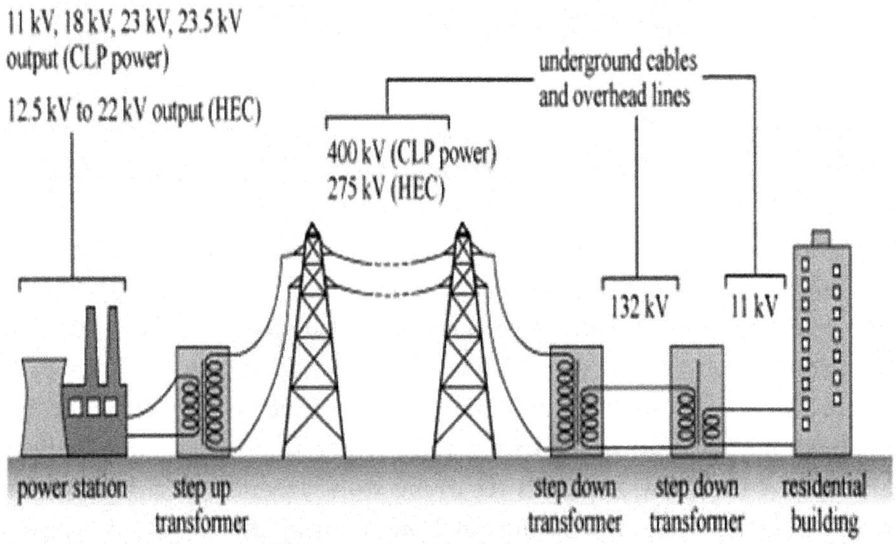

FIGURE 9.8 ASME Industry Safety Standards have provided good illustrations to help workers understand energized power lines.

Wire rope is a popular product used in the mobile crane hoisting and rigging industry. It is strong and flexible and holds up well under heavy lifting. Galvanized steel wire rope is used for many of today's hoisting operations. Many crane companies prefer wire rope for lifting heavy loads because its distinctive design provides multiple steel wires that shape individual strands placed in a helical pattern around a core, giving it flexibility. Its flexibility gives it the ability to handle, grip, and hold during a lift. Different layouts and arrangements of the wire and strand composition can pay big dividends when planning specific hoisting and rigging operations properly. The key is selecting the correct hoist rope for the job. Wire rope is a machine, a real workhorse that lifts the large heavy loads that our mobile cranes handle on the industrial and construction jobsites of today. The more you are informed about the hoist ropes on your crane, the better a resource you will be to the industry you work in and the more knowledgeable a crane operator you will become.

Synthetic rope has grown in popularity in recent years. Synthetic rope is an alternative to steel wire ropes and is made of synthetic fiber. It is abrasion resistant and easy to inspect. The rope is torque neutral so it will not twist and is easy to handle. Synthetic rope provides good pulling power and is light to install and maneuver. Safety is the top priority whether it is wire or synthetic hoist rope. In construction and other industries, we have two reliable options to choose from when selecting hoist rope. It's important that all those working in crane and rigging operations are trained and familiar with both.

10 Hoist Ropes Used in Craning

Choosing the Correct Rope

With a wide selection of hoist ropes available, one decision crane owners will need to make is between wire and synthetic ropes. Several manufacturers produce steel wire rope. First, let's begin with an explanation of what wire rope is.

Wire rope is composed of wires, strands, and a core (Figure 10.1). The basic unit of wire rope is wire, which is processed from selected grades of steel to predetermined physical properties and sizes. A selected number of finished wires are then helically laid together in a uniform geometric pattern to form a strand. The required number of fabricated strands is laid symmetrically with a definite length of lay around a core to form a finished rope.

Wire rope is highly respected in many industries because of its previous accomplishments over many years. Both wire rope and synthetic rope are currently utilized in several different applications. Owners will need to choose what ropes they want to run, based on the crane manufacturer's specifications. Always choose ropes of original capacity and use. Wire rope is designed by the manufacturer of your crane. Always consult your crane manufacturers specifications and you're wire rope dealer's recommendations when purchasing rope to ensure the wire rope you buy will be acceptable.

FIGURE 10.1 Basic construction of a wire rope.

After trial and error using hoist ropes on different jobs and applications, a crane operator may start to build a preference toward one type of rope over another. Owners and crane operators want less abrasion, better fatigue resistance, more strength, and longer wear for their investment. The reality is that no single wire rope design can do everything. Wire ropes offer design property exchanges. It is very difficult to lower both fatigue and abrasion resistance on a wire rope all at the same time. For example, you can increase fatigue resistance by choosing a wire rope with more wires. This gives the wire rope less abrasion resistance because of its greater number of smaller outside wires.

Whatever you choose when replacing wire rope on your mobile crane, it is important to make sure it meets the original capacity and use for the specifications provided by your crane manufacturer. Here are some abrasion and fatigue scenarios to think about:

Fewer Outside Wires per Strand
- It has greater abrasion resistance.
- Resistance increases with larger wires.
- Resistance decreases with smaller wires.

More Outside Wires per Strand
- It has greater fatigue resistance.
- Resistance decreases with larger wires.
- Resistance increases with smaller wires.

How would one go about choosing the best wire rope for the operations being planned for hoisting and rigging operations? Let's consider some important factors before purchasing the wire rope:

- **Strength.** Wire rope in the United States is usually rated in terms of minimum breaking strength shown in number of tons (with a ton being 2,000 lbs). Wire rope strength is shown as minimum breaking force.
- **Fatigue resistance.** Fatigue resistance means fatigue of the wires that make up strands that wrap around the core to form the wire rope.
- **Crushing resistance.** Crushing is the effect of external pressures on a wire rope. This damages the rope by distorting the cross-section shape of the rope or its strands. This could include all

three, cross-section-shaped strands and crushing. Round wire rope has only a single contact location at a cross-over point, which causes very high contact pressures. High contact pressure causes crushing. Crushing also occurs by cross winding, a direct result of improper winding procedures. This problem is often related to an irregular fleet angle, which is the angle between the wire rope and an imaginary line that's perpendicular to the hoist drum. The angle between the center line through the lead sheave and center line of the wire rope leading to the hoist drum is called the *fleet angle*. It is the difference between the angle at which the wire rope is traveling and the center line of the hoist drum. Sheaves that are heavily worn or badly corrugated will accelerate wear and friction, causing individual wires to break down and press together. Wire ropes must be maintained with periodic lubrication to prevent internal corrosion and breakdown.

- **Resistance to metal loss and deformation.** Metal loss means the actual wearing away of metal from the outside wires of the wire rope. It also means deformation that changes the shape of the outer wires of a rope.
- **Resistance to rotation.** When a load is placed on a rope, torque is created within the rope as wires and strands try to straighten out.

Once you have purchased the wire rope of choice based on the manufacturer's requirements and your jobsite's craning needs, it's time to install the rope on your crane. Always check the shipping papers when your wire rope arrives to ensure it is the exact rope you ordered. Installing the wrong rope on your crane will destroy the rope and may damage the sheaves very quickly. Getting the rope properly on your crane can be a challenge, so doing that task correctly the first time will save a bunch of money and a ton of headaches. Therefore, the professionals installing wire rope will recommend the following techniques.

Installing Wire Rope

The preferred technique for installing rope onto a crane is:

1. Unload the rope properly and relieve any twists. Pull the rope off the shipping reel, or unroll it from a shipping coil. If done improperly, you may kink the rope, which will result in permanent damage to it. Next lay the rope on the ground in a direct line with the boom. This helps release any twist in the rope.
2. Attach the rope end to the drum. Pull the rope over the boom point sheave, string it down the length of the boom, and attach the end to the hoist drum. Before making any end attachment, be sure the rope strands are free to adjust. In other words, make sure the end of the rope is not welded together unless it is a specialty product that requires the weld to remain.
3. Wind the rope onto the drum slowly and carefully. At this point, it isn't necessary to provide additional load other than the weight of the rope being pulled across the ground.
4. Spool the first layer tightly. It's essential on smooth-faced drums that the first layer is spooled with wraps tight and close together since the first layer forms the foundation for successive layers. If need be, use a rubber, lead, or brass mallet, but never a steel hammer, to tap the rope into place. Make sure the foundation layer is spooled tightly.
5. Spool multiple layers with sufficient tension. It's very important to apply a tensioning load to the ropes during the rope breaking-in process. If not, the lower layers may be loose enough

that the upper layers become wedged into the lower layers under load, which can seriously damage the rope. Improper tension can result in crushing and "pulling in" of the rope. The tensioning load should range from 2 to 3% of the rope's minimum breaking force.

6. For ropes in multiple parts of line systems, reeve the hoist hook block and boom tip sheaves so the rope spacing is maximized and the traveling hoist hook block hangs straight and level to help ensure hook block stability. Avoid dead-ending the rope at the hoist hook block if possible.
7. Check the rope for twist. With the rope slack, pull enough rope out to allow it to hang in a loop to the right. If the rope hangs without twisting together, there is no twist in the rope. If the rope twists together to the left, the rope has a twist in it.

Selecting the Correct Lay Rope

A glossary of terms:

- **Regular lay.** The wires are twisted in one direction, and the strands are twisted in the opposite direction to form the rope. The wires appear to run parallel to the center line of the rope. Due to the difference in direction between the wires and the strand, regular lay ropes are less likely to unwind or kink. Regular lay ropes are less likely to fail from crushing and distortion because of the shorter length of the exposed outer wires.
- **Lang lay.** The wires and strands spiral in the same direction and appear to run diagonally to the center line of the rope. With the longer length of exposed outer wires, lang lay ropes have more flexibility and greater resistance to abrasion and show less fatigue. However, lang lay is more prone to kinking and twisting and cannot stand abuse from distortion and crushing. Lang lay ropes should have both ends secured to prevent unwinding and should not be on single-part hoist lines or used with swivel end terminals.
- **Alternative lay.** This is where the ropes have three strands right laid and three strands left laid.
- **Wire rope lay.** This term refers to the direction in which the strands are laid, clockwise or counterclockwise. Regular lay rope has the wires and strands laid in opposite directions. Lang lay rope has wires and strands laid in the same direction.
- **Right or left lay.** This refers to the direction in which the strands rotate around the wire rope. With right lay rope, the strands rotate to the right, while left lay has the strands rotate to the left.
- **Wire rope finish.** The term *bright* refers to a wire rope manufactured with no protective coating or finish other than lubricant. Some applications require more corrosion protection than lubricant alone. In these cases, a "galvanized" finish is provided. Consult your supplier for more information on corrosion-resistant wire ropes.

Note

- Strands wound to the *right* around the core, as shown in Figure 10.2, are the majority or nearly all of those used in the construction industry today.
- Strands wound to the left are used in a few special situations such as cable tool drilling and some older tower cranes, as shown in Figure 10.3.

FIGURE 10.2 Right lay ropes are the most common in the construction industry.

FIGURE 10.3 Left lay ropes are used in a few special applications.

The correct lay of rope to be installed on a hoist drum will be determined by the location of the rope's anchorage point and whether the rope spools to the top or to the bottom of the hoist drum. To understand left lays and right lays, first visualize yourself as standing behind the hoist drum looking out. If the rope winds to the top of the drum, clench your fist with the palm down, and extend your index finger. The rope's anchorage point must be on the same side of your hand as your extended thumb. The hand that you use determines whether the rope you use is left or right lay. For hoist drums that spool from the bottom, simply clench your fist with the palm up, and extend the index finger and thumb.

FIGURE 10.4 When removing the rope from the shipping reel or coil, the reel or coil must rotate as the rope unwinds.

When removing the rope from the shipping reel or coil, the reel or coil must rotate as the rope unwinds (Figure 10.4). Any attempt to remove a rope from a stationary reel or coil will almost always result in a kinked rope that may be ruined beyond repair. Mount the reel on a shaft supported by either jacks or blocks. Grasping the free end of the rope, walk away from the reel, which will rotate as the rope unwinds. When winding a rope onto a machine drum or onto another reel, make certain that it bends in the same direction.

Understanding Cable Lays

As noted above, right lay ropes are the most common in construction. The wires wind in one direction, and the strands wind in the opposite direction, as shown in Figure 10.5. Lang lay rope wires in the strands of the rope wind in the same direction, as shown in Figure 10.6.

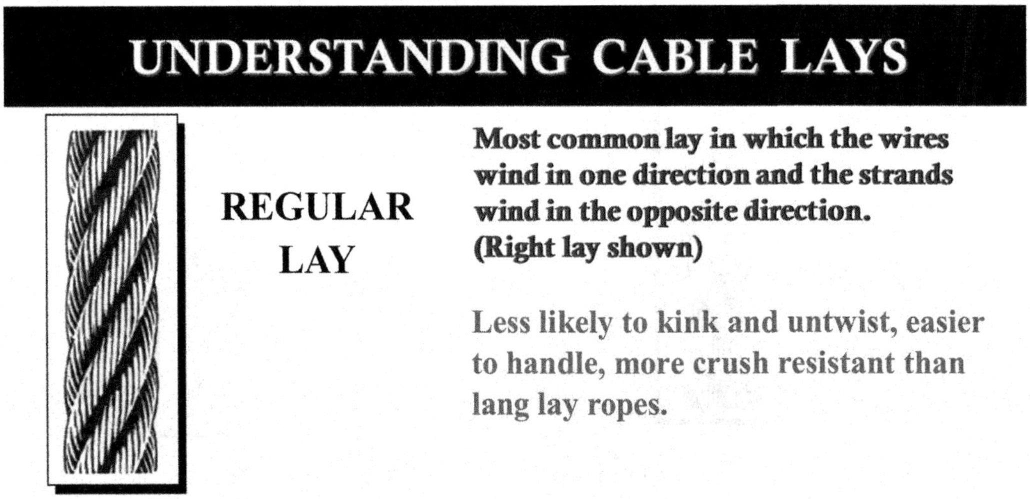

FIGURE 10.5 Regular lay ropes are less likely to kink and untwist. They are easier to handle and are more crush resistant than lang lay ropes.

When to Replace the Wire Ropes on Your Crane

According to the ASME B30.30 safety standard for wire ropes and B30.5 for mobile cranes, there are no special written plans or rules about when you should replace wire rope on your crane.

Proper wire rope installation requires rope to be installed from the top of the reel to the top of the hoist drum (Figure 10.7). Remember *top to top*.

FIGURE 10.6 Lang lay ropes provide increased resistance to abrasion, greater flexibility, and more fatigue resistance than regular lay ropes.

The severity of duty and the working conditions to which the rope has been exposed will be your deciding factors. According to ASME B30.5 Safety Standard for Mobile Cranes, once a hoist rope reaches any of the removal criteria based on the judgment of a qualified person, it may be allowed to operate to the end of that shift but must be replaced before the next shift. The removal criteria required are listed here for your review:

- **Diameter reduction.** Remove when the measured diameter has decreased more than 5% below the nominal diameter at any location along the rope's entire length.

- **Distortion of the rope structure.** Remove any rope with evidence of kinking or crushing where the minimum dimension across the distorted section is 5/6 of the nominal diameter or less or where there is birdcaging, severe doglegs, core protrusions, or any other damage resulting in permanent distortion of the rope structure.
- **Corrosion.** Remove when widespread or localized corrosion is present, as evidenced by surface pitting, magnetic debris coming from the valleys, or any other obvious signs indicating significant rope degradation due to corrosion conditions.
- **Waviness or corkscrew effect.** According to ASME B30.30, excessive waviness of wire rope should be removed from service when the overall diameter has increased to a value greater then 110% of the nominal rope diameter. More information can be found in ASME B30.5.
- **Heat damage.** Remove when you see any apparent damage from any heat source, such as welding, power line strikes, or lightning strikes.
- **High or low strands.** Remove when a strand is higher or lower than half of the strand's diameter above or below the surface of the rope.

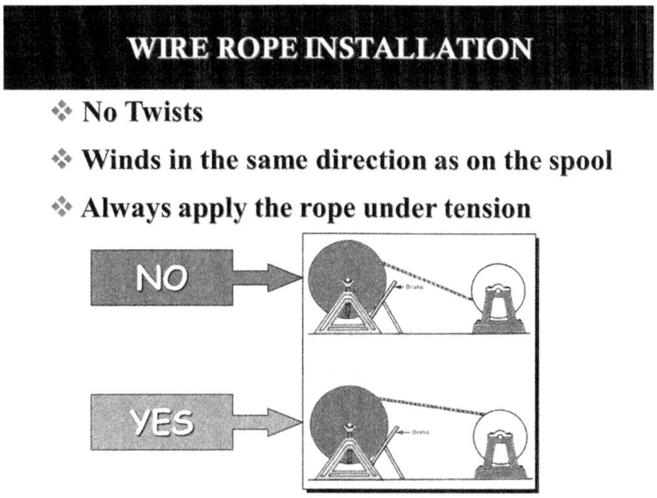

FIGURE 10.7 Proper wire rope installation, is shown here. Always remember *top to top*.

Idle Wire Ropes

Wire rope that has been idle for a period of 1 to 6 months due to shutdown or storage of the machine is subject to the *frequent* inspection requirements as outlined by ASME B30.30 Ropes (Figure 10.8). Wire rope that has been idle for a period of over 6 months due to shutdown or storage of the machine is subject to the *periodic* inspection requirements as outlined by ASME B30.30 Ropes.

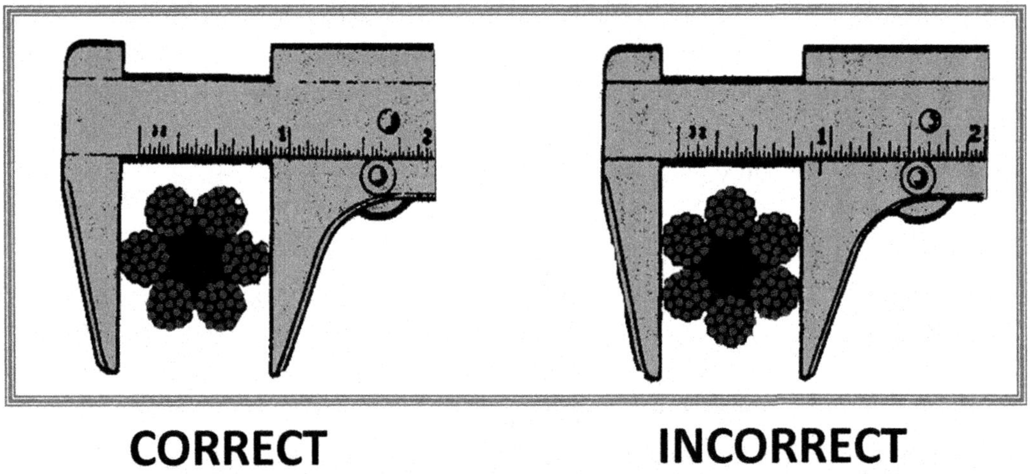

FIGURE 10.8 The correct way to check wear of a wire rope. Always make your measurement on the higher part of the strand rather than the lower as shown here. The diameter of a wire rope is the distance measured between the widest points of the wire rope. Measurements must be made on the outer surface.

Replacement Wire Ropes

Your replacement rope must have a minimum breaking force that is at least as great as the original rope furnished or recommended by the crane manufacturer. Any deviation from the original size, grade, type, or construction must be specified by the wire rope manufacturer, crane manufacturer, or a qualified person.

Inspecting Wire Rope

Wire rope is a machine that, unfortunately, starts to wear out the moment after you've installed it. A highly trained qualified person understands all the dynamics of the wire rope on your crane and must be placed in charge of monitoring this process. The person will use his or her experience along with the ASME B30.5 mobile crane standard and ASME B30.30 ropes safety standards as a guide. If the rope becomes a hazard or meets any one of the removal criteria, it will be time for the qualified person to take the rope out of service. Wire rope that the crane operator expects to use in service must be inspected every day. Although frequent inspections are not required, they are highly recommended. Always check your crane manufacturer's requirements and specifications. Record keeping is required for periodic inspections. Wire rope may need to be inspected more often, depending on the recommendations of a qualified person and whenever the crane is working multiple shifts or duty cycle operations, whenever corrosive conditions exist, and when the rope is reaching the end of its life.

Broken wires are a clear sign that the rope is showing signs of fatigue. Wires outside the crown are a likely sign. Always look for broken wires during your inspection, mainly the areas of the rope that are in contact with the hoist block or auxiliary boom head sheaves or hoist drums when loads are being hoisted.

Other important points to consider are:

- **Heat exposure.** Keep wire ropes away from arc welding. If, for any reason, the rope contacts an electric arc, the entire rope should be replaced. Watch for any other type of heat exposure that has damaged the wire rope.
- **Hoist drums.** Check the condition of drum flanges. Both smooth and grooved hoist drums should be checked for evenness. Check hoist drums to ensure the wire rope is wrapping properly and the correct number of dead wraps remains on the drum when most of the rope is paid out. No fewer than two wraps of wire rope shall remain on the hoist drum at each anchorage of the hoisting drum when the hook is in its extreme low position. The hoist drum flange shall extend a minimum of ½ inch over the top layer of rope when the hoist drum is full at all times. Some manufacturers' specifications may be more stringent than ASME Safety Standards and require three full wraps. Always check and follow your crane manufacturer's requirements.
- **End attachments.** Check anchor points at the hoist drum and at any wedge and socket assemblies for broken wires. These are high-stress areas that can be hard to detect. Also check for signs of corrosion in these areas. Whenever necessary, cut off any damaged ends and reattach any anchors or dead-end attachments. This is also a great time to look over all wedges, socket pins, and other hardware in use.
- **Sheaves.** Check all boom points and auxiliary boom head sheaves for smoothness and groove sizes using a groove gauge. Check for signs of grooves becoming too wide or narrow that could damage the wire ropes on the crane.

On a monthly basis, you should inspect the entire length of the rope. On a daily or frequent basis, you should keep a close eye on the high-stress areas of the wire rope contacting any sheaves or hoist drums in use, particularly when the wire rope is under stress lifting a heavy load. The amount of rope you expect to use during that shift should be inspected daily.

Sheave Sizes

Sheave sizes differ by type of sheave:

- Boom-hoisting sheaves shall have pitch diameters of not less than *15* times the nominal diameter of the rope used.
- Load-hoisting sheaves shall have pitch diameters not less than *18* times the nominal diameter of the rope used.
- Load block sheaves shall have a pitch diameter not less than *16* times the nominal diameter of the rope used.

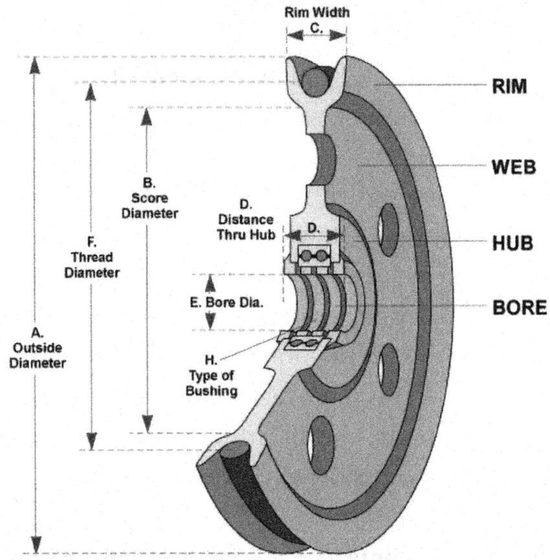

FIGURE 10.9 Nomenclature of a sheeve.

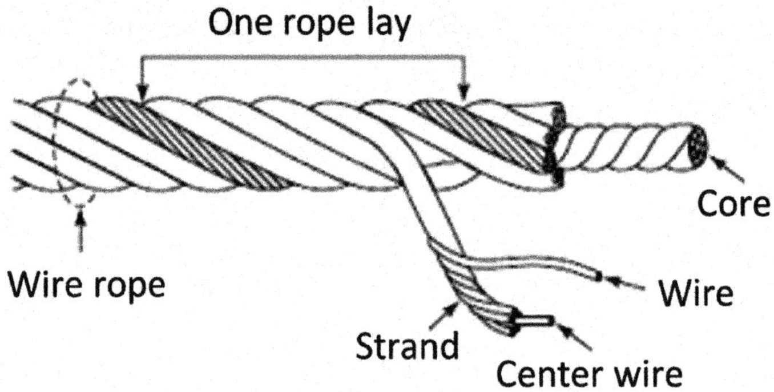

FIGURE 10.10 Wire rope nomenclature.

Grades of Wire Rope

The toughness, strength, abrasion, and corrosion resistance are indicated in the wire's grade designation. These grades measure the tensile strength of the individual wires in long tons and range from the lowest at 100/110 plow steel through 110/120 improved plow steel (IPS), to 120/130 special improved plow steel (SIPS), to a high of approximately 130/140 for extra-improved plow steel (EIPS). The most used of this group is the 110/120 improved plow steel.

Wire Rope Classifications

In addition to the differences in core construction, the number and patterns of the individual wires and the strands themselves will vary, depending on what working characteristics are required (Figure 10.11). Although the number of wires in the strands may vary, the most used ropes are grouped into categories (Figure 10.12):

- **6x19.** This category contains ropes that have 6 strands that are made up of 15 through 26 wires per strand, of which no more than 12 are outside wires.

- **6x37.** This category contains ropes that have 6 strands that are made up of 27 through 49 wires per strand, of which no more than 18 are outside wires.
- **8x19.** This category contains ropes that have 8 strands that are made up of 15 through 26 wires per strand, of which no more than 12 are outside wires.
- **19x7.** This category contains ropes that have 19 strands, where each strand is made up of 7 wires. It is manufactured by covering an inner rope of 7x7 left lang lay construction with 12 strands in right regular lay. This is usually classified as a rotation-resistant rope.

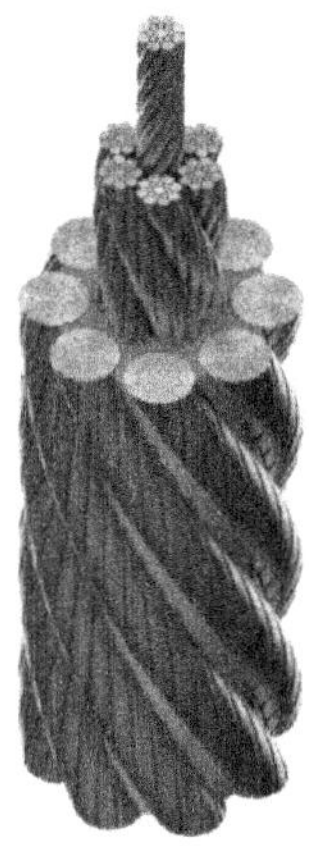

FIGURE 10.11 View of strands and core of a wire rope.

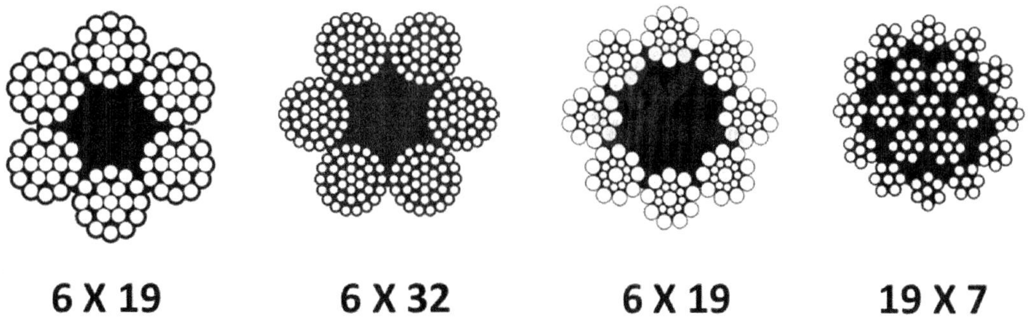

FIGURE 10.12 Views of the four general classifications of wire ropes

Safety Factors

To account for all the stresses placed on wire rope during hoisting operations and to help guard against failure, the actual load placed on the rope should represent only a fraction of its ultimate or breaking strength. This margin of safety, or "safety factor," is used to determine the maximum working load limit (WLL) that can be applied to the rope.

Running Ropes

- Standard running rope: 3.5 to 1
- Rotation-resistant rope: 5.0 to 1
- Standing ropes: 3.0 to 1

Running Ropes When Hoisting Personnel

- Standard running rope: 7.0 to 1
- Rotation-resistant rope: 10.0 to 1

Wire Rope Break-in and Lubrication After Installation

When installing wire rope on a mobile crane, always refer to the manufacturer's recommendations. After installation, the wire rope should be properly broken in as recommended by your wire rope manufacturer and the specifications for your crane.

Keep your wire rope lubricated for a longer service life. Start by cleaning the rope to remove any grime or dirt or any other materials that would prevent lubricants from fully penetrating the wire rope. Develop a schedule of regular maintenance and lubrication for the wire rope based on corrosive conditions and the type of work being done with the crane. Lubricate more often when conditions get rough, like when performing duty cycle operations such as dragline, clamshell, and pile-driving work. Good lubrication applied before high-speed cycling of the wire rope when using the crane for concrete pours using a concrete bucket helps reduce heat and friction. Working around saltwater or when heavy dust is in the air is always a good time to consider lubrication more often. Wire rope dealers can provide specifically formulated lubricants designed to keep hoist ropes in good operating condition (Figure 10.13).

FIGURE 10.13 Properly lubricated wire rope installed on two hoist drums.

Wire ropes come from the factory prelubricated; however, this lubrication won't last long. Wire ropes create a lot of heat and friction as they travel through the hoisting system on the crane. Bending over sheaves and around hoist drums requires ropes to be cared for and lubricated often. Lubrication can be purchased from your wire rope dealer. Make sure the lubricant you buy contains no alkalis or acids. Using motor oil is a bad idea because this type of oil contains properties that can damage the rope.

Consider, whenever possible, carrying out wire rope lubrication and maintenance at the end of the last shift of the week. This task will take two craftworkers to complete—the crane operator and a qualified assistant. The crane operator will fully extend all powered boom sections and spool out all the wire rope on hydraulic cranes to the last three wraps of wire rope on the crane's hoist drum. The crane operator will be at the hoisting controls while the assistant checks the hoist drum (1) to ensure it hasn't moved out of place during regular hoisting operations and (2) to place rags under the drum to absorb any drips or excess lubricant. As the crane operator gently begins to return the hoist line back onto the hoist drum, the assistant will spray lubricant onto the wire rope as it spools back onto the hoist drum. The wire rope will open slightly, allowing the administration of lubricant to penetrate deep down to the core as layer after layer of rope tacks back onto the drum. Letting the lubricant soak into the rope over a weekend allows good penetration and avoids a big oily mess when the crane operator returns to work and extends the boom sections and spools hoist ropes out after returning the crane back to work.

Regardless of how you plan to lubricate the wire rope on your crane, it is highly important you have a plan and carry it out on a frequent basis.

Standard Running Ropes

Wire ropes come in many sizes and types of construction, which determine their breaking strength. Breaking strength can be affected by the overall condition of the rope and how well it has been maintained. Important factors to consider when selecting wire rope on your crane are the wire's breaking strength and working load limit. As noted above, a standard running rope on a mobile crane has a minimum design factor of 3.5. This means the working load for this type of wire rope would be based on a safety factor, wire rope breaking strength divided by the safety, or design factor equals the working load limit of the rope. Wire rope is a complex machine with several different specifications and properties that can affect its performance and service life. A finished wire rope has four basic components that make up its design. They include:

- Wires made from metal that form a strand
- Multiple wire strands laid around a core in a helical pattern
- A fiber or steel core
- Wire rope lubrication

Wire rope is an integral part of a mobile crane system. Proper maintenance, inspection, and use are important in maintaining the wire rope at a safe operating condition. Professional crane operators understand that even a new wire rope properly installed on their crane starts to wear out immediately. It is critical that the crane operator and other members of the crane maintenance program and crew know when the wire rope's condition indicates it is time for the wire rope to be removed from service. Failure to do so in a timely manner could be deadly.

All wire ropes must be thoroughly inspected at regular intervals. The longer a wire rope has been in service, the more often it should be inspected. The severity of the service must determine how frequently the rope is inspected. It is the responsibility of the qualified person inspecting the wire rope to know and follow the inspection criteria for it and the application in question.

Standard running ropes in service on mobile cranes have a safety factor of 3.5 or greater, and if used to hoist personnel, the ropes have a safety factor of 7:1 or greater.

Running Ropes Removal Criteria

Running ropes shall be removed from service when there are more than 6 broken wires in 1 rope lay or 3 broken wires in 1 strand in 1 lay, where a rope lay is the length along the rope in which 1 strand makes a complete revolution around the rope. A rope lay is a unit of measurement that is used when determining if a wire rope should be replaced due to visibly broken wires (Figure 10.14).

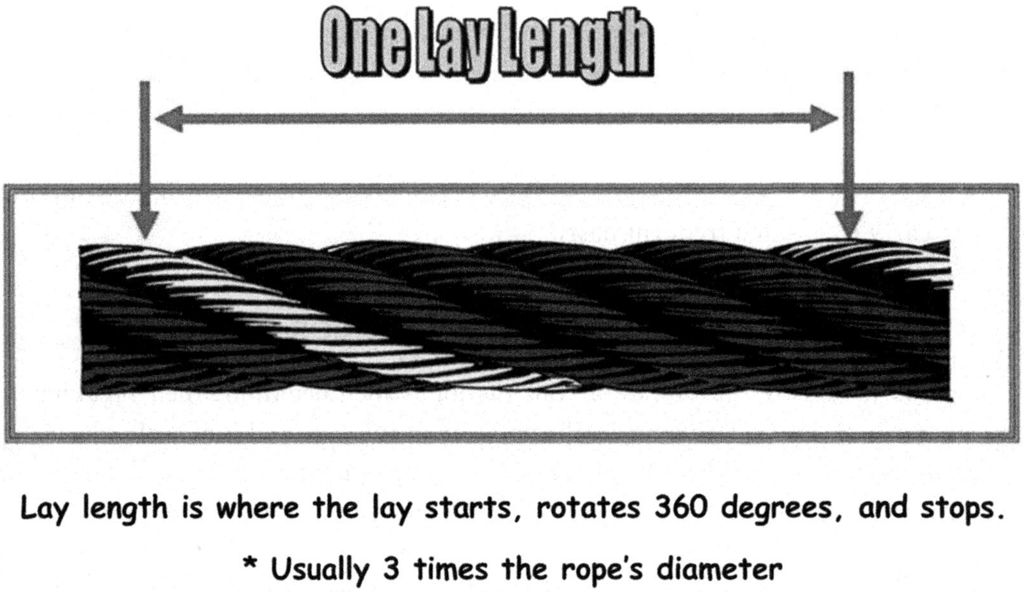

FIGURE 10.14 A rope lay, or lay length, is a unit of measure that is used when determining if a wire rope should be replaced due to visible broken wires.

Standing Ropes

A standing rope is permanently fastened on the mobile crane and used as a guy. A standing rope is a fixed-length rope forming part of the boom suspension system; it is also called a *hog line*, *boom stay standing line*, *stay rope*, or *pendant*.

Wire rope defections to watch for include:

- Broken wires due to fatigue
- A reduction in rope diameter due to failure of the rope core
- Corroded wire rope, which leads to loss of metallic cross section
- Wave formation in the rope
- An increase in rope diameter

Fracture, wear, and corrosion are the main damage types for wire ropes. Unlike fracture, both wear and corrosion are gradual processes that interact with each other, inducing failure of wire ropes.

As noted above, standing ropes are often called *boom pendants*. Boom pendant ropes operate within pairs as a means of connecting the boom hoist rope to the end of the boom, allowing various lengths of boom to be installed on the lattice boom crane. Jib forestay and backstay pendant lines are included. A jib is an extension of a boom cable being mounted with its own forestay pendant lines connected from the jib tip to the jib mast. The jib mast is connected to the boom tip. The jib backstay pendant is manually adjustable to change the angle of the jib. Boom standing ropes, or pendants, and jib forestay and backstay pendants must be inspected by a qualified person. Standing ropes used on mobile cranes have a safety factor of 3.0 or greater.

Criteria for removing standing ropes include:

- When there are 3 or more broken wires in 1 rope lay anywhere in the wire rope or 2 or more broken wires at the end connections (Figure 10.15).
- When there is severe corrosion as evident by pitting.
- When there is pitting of the wire rope. Socketing shall be done in the manner specified by the manufacturer to detect pitting of the wire rope. Watch for broken wires in this area. If found, the assembly must be replaced.

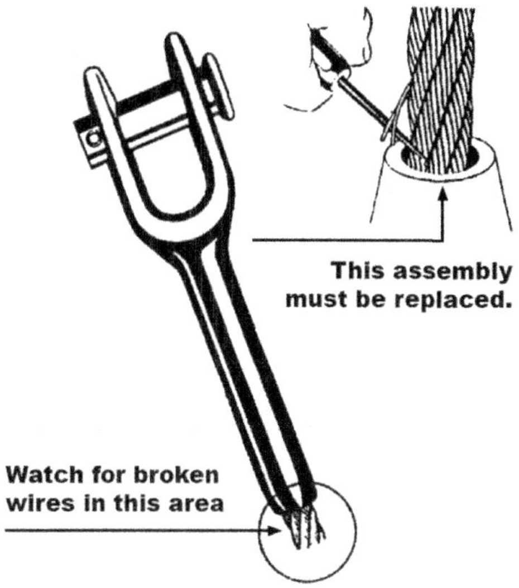

FIGURE 10.15 If two or more broken wires are found in this end connection, the assembly must be replaced.

Rotation-Resistant Rope

When a load is placed on a hoisting rope, torque energy is created inside the rope as the wires and strands begin to straighten out. Hoist ropes are designed to have load torque.

Rotation-resistant ropes are designed so that the torque energy created by the wire rope core acts in the opposite direction of the torque created by the outer strands (Figure 10.16). The outer strands create more torque than core strands, so rotation occurs in a direction that unlays the outer strands whenever the rope is allowed to rotate while it is under load. Rotation-resistant rope accomplishes better results when it's not allowed to rotate. When a rope is allowed to rotate, it will cause internal wear to the rope that will be very difficult to see when performing an inspection. Loss of strength will occur, and there will be a possible load unbalance due to unequal loading in the rope.

FIGURE 10.16 A view of a typical rotation-resistant rope design.

Rotation-Resistant Rope Removal Criteria

Many of the areas for removal criteria for rotation-resistant ropes are more stringent than the criteria for standard running rope. Two of the more stringent areas to review are:

- Abnormal loss of rope diameter
- Unconventional lengthening of wire rope lay or protrusion of wires between the outer strands. Both are cause for prompt removal of the wire rope from service.

Some kinds of wire rope, such as lang lay ropes, are more vulnerable to rotation when under load. Rotation-resistant wire rope is fabricated to resist spinning, twisting, and rotation. This type of rope is usually used as a single line. However, if the manufacturer allows, the rope may be used when reeving up multiple parts of a line.

Other reasons to remove rotation-resistant rope are:

FIGURE 10.17 Rotation-resistant ropes require a higher safety factor because they are more likely to become damaged internally without the damage being detected.

1. When you see two randomly distributed crown wire breaks in six rope diameters, or four randomly distributed crown wire breaks in 30 rope diameters. Rotation-resistant ropes, because of their elevated vulnerability to damage, increase degradation when working on equipment with design limitations. The internal worsening of rotation-resistant ropes may not be observable. Rotation-resistant wire ropes require a higher safety factor because they are more likely to become damaged internally without the damage being detected (Figure 10.17). Rotation-resistant wire ropes used on mobile cranes have a safety factor of five or greater, and if being used to hoist personnel, have a safety factor of ten or greater.
2. When 1 outer wire is broken at the point of contact with the core of the rope that has worked its way out of the rope structure and protrudes or loops out from the rope structure.
3. When there is an independent wire rope core protrusion between the outer strands.
4. When there is any kinking, crushing, birdcaging, or other damage resulting in distortion of the rope structure.
5. When there is any apparent damage from any heat source, including, but not limited to, welding, power lines, and lightning.
6. When there is a reduction from the nominal diameter of more than 5%.

Rotation-resistant ropes must be handled differently than other types of wire ropes. The operational properties of rotation-resistant ropes are the result of the lay of the outer strands to the inner strands. The outer strands are not pre-formed and must be tightly restrained when the rope is cut to maintain the rope's operational properties. The welded ends provided on both ends of the rope should not be removed.

If you must cut a rotation-resistant rope, the following steps must be carefully observed:

1. Before cutting the rope, make three separate bands with seizing on each side where the cut is to be made (Figures 10.18 and 10.19). A total of six bands of seizing for each cut will be needed. Each band of seizing is to be tightly wrapped and have a minimum length of 1½ times the rope diameter. The two bands of seizing closest to the cut should be located at a distance equal to one rope diameter away from the cut. The four remaining bands of seizing should be evenly spaced at a distance equal to three rope diameters.
2. If a welder is available, the cut should be made with an abrasive saw. Immediately after the cut, both ends of the rope are to be cap-welded so that all inner and outer strands are welded

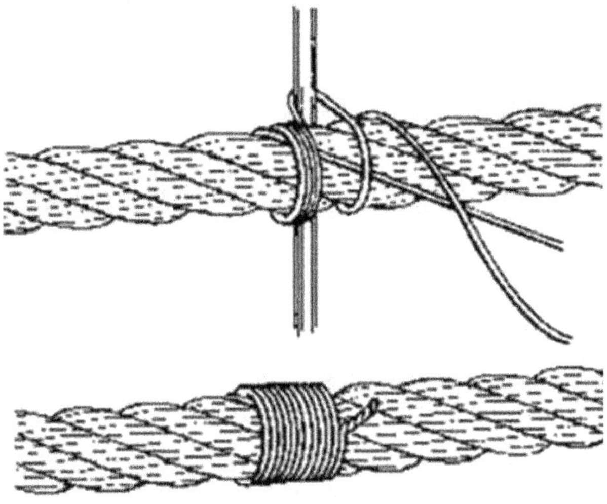

FIGURE 10.18 Proper wire rope seizing methods.

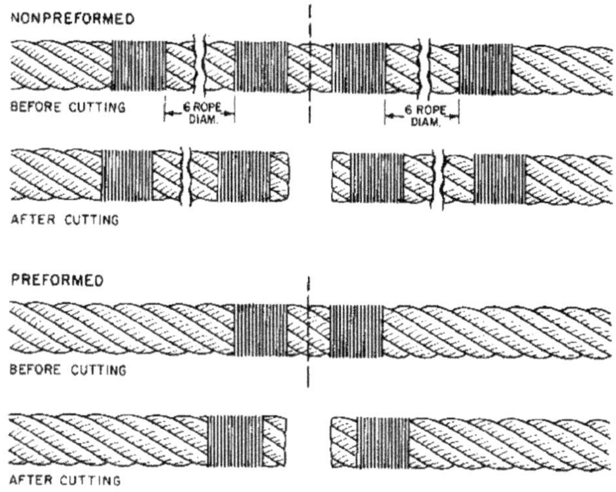

FIGURE 10.19 Steps to properly seize rope before cutting.

together, preventing any movement between them. *Note:* The outer strands must not be able to move with respect to the inner strands. The weld must not exceed the diameter of the rope.

If a welder is not available, the cut is to be made with an acetylene torch. The cut is to be made in such a way that both ends of the rope are completely fused so that all inner and outer strands are bonded together, preventing any movement between strands. It is important that the outer strands not be allowed to move with respect to the inner strands. The weld must not exceed the diameter of the rope.

3. Once the cuts and welding have been completed, the seizing bands are to be left in place if possible.

Synthetic Crane Hoist Line

Synthetic rope used for mobile crane hoist line operations is being utilized more than ever before (Figure 10.20). Both manufacturers and users say that synthetic ropes are easy to handle, much lighter and easier to work with than wire rope, and easy to inspect. Other benefits of the rope are its high strength-to-weight ratio, its bend fatigue durability, and its ease to spool onto the crane. Statistics show strength similar to wire rope hoist lines, with an 80% weight reduction. Synthetic ropes can reduce wear on hoist drums and reeving components due to less surface pressure and friction. This type of rope has other advantages in comparison with wire rope because it does not fray or produce sharp barbs that cause hand injuries and does not produce defects like kinking and birdcaging.

FIGURE 10.20 Synthetic hoist rope installed on a mobile crane.

Other reported benefits include:

- Minimizes wear on sheaves and drums
- Is less likely to have rope drum crushing
- Diminishes hand injuries from sharp broken wires
- Decreases load spin and cabling
- Presents less difficulties during installation and reeving multiple parts of line
- Has a 5-to-1 safety factor

Synthetic ropes offer an alternative to traditional wire ropes. However, regardless of what type of rope is on your mobile crane, it's only a matter of time before it shows the signs of everyday use of hoisting heavy loads running over sheaves and around hoist drums. All rope, regardless of how good or durable, will wear out over time. The most customary error when using hoist ropes is to disregard signs of wear and damage that occurred during use on the job. All ropes should receive a thorough, frequent inspection. Rope failure can lead to serious hoisting and rigging accidents, property damage, personnel injuries, and loss of life.

When inspecting synthetic crane hoist lines, watch for these defects:

- Cut strands
- Gouges and badly abraded spots
- Discoloration
- Seriously worn surface yams
- Weak or brittle filaments or fibers
- Inconsistent diameter
- Signs of melting or glazing

Other considerations, observations, and concerns include:

- Signs of chemical contamination
- Visible sheen
- Stiffness
- Fused fibers
- Strands pulled away from the rest of the rope

It is very important that the synthetic rope used on your crane is in good condition and meets the operating requirements of the manufacturer of the crane. Always select the correct rope for the job at hand, and be sure that it meets all safety requirements.

Mobile Crane Wire Rope Terminations

The type, condition, and correct installation of any wire rope termination is important to the safety of any lifting operation. Ideally, to attain maximum rope strength, the termination should not bend or distort the rope in any way. Swaged or spelter sockets used on crane pendant lines are great examples of a good termination.

As a crane operator, lift director, rigger, or other member of the ground crew and lifting team, it is important to know and understand the various types of end terminations that can be used at the end of a length of wire rope on a mobile crane. It is very beneficial that members of this team understand all the various types of wire rope on the crane and their specifications and know when to remove the hoisting rope from service.

Common Wire Rope Fittings

Wedge sockets are fittings used to temporarily terminate the end of a wire rope. Wedge sockets are very popular in the craning industry. They are easy to install and fast to dismantle. Two types of wedge sockets are offered: the standard type and a special type with a wedge that is designed to protrude from the nose of the socket, allowing a wire rope clip to be attached through the wedge to the dead end of the rope. Wedge socket efficiency is 75 to 80% of rope breaking strength. Wedge sockets secure the rope to the end attachment by passing it around a grooved wedge and then pulling it down under load into the bowl of the attachment to seat it. When properly applied and maintained, these terminations are 100% efficient. When installing a wire rope clip, it is attached through the wedge to the dead end of the wire rope. A tail length is required based on the type of wire rope as listed here. When installing the live end, the rope must be in line with the pin hole of the socket assembly.

Tail Length Requirement

- For 6- to 8-strand rope, a minimum of 6 rope diameters and not less than 6 inches
- For rotation-resistant wire ropes, a minimum of 20 rope diameters and not less than 6 inches

Always refer to the ASME B30.5 safety standard for more guidance. Several methods for securing the dead end of a rope or tail to prevent it from slipping or pulling through under load have been approved and can be reviewed in the safety standard. Although the methods shown in Figure 10.21 may or may not be practiced for all crane hoisting operations, one point is clear: It is no longer acceptable to secure the dead end of the rope to the live end by means of a U-bolt clip because damage can occur to the live end of the rope, weakening its capabilities.

The most common wire rope installation used in crane operations is the wedge and socket assembly. These assemblies are used on hoist lines or running ropes where the dead end terminated must be changed frequently. In good condition and correctly installed, they will maintain approximately 70% of the rope's minimum catalog breaking strength. Before installing the wedge, make sure that:

1. The wedge and socket are the same size and from the same manufacturer.
2. There is no excessive wear or cracks.
3. There is no sign of welding or brazing.
4. No modified substitute parts are used.

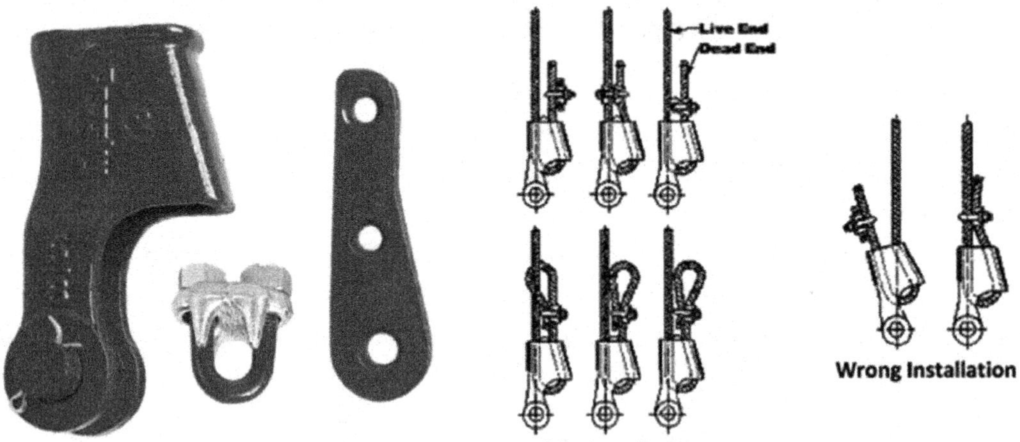

FIGURE 10.21 Several methods for securing the dead end or tail from slipping or pulling through under load have been approved. Displayed here are several correct applications. Correct installation will be based on type of wire rope being used and its diameter. Review your cranes operating manual for specific guidelines.

When installing running ropes, cut the rope clean, leaving the ends to move freely (Figure 10.22).

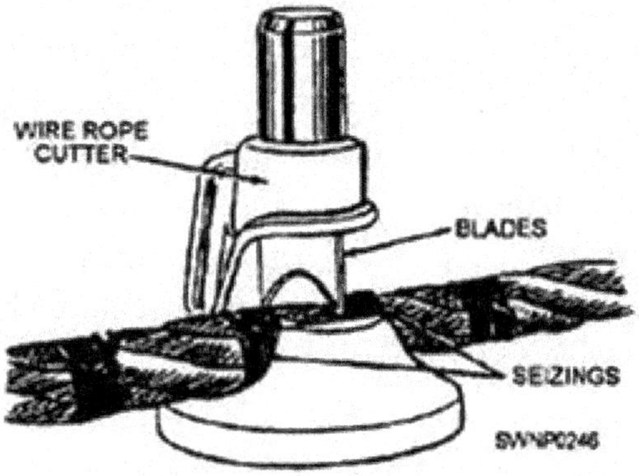

FIGURE 10.22 Cut running ropes clean, leaving the ends to move freely.

Wire Rope Clip Terminations

When wire rope U-bolts or clips are used to create the end fitting, it is highly important that the proper procedures be followed, because at the very best, this type of termination provides only 80 to 90% efficiency of the rope's minimum catalog breaking strength, depending on the size of the wire rope used. When installing wire rope U-bolts or clips in the configuration shown in Figure 10.22, make sure that:

- The proper amount of turnback is used, based on the wire rope diameter.
- Sling eyes are formed around thimbles.
- Only the correct type of clip is used, made of lifting-grade drop forged steel.
- U-bolts are properly spaced and torqued, based on rope diameter as shown in Figure 10.23.

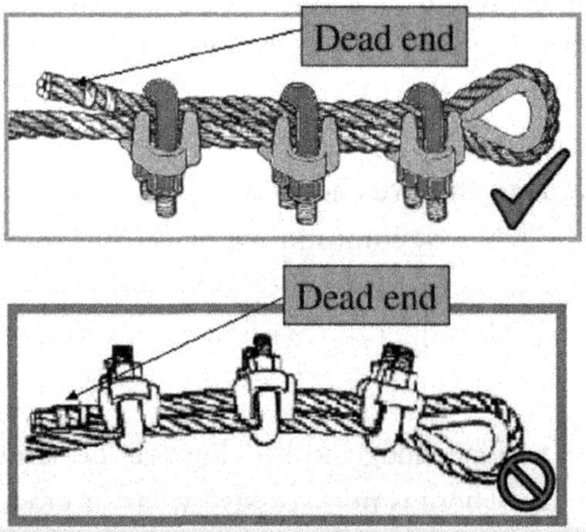

FIGURE 10.23 Make sure you closely follow the manufacturer's specifications when installing wire rope U-bolts or clips.

Have you ever heard the old saying, "Never saddle a dead horse"? This means when installing a U-bolt, always install the saddle part of the clip on the live end of the rope and the nuts and U-bolt on the dead end. The statement "Never saddle a dead horse" has been used throughout the craning and rigging industry for decades. The phrasing helps everyone installing U-bolts to remember the correct way to install them. The *dead end* is the short portion that is turned back to form the tail. The U-bolts are referred to as a *termination*. This complete unit is known as the *assembly turnback*. Always wrap or seize the end of the turnback tail to prevent the movement of any wires or strands. Correct installation procedures, measurements for U-bolts, and torque specifications will be provided when you purchase these components from your wire rope or rigging provider. Remember to use only drop forged steel U-bolts for making these types of terminations. A cheaper, light-duty, malleable U-bolt clip can be found in local hardware stores. Never use malleable clips for lifting terminations; they are made for securement, not for lifting.

Boom Hoist Wire Ropes

Mobile cranes manufactured in North America in the 1950s, '60s, and '70s were truly the workhorses of America. These cranes were instrumental in building the highway systems, bridges, power plants, and dams we all benefit from today. The lattice boom–type machines of this time period were known as *conventional cranes*. The mechanics of these machines were much different from the workings of the modern lattice boom cranes of today. The conventional cranes of the past worked with torque converters, clutches, breaks, and frictions to hoist heavy loads. Lattice boom cranes raised and lowered their booms by means of a gantry system, boom hoist drum, and wire rope.

Today hydraulic boom cylinders raise the booms of the modern lattice boom cranes. However, there are plenty of the older, conventional-type cranes still in service throughout North America. This means crane operators and maintenance personnel must understand wire rope, particularly the rope used on the boom hoist—how to inspect it, maintain it, and lubricate it, and also when to take it out of service. Boom hoist ropes run at a 3.5:1 design factor and 15:1 bending ratios that will wear out much faster than other hoist line wire ropes on the crane operating at a 5:1 design factor and 18:1 bending ratio. When performing high cycling operations such as duty cycle work with a conventional crane that is reeved with 18 parts of boom hoist wire rope, lifting and lowering of just one load can cause sections of the wire rope to bend as many as 32 times. Other high-stress activities include raising the lattice boom from the ground after boom assembly operations due to high crushing stresses at the boom hoist drum, primarily at the crossover and step-up points on the drum. Broken wires can cause considerable damage and problems with boom hoist ropes.

The way a crane is being used will have a great influence on whether the wire rope will deteriorate on the drum or within the rope reeving. Some parts of the rope will be difficult to inspect, and other areas very easy, depending on the length of rope being looked at. One thing that is important: Boom hoist wire rope needs to be inspected and lubricated often. Criteria for removing boom hoist ropes from service are listed here for your review:

- ➤ Standard running rope is three broken wires in one strand or six broken wires in one lay.
- ➤ Rotation-resistant rope is two broken wires in six rope diameters or four broken wires in 30 rope diameters:
  - One wire broken in a valley
  - Kinking, crushing, birdcaging, or any other damage resulting in distortion of the rope structure
  - Any heat damage
  - Corrosion of the wire rope

Be aware of any boom hoist wire rope on your crane and how to properly maintain it and inspect it and when to take it out of service. Conventional lattice boom cranes are still widely used in the craning industry today and will continue to be used for years to come as long as they are properly taken care of.

Chapter 10 Review Quiz

1. Running ropes shall be removed from service when there are
   a. three broken wires in any one lay length or 30 rope diameters
   b. six randomly distributed broken wires within one lay length or three broken wires in one strand in one lay
   c. six or more broken strands in any three individual wires or two rope lays
   d. more than six broken wires in a rope's length or in two lay lengths
2. The frequency of wire rope lubrication will depend on
   a. the rope diameter
   b. the wire rope design factors
   c. the severity of duty and corrosive conditions
   d. the age of the rope
3. When replacing wire rope on your mobile crane, it is important to make sure that
   a. it is regular lay rope
   b. it is the right lay rope
   c. it is the original capacity and use
   d. it is rotation resistant
4. The design factor for running ropes that wind on drums or travel over sheaves shall not be less than
   a. 3.0:1
   b. 3.5:1
   c. 5.5:1
   d. 6.0:1
5. A wedge socket assembly must be assembled with the load line pulling in a straight line with the
   a. wedge
   b. shank
   c. pinhole
   d. hoist block
6. The correct method for installing a new wire rope from a reel to the hoist drum is to
   a. unspool the rope from the top of the reel to the bottom of the hoist drum
   b. unspool the rope from the top of the reel to the top of the hoist drum
   c. unspool the rope from the bottom of the reel to the top of the hoist drum
   d. remove all the rope from the reel so it won't make any difference
7. The design factor for live or rotation-resistant ropes that travel over sheaves shall be
   a. 3.5 or greater
   b. 5.0 or greater
   c. 4.0 or greater
   d. 2.0 or greater

8. Rotation-resistant wire rope requires a higher safety factor because
   a. it is used frequently with a swivel
   b. it is more likely to become damaged internally without being detected
   c. it is frequently used when there is repeated shock loading
   d. it is frequently used when there is repeated dynamic loading

9. According to the ASME B30.5 mobile crane safety standard, wire rope clips of the single-saddle U-bolt–type clip shall be
   a. cast iron
   b. malleable cast
   c. drop forged steel
   d. chrome-plated cast

10. Wire rope clips attached with U-bolts shall have
   a. the U-bolt over the dead end of the rope and the live rope resting in the saddle
   b. the U-bolt over the live end of the rope and the dead rope resting in the saddle
   c. the U-bolt over the dead end and a loop connecting the live end
   d. the U-bolt over the live end and a loop connecting the dead end

11. According to the ASME B30.5 mobile crane safety standard, the minimum that the drum flange shall extend over the top layer of rope at all times is
   a. ¼ inch
   b. ½ inch
   c. ¾ inch
   d. 1 inch

12. According to the ASME B30.5 mobile crane safety standard, the winch drums shall provide a first-layer rope pitch diameter of not less than
   a. 14 times the nominal diameter of the rope used
   b. 16 times the nominal diameter of the rope used
   c. 18 times the nominal diameter of the rope used
   d. 20 times the nominal diameter of the rope used

13. According to the ASME B30.5 mobile crane safety standard, all hook and ball assemblies shall be labeled with their
   a. effective weights
   b. diameter
   c. rated capacity and weight
   d. circumference

14. According to the ASME B30.5 mobile crane safety standard, the design factor for boom pendants or standing ropes shall not be less than
   a. 2.0
   b. 3.0
   c. 4.0
   d. 5.0

15. According to the ASME B30.5 mobile crane safety standard, the design factor for live or running ropes that wind on drums or travel over sheaves shall not be less than

- **a.** 3.5
- **b.** 4.0
- **c.** 4.5
- **d.** 5.0

16. According to the ASME B30.5 mobile crane safety standard, for supporting rated loads, rotation-resistant ropes shall have a design factor of

- **a.** 3 or greater
- **b.** 4 or greater
- **c.** 5 or greater
- **d.** 6 or greater

17. Wire rope inspection on your crane should

- **a.** occur daily
- **b.** occur weekly
- **c.** occur monthly
- **d.** be conducted by a certified crane operator

18. A rope lay is described as the

- **a.** axial distance along the rope in which one strand makes one complete turn around the rope
- **b.** axial distance along the strand in which the strand makes one complete turn around the rope
- **c.** axial distance along the rope in which the strand makes half a turn around the core
- **d.** axial distance along the core in which the rope makes a full revolution around the strand

19. Standing or pendant ropes should be removed from service when there are

- **a.** three broken wires in any one lay length or one broken at the end connection
- **b.** five randomly distributed broken wires within a lay length or four at the end connection
- **c.** three or more broken wires in one rope lay anywhere in the wire rope or two or more broken wires between end connections
- **d.** more than six broken wires between end fittings or three in a lay length

20. According to the ASME B30.5 mobile crane safety standard, how many full wraps of rope shall remain on the drum when the hook is in the extreme low position and the boom is at maximum elevation and extension?

- **a.** One full wrap
- **b.** Two full wraps
- **c.** Three full wraps
- **d.** Four full wraps

21. Wire rope inspections on your crane need not be conducted at equal calendar intervals and should be more frequent

- **a.** as the crane ages
- **b.** during poor weather conditions
- **c.** as the rope approaches the end of its life
- **d.** whenever extra time allows

22. You are installing rotation-resistant wire rope and a 7.5-ton overhaul ball on your 100-ton truck crane. In this configuration, it is permissible to use a swivel

- **a.** if the master mechanic approves
- **b.** if the lift director approves
- **c.** if the certified rigger approves
- **d.** if the wire rope manufacturer approves

Chapter Summary—and Then Some

Clearly, there is so much to know and understand about hoist ropes. It is a specialized area for sure. Those who understand how to use the proper hoist rope, how to maintain it, and when to take it out of service are an asset to any crane hoisting and rigging operation. Much can be gained with this knowledge in the way of both production and safety. Hoist ropes are the lifelines of a mobile crane that hold the weight of the heavy loads we lift and place. Remember, no single hoist rope can do it all, and so it's important you know what rope to use and when. Characteristics of specific rope will influence the type of rope you use and, when based on strength, the fatigue resistance, crushing resistance, metal loss resistance, and deformation as well as resistance to rotation. Crane operators and other members of the lifting team who are experts in this area are truly a valuable and important part of the entire hoisting and rigging team. This type of knowledge is what separates the good crane operators and riggers from the great ones. Learn all you can about hoist ropes and other types of ropes on your crane.

Signal persons are often the eyes and ears of the crane operator. A qualified signal person provides hand signals approved by OSHA and the mobile crane safety standard ASME B30.5. Simultaneously with the lifting team, it is essential that the crane operator and the signal person, as well as all those involved in the lift, work in conjunction to keep the project running smoothly and on schedule. All crane operators, signal persons, and ground crew support members must be knowledgeable about standard hand signals. Approved mobile crane signals should be routine to crew members and should become second nature to the daily hoisting and rigging operations they perform regularly. Signal persons must provide clear, clean, and correct signals to the crane operator in order to make load movements. The qualified and highly trained signal person and his or her ability to give accurate hand signals are of vital importance on the jobsite.

11 Mobile Crane Communications

Mobile crane hand signals are a systematic set of hand and arm positions and movements clearly understood by the crane operator and the signal person. Hand signals, when given properly, allow the signal person and the crane operator to communicate over significant distances. When the crane operator and signal person work together, loads can be moved quickly and safely on the jobsite. Crane signal regulations provided by OSHA and safety standards published by ASME B30.5 stipulate that the qualified signal person and the certified crane operator must communicate by either discernible or audible signals at all times. Cranes are a very effective way of moving heavy loads on industrial and construction projects when a strong communication system is in place.

Properly delivered hand signals are an important part of safe crane operations. Signal persons must attend training to learn and memorize all the basic mobile crane hand signals. A qualified signal person is then capable to safely direct crane operators to raise or lower the boom, including hoist lines, load handling devices, and the load itself. The signal person must know how to direct the crane operator to swing the crane and move the load horizontally. They must also know how to safely stop the crane and understand how to eliminate any sway or swing on the crane, stopping any pendulum effect where the load moves independently of the crane. This is a skill used by both the crane operator and the signal person, and is often referred to as *covering the hook*, as shown in the drawing in Figure 11.1.

Communication between the crane operator and the signal person must be kept intact constantly during all movements of a lift. If at any time during the lift communication stops, the crane operator must cease all hoisting activities until the problem is resolved and communication is restored. All persons, prior to giving any signals to the crane operator, must be tested by a designated person to ensure their qualifications. The signal person must show that he or she understands the basic functions of the crane and its charted capacities. The signal person must be able to demonstrate hand signals when they are to be used as well as voice signals whenever they are chosen. Hand signals shall be those shown in Figure 11.2.

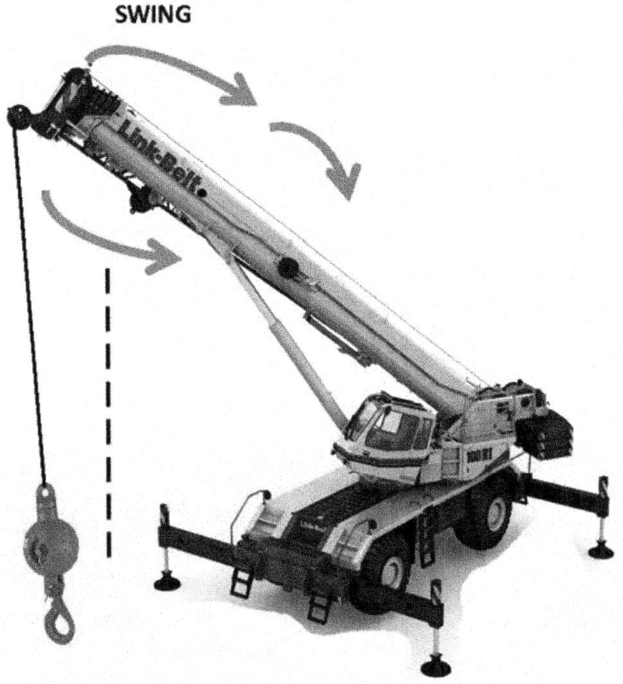

FIGURE 11.1 Crane operators and signal persons must know how to "catch the swing," to safely stop the crane and cover the hook.

Hand signals must be properly posted on the crane and throughout the jobsite, such as in job shacks, lunchrooms, or mechanic trailers. If voice signals are chosen for hoisting operations, the signals must be talked over and agreed on by the signal person and the crane operator, including the lift director. Any equipment used for communication should be tested before any lifts take place. Good-quality communication equipment is a great idea, and if batteries are the power source, backup charged batteries should be available at all times as well as another additional set. Before the lift begins, the crane operator and signal person should start with a simple greeting to ensure all equipment is in good working order. Start by calling for the crane operator by name. The crane operator should respond with the signal person's name. All directions given to the crane operator must be given from the crane operator's direction perspective. According to the ASME B30.5 safety standard, each series of voice signals shall contain three elements stated in order, as shown here:

1. Function and direction
2. Distance and/or speed
3. Function stop

The directive *slowly* can be used at any time with any signal. *Slowly* is always at the end of a command. For example, "Boom up slowly" or "Hoist down slowly."

For hoisting operations using voice signals, the lift director must consider the difficulty of the lift, the charted capacities, and the capabilities of the crane to be used, as well as the skill, knowledge, and experience of all those involved in the lift, including the crane operator and the signal person. For special operations, additions to, or modifications of, the standard signals may be needed. If so, the lift director, crane operator, and signal person must agree on these signals in advance of any hoisting oper-

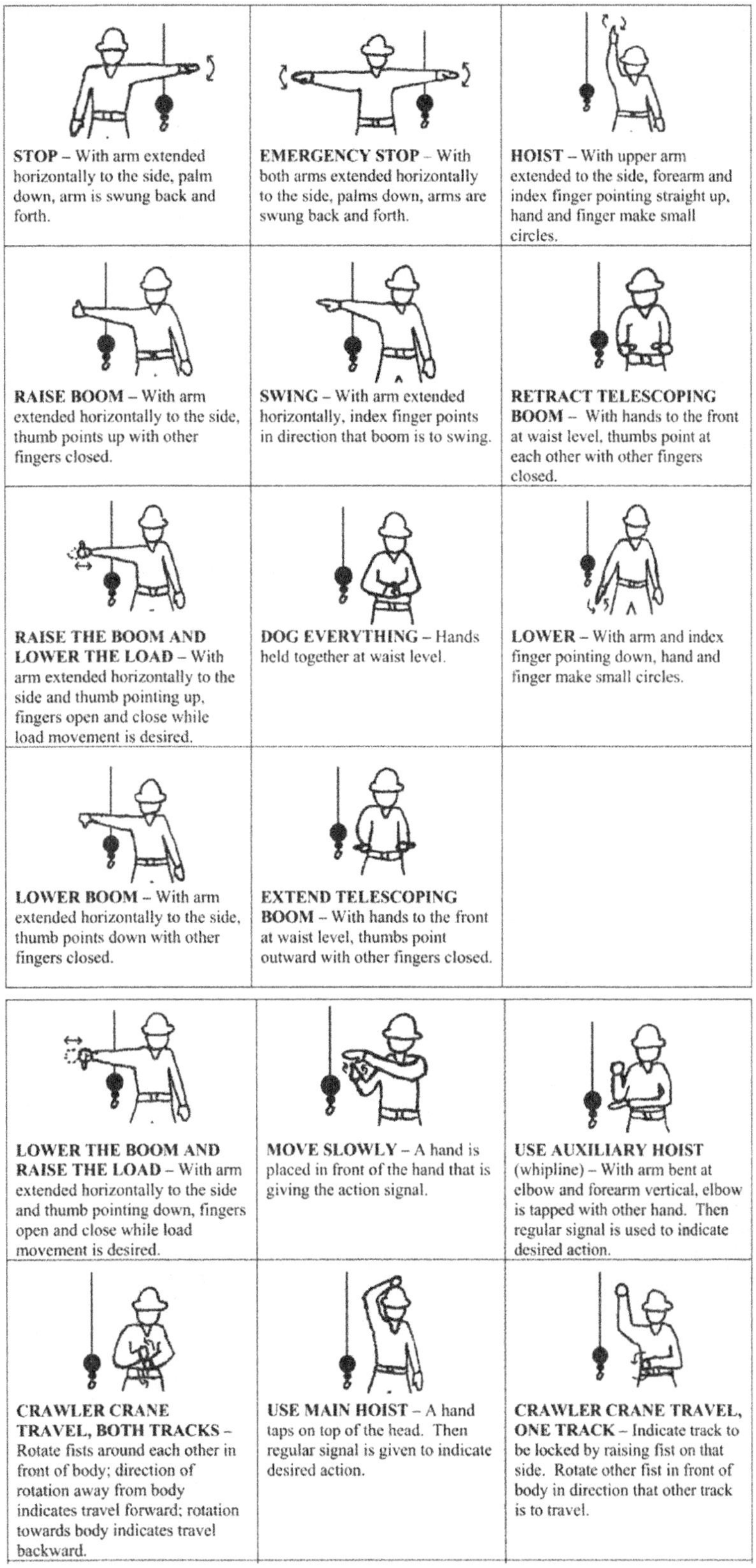

FIGURE 11.2 Standard signals to the crane operator must be those approved by ASME or OSHA.

ations, making sure new signals won't be in contention with regular standard signals. Function names should be the same as those illustrated in the ASME B30.5 mobile crane safety standard.

As an example, say you are a qualified signal person. You are about to give voice signals to a crane operator to swing the crane, remembering all directions will be from the operator's perspective. This is how that would sound:

"Swing left 75 feet, 50 feet, 25 feet, 15 feet. Slow down, 10 feet. Slow, slow 5 feet, swing stop."

Now you want to have the crane operator lower a load:

"Lower load 150 feet, 125 feet, 100 feet, 75 feet, 50 feet, 40 feet, 30 feet, 20 feet. Slow down, 10 feet. Slow, slow 5 feet. Slow, slow, lower stop."

It is important for the signal person to be detailed when giving the crane operator a description of the lift when the operator cannot see. This is known as *working in the blind*. Your hope is to turn the lights on in a dark room for the crane operator and paint a picture for the crane operator of what's going on.

Figure 11.3 illustrates the correct way a function such as "Hoist up" would be given as a voice signal in comparison with an OSHA standard hand signal. Once a function has started, the signal person should maintain constant communication with the crane operator and should avoid un-keying the radio microphone once the load is in motion (Figure 11.4). When the signal person wants to end a function, they must communicate to the crane operator function to stop to make sure the crane operator stops all movement, as shown in the Stop box in Figure 11.2.

From time to time, you must move cranes around on industrial or construction sites. Cranes, much like other kinds of heavy equipment, can be very large and have many blind spots for the driver moving the machine. Every year, ground personnel are run over and seriously injured or killed. For this reason, audible travel signals are provided to warn ground personnel that movement of the crane is about to take place. When moving the crane, these signals shall be used to help warn workers:

FIGURE 11.3 Once a function has started, the signal person should maintain constant communication with the crane operator and should avoid un-keying the radio microphone once the load is in motion.

FIGURE 11.4 The signal person must be alert and prepared for the job. They must always use proper ASME 30.5 hand and voice signals.

- Stop—one short audible signal
- Go ahead—two short audible signals
- Back up—three short audible signals

An audible emergency signal can be determined for a jobsite if agreed upon in advance. All emergency signals can be given by anyone. An emergency audible signal could be multiple short signals or a continuous signal.

The employer of the signal person must ensure, prior to giving any signals, that person meets qualification requirements. This can be accomplished by a third-party qualified evaluator or through a company qualified employer plan in accordance with ASME B30.5 mobile crane safety standards and OSHA regulations. Each signal person must meet these qualification requirements. Other important guidelines include:

- Know and understand the type of signals used for the crane hoisting the load (Figure 11.5).
- Be proficient in the application of the type of signals used.
- Have a basic understanding of the crane used and its limitations. It is important to understand the crane dynamics involved when swinging and stopping loads as well as boom deflection from lifting loads.
- Successfully complete both an oral and written examination as well as a practical test in accordance with applicable OSHA regulations or ASME safety standards.

FIGURE 11.5 A good signal person is also a qualified rigger who keeps an extra eye on the load. Here the signal person is giving the appropriate signal for hoist.

Other important skills and knowledge include:

- The signal person must be responsible and qualified.
- The signal person must be familiar with the work being done.
- The signal person must stay in view of the crane operator whenever possible.

- The signal person must use only proper signals for the crane in operation.
- Special signals must be agreed upon in advance of the lift and clearly understood between the crane operator and the signal person.
- The signal person must be clear and concise with signals, holding his hands and arms away from his body.
- All signals must be given in the crane operator's direction perspective.
- The crane operator shall receive signals from one designated person at a time.
- The crane operator shall receive an emergency stop signal from anyone.
- The signal person is knowledgeable about and understands the limitations of the crane.
- Whenever possible, the signal person must avoid holding tag lines while giving signals to the crane operator (Figure 11.6).

FIGURE 11.6 When using the radio to deliver voice signals, a good signal person avoids chewing gum or speaking with food in the mouth. A good signal person will speak clearly at all times.

- A good signal person is also a knowledgeable rigger.
- The signal person must ensure the boom point of the crane is directly over the center of gravity of the load before making the lift.
- The qualified signal person checks that the rigging is adequate and securely fastened before hoisting the load.
- The signal person makes sure all personnel are clear of any pinch points.
- When using the radio to deliver voice signals, a good signal person avoids chewing gum or speaking with food in the mouth. A good signal person will speak clearly at all times.

- When using radios, the crane operator must have good-quality, hands-free equipment; and fully charged batteries with backup units should always be available.
- Prior to beginning lifting operations, any voice signals shall be discussed and agreed upon in advance.
- The devices to transmit signals must be tested onsite to ensure the transmission is clear and reliable.
- The crane operator's reception of voice radio signals must be a hands-free system.
- The signal person must never direct a load over another person or other ground crew personnel.
- The signal person must never give more signals than a crane operator can perform.
- The signal person must always give signals to the crane operator's right- or left-hand side.

Signals using radios, phones, or other means of electronic transmission must be given through a dedicated channel except when more than one crane is being used. One or more signal persons may share a dedicated channel to coordinate hoisting operations. A dedicated channel must also be shared when a crane is operated on or near railroad tracks and the actions of the crane operator need to be coordinated with the operations of other equipment such as trains.

Chapter 11 Review Quiz

1. The standard OSHA hand signal shown here is
   a. lower the boom and raise the load
   b. raise the boom and lower the load
   c. move down and raise the load
   d. move up and lower the load

2. The standard OSHA hand signal shown here is
   a. boom outward
   b. telescope outward
   c. reach out boom
   d. extend telescoping boom

3. The standard OSHA hand signal shown here is
   a. travel to right
   b. crawler crane travel, one track
   c. travel to left
   d. travel and turn

4. The standard OSHA hand signal shown here is
   a. raise boom
   b. lift boom
   c. telescope up
   d. boom upward

5. The standard OSHA hand signal shown here is
   a. travel backward
   b. crawler crane travel, both tracks
   c. travel forward
   d. move toward me

6. The standard OSHA hand signal shown here is
   a. hoist quickly
   b. raise helicopter
   c. move slowly
   d. raise boom and load

7. The standard OSHA hand signal shown here is
   a. emergency stop
   b. stop
   c. move load this way
   d. swing

8. The standard OSHA hand signal shown here is

- **a.** travel one track
- **b.** use auxiliary hoist
- **c.** use main hoist
- **d.** use one part of line

9. The standard OSHA hand signal shown here is

- **a.** stop travel
- **b.** hold everything
- **c.** dog everything
- **d.** stop

10. The standard OSHA hand signal shown here is

- **a.** raise the boom and lower the load
- **b.** raise boom slowly
- **c.** come up quickly
- **d.** hoist

11. The standard hand signal shown here is

- **a.** raise the boom and lower the load
- **b.** raise the boom and hold the load
- **c.** telescope up and lower the load
- **d.** telescope up and hold the load

12. The standard OSHA hand signal implemented here is

- **a.** telescope
- **b.** telescope out
- **c.** move out
- **d.** extend boom

13. The standard OSHA hand signal shown here is

- **a.** swing over here
- **b.** stop
- **c.** swing right
- **d.** emergency stop

14. The standard OSHA hand signal shown here is

- **a.** use overhaul ball
- **b.** use whip line
- **c.** use main hoist
- **d.** use headache ball

15. The standard OSHA hand signal shown here is

- **a.** boom up
- **b.** boom up and lower load
- **c.** retract cylinders
- **d.** retract boom

16. The standard OSHA hand signal shown here is

- **a.** lower
- **b.** rotate load
- **c.** telescope downward
- **d.** telescope down and lower load

17. The standard OSHA hand signal depicted here is

- **a.** stop
- **b.** stop hoist
- **c.** stop both hoist
- **d.** emergency stop

18. The crane operator should move a load only after receiving a signal from
- **a.** the lift director
- **b.** the foreman or supervisor
- **c.** the designated signal person
- **d.** the rigger

19. When moving the crane, the following shall be used for stop:
- **a.** multiple short audible signals
- **b.** three short audible signals
- **c.** two short audible signals
- **d.** one short audible signal

20. When moving the crane, the following shall be used for backing up:
- **a.** multiple short audible signals
- **b.** three short audible signals
- **c.** two short audible signals
- **d.** one short audible signal

21. According to the ASME B30.5 crane safety standard, all directions given to the crane operator shall be given from
- **a.** the operator's direction perspective
- **b.** the signal person's direction perspective
- **c.** the lift director's recommendation
- **d.** the site supervisor's recommendation

22. You are communicating during a lift by voice signals. According to ASME B305, each series of voice signals shall contain three elements. Which voice command should be given first?

a. Distance and/or speed

b. Function and/or speed

c. Distance and/or function

d. Function and/or direction

Chapter Summary—and Then Some

A qualified signal person who takes the job seriously is truly an asset to any hoisting and rigging operation. The signal person that shows up with reflective safety attire and takes the operation to heart will lower the anxiety of the crane operator and the entire lifting team (Figure 11.7). Positive body language is used to enhance the signals. When a signal person gives clear, concise signals in a very professional way, the whole crew takes notice. Morale will be elevated, safety is at its best, and production will be high. The leadership position of the qualified signal person is given to those who are skilled and knowledgeable and take the job very seriously. After all, the qualified signal person truly is the eyes and ears of the crane operator. This person represents the rest of the riggers and support team on the ground. When this entire group works collectively, accidents will be reduced and lives will be saved.

FIGURE 11.7 The professional signal person wears bright reflective clothing and gives clean, crisp international hand signals.

OSHA requires that all man baskets be designed by a qualified engineer. All work platforms must be capable of supporting their own weight plus five times the maximum intended load. The man basket must clearly indicate load bearing capacity on the basket. It must be surrounded by a guardrail. All access gates on the man basket should open inward and require safety latches. Before each use of a man basket, the work platform must be tested for safety.

12 Personnel Lifting Systems

OSHA prohibits hoisting of personnel on a suspended platform by a crane except when no safe alternative is possible (Figure 12.1). Steel erection operations are special because conventional means of access to the work area is not always available. When conventional means of access to the work, such as scaffolds and ladders, are considered to be unsafe, personnel hoisting operations using a mobile crane that comply with strict provisions of the OSHA 1926.1413 safety standard for hoisting personnel are allowed.

Because using mobile cranes to hoist personnel poses a serious risk to the workers being lifted, any mobile cranes that hoist workers must comply with very strict rules (Figure 12.2). Mobile cranes may be used to hoist personnel in a suspended platform for working at heights when it is the least hazardous way to complete the job. These important rules must be considered as described in the ASME B30.23 safety standard:

- The suspended platform shall have a minimum design factor of 5 based on the stated platform rating.

FIGURE 12.1 OSHA allows lifting personnel only when there is no other practical way to get the work done. Lifting personnel with a crane is always risky. Whenever possible, use an aerial man lift or scissor lift.

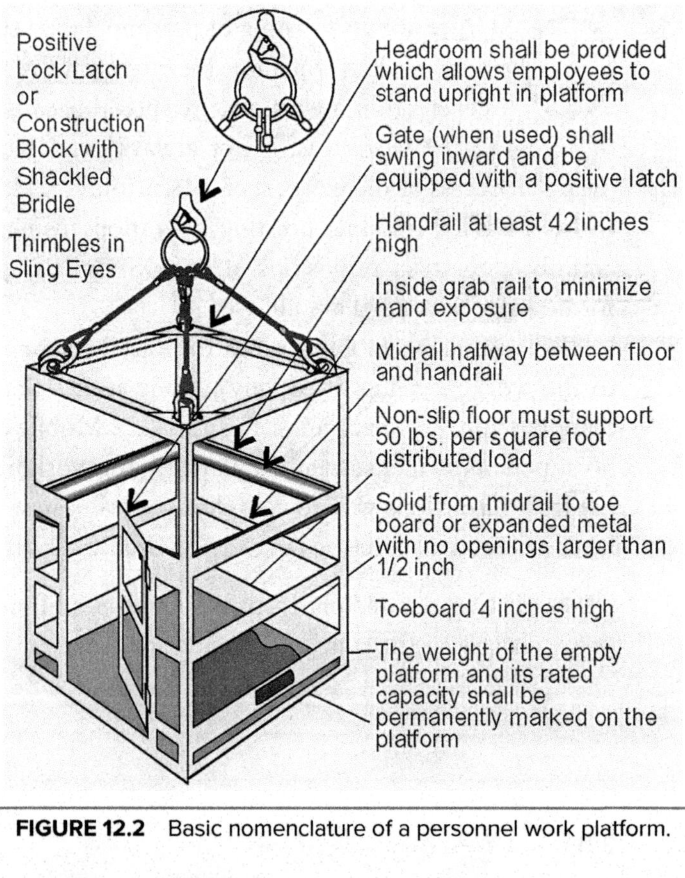

FIGURE 12.2 Basic nomenclature of a personnel work platform.

- Each load-supporting component shall have a minimum design factor of 5 based on the maximum intended resultant stress on that component.
  - One-leg system—design factor of 7
  - Two- or more leg system—design factor of 5 for each leg with only two legs under stress
- Platforms and suspended systems shall be designed to minimize hazards to the occupants:
  - A minimum platform rating of 300 pounds.
  - Guardrail protection consists of a top rail, intermediate rail, toeboard, and lower barrier. Each barrier shall be capable of withstanding a concentrated load of 300 pounds applied at any point in an outward or downward direction.
  - Anchorage points shall be suitable for attaching personnel fall protection device lanyards and capable of withstanding, without failure, the impact load created by falling loads.
  - Falling loads shall be calculated using 500 pounds as a single-occupant weight, which is then multiplied by the number of occupants that are required to attach to the anchorage point.
  - Flooring that has a slip-resistant surface to facilitate the free drainage of liquids and that will withstand the forces developed under proof test loading shall be provided.
  - An identification plate shall be provided and displayed to allow easy viewing from both the interior while hoisted and the exterior while not hoisted.
- The identification plate should display the manufacturer's name and address, platform's rating, identification number, weight of empty platform, and date of fabrication. Required hoist equipment features while handling personnel include:
  - Live boom or live load line operation shall be made inoperable when hoisting personnel.
  - A boom angle indicator shall be visible to the operator.
  - Hoisting equipment with a telescoping boom shall have a boom length indicator readable from the operator's station.
  - Hooks used for attachment of a personnel lifting platform shall be of a type that can be positively locked closed.
  - Hoisting equipment shall have automatic brakes such that when the equipment operating controls are released, the motions are brought to rest.
  - A holding device such as a check valve shall be provided to prevent uncontrolled movement of the hoisting equipment, outriggers, or stabilizers, in the event of a system failure.
  - Pendant-supported jib-type boom extensions without positive stops shall not be used to lift personnel.
- All equipment used in the lifting of personnel shall be inspected, tested, and maintained to protect against failure during lifting operations:
  - The platform manufacturer shall furnish complete inspection criteria for platform users. The criteria shall address all inspection frequencies and shall cover the platform, rigging components, fasteners, and all safety features and attachments.
  - The platform manufacturer shall receive an initial inspection prior to initial use and at each new jobsite by a qualified person.
  - The platform suspension system, attachment points, and any motion controls shall be inspected at least once each working day by a designated person (Figure 12.3).

FIGURE 12.3 The suspended platform must be inspected by a designated person at least once a day.

- At each new jobsite prior to lifting people in the personnel platform, the platform and rigging shall be proof-tested to 125% of the platform's rating. With the proof test load evenly distributed in the platform, the platform shall be hoisted and held in a suspended position for not less than five minutes. After proof-testing, the platform shall be inspected by a qualified person.
- Jobsite management shall not allow or require any operator to lift personnel when the operator does not feel physically or mentally fit or has been working more than 10 hours prior to the start of the lift.
- The individuals assigned as ground crew shall be physically and mentally capable of performing inspections and operations associated with the lift and be trained in the specific requirements of their tasks associated with the lift.
- Occupants of the platform shall attend a pre-lift briefing, as required, with the ground crew, crane operator, and lift supervisor.
- Occupants shall wear a personnel fall protection device and lanyard, fastened to a provided anchorage point, at all times, unless special work requirements dictate otherwise.
- Personnel platforms are not to be used in winds in excess of 20 mph at the hoisted platform height or in electrical storms, snow, ice, sleet, or other adverse weather conditions that could affect the safety of personnel.

  - Personnel platforms are to be used for personnel and their tools and sufficient materials to do their work. They should not be used solely for transporting bulk materials.
  - When personnel lifts are conducted over water, personnel flotation shall be provided and required to be worn. The fall protection device shall be appropriately attached, with quick-release features, while personnel are lifted over land and detached while personnel are lifted over water.
  - A boat with appropriate rescue personnel shall be available at all times during a personnel lift over water.
- The hoisting equipment operator shall consult the hoist equipment and platform manufacturers for specific instructions on the equipment operations:
  - The operator shall inspect the crane setup area before the personnel lift and report observations to the personnel lift supervisor.
  - The operator shall inspect the area for potential hazards such as excessive load, radius, overhead obstructions and electric transmission lines, hazardous location ground conditions, wind, weather, unstable conditions, and any potentially hazardous locations.
  - The crane must be operated with outriggers or stabilizers extended, blocked, and locked.
  - The crane is set up and maintained within 1% of level during all personnel lifts.
  - Never allow the total weight of the lifted load, including rigging, platform, personal tools, and materials, to exceed 50% of the crane's charted capacity.
  - A qualified person must perform a trial lift prior to lifting personnel with the platform on each shift and after any change of setup location, crane equipment configuration, or operator.
  - The operation should remain in continuous sight or in communication with the operator and signal person.
  - Hand signals to the operator shall be in accordance with the ASME B30.5 mobile crane safety standard.
  - If communication with the operator and the platform occupants is disrupted, all operations shall be stopped until communication is re-established.
- When lifting personnel are near electric power lines, special procedures must be followed, depending on the conditions and the situation in each case. OSHA regulations and ASME mobile crane safety standards must be consulted prior to any lift.

Chapter 12 Review Quiz

1. An important safety procedure to have in place when lifting personnel in a platform suspension system over water is to
   a. have extra flotation devices
   b. have an emergency medical technician in a platform
   c. have an ambulance on standby
   d. have a boat with rescue personnel available

2. When lifting personnel with a crane, the outriggers or stabilizers must
   a. be at least 50% extended
   b. be at least 75% extended
   c. be provided with adequate blocking
   d. be fully extended, blocked, and locked

3. When lifting personnel with a crane, the operator should remain in continuous sight or in communication, and hand signals to the operator should be in accordance with
   a. OSHA 1926.550
   b. ASME B30.20
   c. ASME B30.7
   d. ASME B30.5

4. According to federal requirements, OSHA permits the hoisting of personnel by an approved lifting system when
   a. the crane is over 50-ton capacity
   b. the crane is properly set up
   c. there is no forklift available
   d. there is no less hazardous or practical way

5. When lifting personnel in an approved lifting system, the total weight of the loaded platform must not exceed
   a. 20% of the crane's charted capacity
   b. 30% of the crane's charted capacity
   c. 40% of the crane's charted capacity
   d. 50% of the crane's charted capacity

6. Before being put into use, personnel platforms are to be loaded to 125% of their rated capacity and lifted and held for
   a. 5 minutes
   b. 10 minutes
   c. 15 minutes
   d. 20 minutes

7. According to the ASME B30.23 standard, when personnel lifts are conducted over water, personnel shall
   a. attend a company safety meeting
   b. be certified for overwater operation
   c. wear a highly visible safety vest
   d. wear an approved flotation device

8. According to the ASME B30.23 safety standard, while in the platform, personnel fall protection devices with quick-release features shall be provided and required to be worn *except*
   a. when no other practical means are possible
   b. when working at heights less than stipulated by federal requirements
   c. when being lifted over water
   d. when a certified lift director authorizes

9. When occupying an approved lifting device, personnel are not to be lifted in winds in excess of
   a. 10 mph
   b. 15 mph
   c. 20 mph
   d. 25 mph

10. The personnel lifting system will have an identification plate that shall be provided and displayed to allow easy viewing from
   a. the operator's station while hoisted and not hoisted
   b. the interior while hoisted and not hoisted
   c. the exterior while hoisted and not hoisted
   d. both the interior while hoisted and the exterior while not hoisted

Summary—and Then Some

It is critical that personnel who hoist persons in a platform meet a minimum set of guidelines. This must apply to the crane operator as well as those in the platform and must include all those involved in the operations (Figure 12.4). The crane operator must be trained as well as qualified and certified. Workers in the platform must be trained and know how to effectively communicate with the crane operator to provide directions for the movement of the platform. A backup communication system should always be in place before and during all operations. The crane operator and personnel in the platform must communicate in the same language. Workers should always wear the appropriate personal protective equipment for the operation. Additional safety considerations may be required when the platform is used in special environments. Always check the requirements in the jurisdiction you are operating in.

FIGURE 12.4 Occupants shall wear a personal fall protection device and lanyard, fastened to a provided anchorage point at all times unless special work requirements dictate otherwise.

All those who are involved with mobile cranes and heavy lifting should have the most recent edition of the ASME safety standard for hooks, B30.10, in their library. The standard includes two chapters, 10-1 and 10-2, that address the different types of hooks used in our industry. The two chapters provide information regarding components, proof testing, fabrication, and modification. It is important for all those who operate or use cranes to read and fully understand this safety standard so the hooks you are using on or off the crane are being used, inspected, and maintained correctly and within the guidelines of industry safety standards as well as the manufacturer's specifications.

Crane hooks are classified by the materials used to manufacture them. They are made in a variety of designs to meet specific needs and rated to lift loads of different types, sizes, and weights. The correct hook size for your operation is determined by weight, size, and working environment and by the type of crane involved to hoist the load. Always follow the manufacturer's recommendations and safety standards for inspection and maintenance requirements.

Crane hooks are important components of any heavy lifting operations. A crane uses one type of hook to lift, move, and place heavy materials on a jobsite. Another type of hook may be used as an attachment on a set of slings or to tie down and secure a load. It is critical to know the difference between a hook that is load bearing and rated to lift heavy loads and another that is specifically designed to pull or tie down loads. Hooks used on cranes are rated to be used for hoisting hooks on load blocks and overhaul balls. Other hooks are designed for pulling purposes or for tying down or securing a load. They must not be used for hoisting loads with a crane.

13 Crane Hook Types, Applications, and Inspection

In this chapter, we first cover overhead lifting hooks and non-overhead lifting hooks. We then explore hook applications and go over recommendations for hook inspections.

Overhead Lifting Hooks

Not all rigging and lifting hooks are appropriate for overhead lifting; only alloy hooks should be used in overhead hoisting activities. When choosing an overhead lifting hook, it's important to consider the application you will be using it for. If you are lifting a plate, you may need one type of hook, while lifting an engine out of a bulldozer may require another.

Types of overhead lifting hooks include:

- **Clevlok hooks.** Clevlok alloy hooks are typically used for overhead lifting applications (Figure 13.1). This line of hooks offers easy installation in the shop or on the jobsite. These hooks are 100% proof-tested at the factory, requiring no additional testing once installed.

- **Eye hooks.** Eye hooks are an excellent choice when used with mechanical couplers (Figure 13.2). These hooks are designed for overhead hoisting and can be used in place of Clevlok hooks. Overhead lifting eye hooks are 100% proof-tested at the factory, so they require no additional testing once put to work.
- **S hooks.** S hooks are built and designed for special lifting requisitions (Figure 13.3). S hooks are 100% proof-tested at the factory and can be used for various applications where a wide throat opening is required to make the lift.
- **Plate hooks.** Plate hooks are designed for lifting plate material, like steel, in vertical and horizontal positions (Figure 13.4). Plate hooks should be used in pairs, and careful attention should be paid to sling angles when determining the working load limit of the rigging used.
- **Sorting hooks.** Sorting hooks are designed to lift and move material with long, narrow throat openings (Figure 13.5). Sorting hooks are 100% proof-tested and are available with or without handles.

FIGURE 13.1 Clevlok hooks, like this one, are typically used for overhead lifting.

FIGURE 13.2 Eye hooks, like the one shown here, are an excellent choice when used with mechanical couplers.

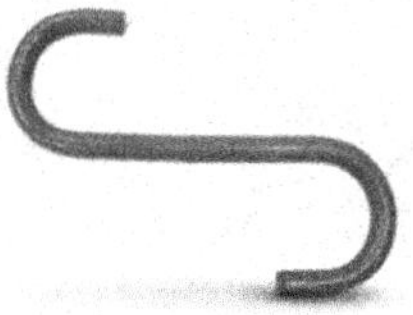

FIGURE 13.3 S hooks, like this one, are built and designed for special lifting applications.

FIGURE 13.4 Plate hooks, like the one pictured here, are designed for lifting plate materials.

FIGURE 13.5 Sorting hooks, like the one shown here, are designed to lift and move materials with long, narrow throat openings.

Non-Overhead Lifting Hooks

Non-overhead lifting hooks are designed for pulling or load security. These hooks do not have the same requirements as those used for overhead lifting. Non-overhead lifting hooks are available in Grades 30, 40, and 70. Grade 80 hooks are not suitable for overhead lifting and are marked T-80. They should only be used for load security, such as tying down loads.

Types of non-overhead lifting hooks include:

- **Clevis hooks.** Clevis hooks are not designed for overhead lifting. They are most often used for load securement with tie-down chains (Figure 13.6). Clevis hooks can be used in combination with various grades of chain including Grades 30, 40, 70, and 80. These offer a U-shaped attachment point with a pin to secure chains. They can be used with other rigging applications as well. Different grades of clevis hooks have different working load

FIGURE 13.6 An image of a Clevis hook. Clevis hooks are used for load securement with tie-down chains.

FIGURE 13.7 Eye hooks, like the one shown here, are used for load securement with tie-down chains.

limits. For that reason, you must make sure you use the correct hook grade and size for your application and the weight of the load to be hoisted.

- **Eye hooks, or non-cradle grab and slip hooks.** Eye hooks, including standard eye hooks, are not designed for overhead lifting but instead are most often used for load securement with tie-down chains (Figure 13.7). Eye hooks are used in combination with various grades of chain, including Grades 30, 43, 70, and 80. Eye hooks offer a simple circular attachment point for rigging applications. Different grades of eye hooks have different working load limits, so you must make sure you use the correct hook grade and size for your operation.

Rigging Hook and Lifting Safety Steps

Following are essential safety steps:

- Care should be exercised during use, so the crane hook is not abused or damaged.
- Do not exceed the working load limit or shock-load the rigging, including any load handling devices. Loads applied rapidly or dropped freely can result in serious overloading of the hook as well as other components used in the lift.
- Hooks should not be subjected to bending, exposed to sharp objects, or tip-loaded unless specified by the manufacturer. Hooks must not be loaded in a fashion inconsistent with their design.
- Hooks attached to rigging should be selected to match the size and working load limit of the rigging application used.
- Avoid exposure to corrosive conditions or high temperatures that could affect the thermal treatment and strength of the hook. Hooks can be used from –40 degrees Fahrenheit to 400 degrees Fahrenheit without reduction of the working load limit.
- When rigging and lifting a load, it is important that the hook is correctly positioned where the weight of the object lifted is applied in the bowl or saddle of the hook. Side, back, and tip loading conditions can reduce the hook's lifting capacity and cause damage to the hook (Figure 13.8).
- The load must be centered in the bowl or saddle of the hook to prevent point loading or side pulling.
- Make sure the load, including all rigging and load handling devices applied to the hook, does not exceed its working load limit.
- When using multi-leg slings, they must be placed in the bowl or saddle of the hook. The maximum angle between all sling legs must not exceed 90 degrees or the manufacturer's recommendations.
- The strength of hooks can be affected by chemically active environments such as caustic or acid materials. The manufacturer should be consulted before using hooks in these environments.

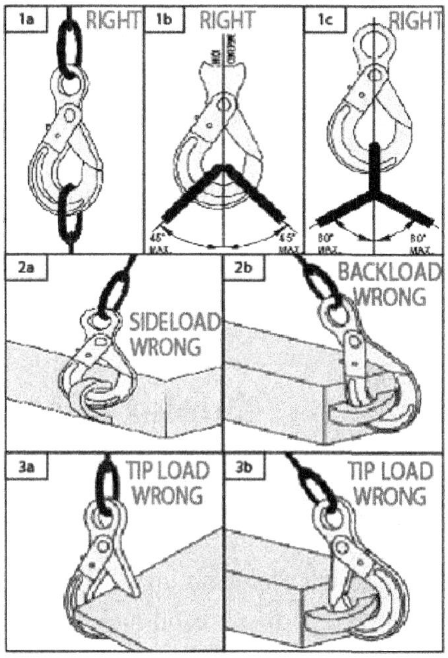

FIGURE 13.8 Side, back, and tip loading will reduce the hook's lifting capacity.

- All hooks that are designed to have a safety latch should have a latch installed.
- Rated shackles, links, or rings should be applied when more than two sling legs are placed in a hook saddle for angles greater than 45 degrees in respect to the centerline of the hook (Figure 13.9).

FIGURE 13.9 Use rated links and rings when more than one sling is used.

- Self-locking hooks must be locked during all lifts.
- Crane hooks shall be equipped with a safety latch unless the application makes the use of a latch impractical.
- When the hook is equipped with a safety latch, the load must not restrict the closure of the latch.
- Arms, hands, fingers, and other body parts must be kept away from pinch points between the load and the hook.

Duplex (Sister) and Quad Hook Applications

The safety applications for the use of a duplex, or sister, hook are the same as for the use of a single hook. Here are some additional points to be aware of:

- When determining allowable sling angles for duplex hooks, you should consult the manufacturer for its recommendations.
- You must load each side of a duplex hook equally on both sides unless the hook is manufactured and approved for single-point loading. You must use great care to avoid any rotation of the load that could cause shifting of any payload in the rigging apparatus (Figure 13.10).
- Duplex hooks shall not be used in an application to place a side load, back load, or tip load on the hook.
- A qualified person must be consulted if you plan to hoist a load with unbalanced prong loading.

FIGURE 13.10 A duplex, or sister hook. You must load each side of this hook equally.

FIGURE 13.11 Shown is a special forged lifting hook with a working load limit of 1,550 tons.

- If you plan to load a duplex hook at the pinhole of the two saddles, the load applied shall not exceed the working load limit that would normally be shared between the two saddles (Figure 13.11). In addition, you must not overload any load handling equipment used in the lift.
- Duplex or quad load hooks shall be loaded with all prongs rigged to balance the prong loading forces. Contact the manufacturer for consultation if any unbalanced prong hoisting is planned during the hoisting operation.

Crane Hook Inspections

For crane hooks in regular service, manufacturers' recommendations and the ASME B30.10 crane hook safety standards provide inspection requirements as well as record-keeping requirements (Figure 13.12). Hook inspections shall be administrated by a designated person. However, if deficiencies are found in that inspection, a qualified person must be consulted to determine if a hazard exists. Specialized equipment, such as load handling devices, may require more stringent inspection by the manufacturer or safety standard. Hook inspection procedures are listed here for your review.

FIGURE 13.12 Crane hook inspection should be done periodically in accordance with the ASME B30.12 hook safety standard.

Initial Inspections

Prior to use, crane hooks that are new, altered, or modified, as well as repaired crane hooks, shall be inspected to verify compliance with the ASME B30.10 hook safety standard and any manufacturer's requirements.

Frequent Inspections

Frequent inspections include observations of the hook being used during hoisting as well as visual inspection to identify any issues or removal criteria outlined in the ASME B30.10 hook safety standard or any manufacturer's requirements.

For semi-permanent and inaccessible locations where frequent inspections are not feasible, a qualified person will determine the frequency of periodic in-section requirements to satisfy the ASME B30.10 hook inspection requirements (Figure 13.13). Inspection intervals should be based on the criteria listed here or the manufacturer's recommendations:

- The frequency of use of the hook
- The severity of the service conditions the hook experiences during daily operations
- The nature of the load handling activities
- The practical knowledge gained on the service life of the hook used in similar hoisting and rigging operations

Guidelines for frequent inspections include:

- Normal service requires a monthly inspection.
- Heavy service requires a weekly to monthly inspection.
- Severe service requires a daily to weekly inspection.

FIGURE 13.13 Crane hooks should receive a frequent inspection and be part of your checklist.

Periodic Inspections

A complete and thorough inspection of the hook shall be performed (Figure 13.14). Disassembly of the hook may be required to perform a complete examination and identify conditions based on the removal criteria outlined in the ASME B30.10 hook inspection requirements. Periodic inspections shall be performed at a minimum interval of 12 months. Periodic inspection intervals should be based on these conditions and criteria:

- The frequency of hook use
- The severity of the service conditions the hook experiences in daily operations
- The nature of the load handling activities
- The understanding gained on the service life of a hook used in similar operations

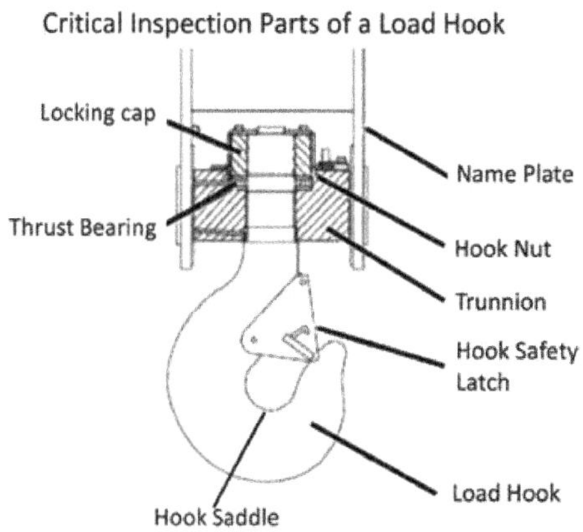

FIGURE 13.14 Components to consider when performing frequent and periodic inspections.

FIGURE 13.15 When reeving multiple parts of line for a hook block, manufacturer's instructions must be followed.

Rigging Hook Inspection, Maintenance, and Removal

Rigging hook inspection, maintenance, and removal criteria include the following:

- Discard hooks that are worn more than 10% of the original dimension or are worn beyond a specific dimension or tolerance as provided in wear allowance tables, charts, or diagrams.
- Discard hooks that have an increase in throat or slot opening more than 5% of the original opening not to exceed ¼ inch.
- Discard hooks with any visibly apparent bend or twist from the plane of the unbent hook. Replace load pins that are permanently distorted.
- Replacement parts, such as load pins for clevis hooks, should be equal to the original manufacturer's specifications.
- A safety hook latch that is inoperative or missing shall be repaired or replaced. If a hook has a latch that does not bridge the throat opening, remove it from service until the latch is replaced or repaired and the hook is examined for deformation, with special attention to the throat opening.
- A designated person shall repair cracks, nicks, and gouges by grinding longitudinally, following the contour of the hook, provided no dimension is reduced more than 10%—or as recommended by the manufacturer—of its original size and shape. All other repairs shall be performed by the manufacturer or a qualified person.
- Missing or illegible hook manufacturer identification.
- Missing or illegible rate load identification.

- Any self-locking hook that does not lock properly.
- Evidence of heat exposure or unauthorized welding.
- Evidence of unauthorized alterations such as drilling, machining, grinding, or other modifications.

Hooks having any of the deficiencies listed shall be removed from service unless a qualified person approves their continued use and initiates corrective action. Hooks approved for continued use shall be subject to periodic inspection.

Chapter 13 Review Quiz

1. All lifting hooks should have which of the following to retain slings and lifting attachments?
   a. Saddle
   b. Keeper
   c. Safety latch
   d. Shackle
2. Lifting hooks shall be removed from service when the throat opening exceeds what percent?
   a. 5%
   b. 10%
   c. 15%
   d. 20%
3. Crane hooks shall be equipped with a safety latch
   a. during all lifting applications
   b. unless the application makes the use of a latch impractical
   c. whenever rigging is applied
   d. when lifting loads over ground personnel
4. Safety latches shall close
   a. over wire rope slings
   b. over chain slings
   c. over synthetic slings
   d. over the entire throat opening
5. Crane hooks are designed and manufactured to lift specific loads. Load hook block assemblies shall
   a. be load-tested weekly
   b. be load-tested monthly
   c. receive a yearly inspection
   d. be labeled with rate capacity and weight
6. Crane hooks shall be removed from service when there is
   a. a 5% visible bend or twist from the plane of the unbent hook
   b. a 10% visible bend or twist from the plane of the unbent hook
   c. a 15% visible bend or twist from the plane of the unbent hook
   d. any visible bend or twist from the plane of the unbent hook
7. Crane hooks shall be removed from service
   a. if wear exceeds 4% of the original section dimension
   b. if wear exceeds 6% of the original section dimension
   c. if wear exceeds 8% of the original section dimension
   d. if wear exceeds 10% of the original section dimension

8. Crane hooks having any deficiencies shall be removed from service unless
- **a.** an appointed person approves their continued use and initiates corrective actions
- **b.** a qualified person approves their continued use and initiates corrective actions
- **c.** a hook manufacturer representative provides authorization
- **d.** a crane manufacturer representative provides authorization

9. Crane hooks approved for continued use shall be subjected to
- **a.** periodic inspection
- **b.** frequent inspection
- **c.** an initial inspection
- **d.** a Magnaflux inspection

10. Crane hooks shall be removed from service when
- **a.** an increase in throat opening exceeds 2%
- **b.** an increase in throat opening exceeds 3%
- **c.** an increase in throat opening exceeds 5%
- **d.** an increase in throat opening exceeds 7%

Chapter 13 Summary—and Then Some

Crane hooks are made of drop forged alloy steel Grades 80 through 100 and are made in accordance with strict guidelines. Crane hooks are strong and don't have problems very often; however, they must be examined with a critical eye. The part of the hook that is most susceptible to stress, cracks, and damage is the part you cannot see. The hook shank is the portion that is encased inside the bottom block. It is important to watch the hook closely because if one does fail, it can be catastrophic. If you have concerns, crane inspectors and rigging dealers can provide a dye or Magnaflux test to look for cracks. A newer technique, called a *magnetic rubber test*, can also be administered. It is important that you develop a compliance program and inspect the hooks in accordance with ASME B30.10 safety standard and any manufacturer's requirements. It is also very important to make sure everyone in the ground crew, as well as the crane operator and lift director, is properly trained about crane hooks and all rigging use, applications, and inspections. Doing it right just may save a life.

As with all activities involving cranes, the effective management of safety during the lifting and movement of a load during pick and carry operations can only succeed if all those involved are clear about their roles and responsibilities. It is essential for all those undertaking roles during a pick and carry operation with a crane to be competent. The crane operator, as well as everyone else involved in the operation, must have relevant up-to-date training. All team members must have the qualifications and certifications, including the experience appropriate for the pick and carry operations for which they are responsible.

14 Pick and Carry Operations

When you operate cranes without the outriggers deployed, you are doing what is referred to as "operating on rubber." Whenever operating on rubber tires, following the correct procedures and operating limitations specified by the crane manufacturer cannot be emphasized enough. Just because a crane has rubber tires does not necessarily mean it has rubber pick and carry capabilities. When on rubber, if pick and carry operations are allowed, the manufacturer will provide an on-rubber load chart that must be used along with other special specifications. All-terrain rubber-tired cranes are often the tool for conducting pick and carry operations (Figure 14.1).

Where the load will be picked from, what the travel route will be, and where the load will be placed are very important assessments that must be made in the planning process before the load ever leaves the ground. Many factors need to be considered when planning a pick and carry operation. Crane stability depends on the integrity of the ground on which it's going to travel. To reduce the risks of accidents during on-rubber lifts, planning activities for pick and carry operations shall be carried out by competent persons to assess the capability of the ground throughout the travel route to withstand the loads and ground pressures imposed by the crane along with the weight of the load (Figure 14.2).

Because so many factors affect a mobile crane's rated capacity when operating on rubber tires, a pre-lift checklist must be developed, and a thorough lift plan including travel route must be in place. A lift director must be appointed to supervise the operation. He will confirm all requirements of the operation are in place, including any required tests or inspections as well as all rigging components to be used. The lift director will hold a pre-lift meeting to discuss the lift plan and the roles of all personnel involved in the operation.

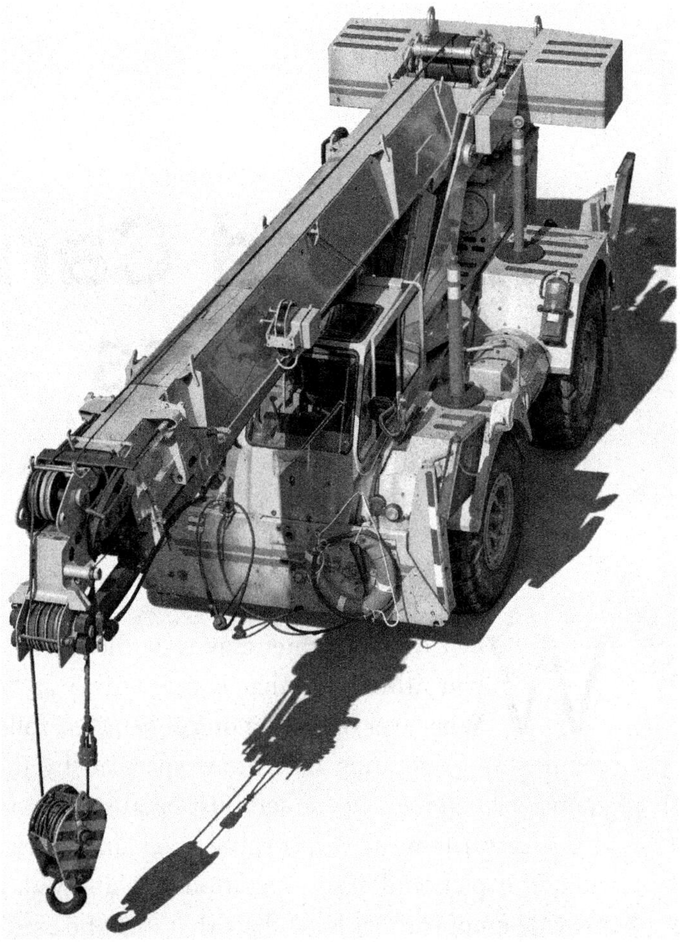

FIGURE 14.1 An all-terrain rubber-tired crane. All-terrain rubber-tired cranes are often the tool for conducting pick and carry operations. This figure shows the crane operator is keeping the load low to the ground with a short boom length.

Statistics indicate that the majority of craning accidents occur when the crane is either improperly set up prior to making a lift or traveling on the jobsite with or without a suspended load. Lifting, carrying, and placing a load while the crane is mobile and unstable requires skill and caution:

- Pick and carry operations must follow the crane manufacturer's specifications.
  - Pay special attention to any decrease or de-rating of the on-rubber load chart.
- Pick and carry operations require that particular attention be paid to the ground conditions.
  - The ground should be level and capable of supporting the crane and the suspended load.
  - Plan the path of travel before operating the crane, and develop a lift plan.
  - In cases of poor ground conditions, it may be necessary to improve the ground conditions.

FIGURE 14.2 Pick and carry operations are for the experienced crane operator. Specialized cranes like this one are purpose-built to handle conditions often involved with pick and carry operations.

- Crane operations, as well as all support personnel, must be experienced and have specific training, such as lifting on rubber during pick and carry operations.
  - Pick and carry operations shall not take place until the crane operator and the crew have read and fully understand the manufacturer's limitations and specifications.
- During pick and carry operations with a rough-terrain crane, the boom length should be as short as possible, and the load should be as close to the ground as possible to limit load movement.
  - The crane should travel at a slow, even pace comfortable for on-ground support personnel and be capable of keeping inertia of the crane and the load its carrying to a minimum. Rough-terrain cranes are designed to travel on firm, level ground capable of supporting the crane and the suspended load.
- When moving the load, tag lines should be considered to prevent load swing that can cause side loading on the boom.
  - Tag lines should be held by an experienced spotter.
  - Securing the load back to the crane is another consideration for some operations.
- A side slope inclinometer is another very useful instrument to consider using as a guide to ensure that the manufacturer's travel specifications regarding fore-and-aft limitations are being followed.
  - Remember, an inclinometer is only accurate when the crane is moving straight ahead with no load.

- ➤ When the crane is required to travel uphill, keep the load close to the crane to keep the operating radius at a minimum.
  - – When traveling down a hill, consider backing down with the load close to the crane at a very slow speed.
- ➤ Crane manufacturers may have important instructions and limitations regarding the rated capacity of a crane's fore-and-aft slope that are often overlooked. In some cases, manufacturers may require you to de-rate the load chart, depending on the ground conditions or other special requirements (Figure 14.3).
  - – Whatever the method, always apply and lock the crane's house swing brake.

FIGURE 14.3 Pick and carry operations often require the load to be carried close to the crane as shown here. Keep the inertia of the load and crane to a minimum, while traveling with a load.

- ➤ Always use qualified signal persons and spotters when directing the operation on the ground.
  - – Keep any ground crew in sight of the crane operator whenever possible.
  - – The operator should be able to observe the load, rigging, and overhaul ball or block whenever operations will allow.
  - – Never allow any support ground crew, any riggers, or the signal person between the crane and the load or any other area a crushing or struck-by-load hazard exists.
- ➤ Always inspect the tires on the cranes to ensure they are in good condition.
  - – Check for correct air pressure before being put into service.
  - – Observe tire lugs and tread for good condition.
  - – Carefully observe the side walls of the tires for any server cuts or bulges.

In regard to crawler cranes and mobile cranes:

- Traveling a crawler crane (Figure 14.4) with a suspended load can be particularly hazardous due to its fixed boom length and design. The rigidness of the carrier and its inability to compensate for slightly rough or uneven ground, as well as the crane's difficulty in producing smooth travel and turning motions, are sometimes problems. Extra caution must be taken when crawler cranes are used during pick and carry operations:
  - Use extra care when carrying loads over the side of the crawler crane.
  - Use extra spotters and signal persons when traveling with a load over the side quadrant of a crawler crane.
- Mobile cranes are designed to travel on firm, level ground within 1% gradient.
  - A mobile crane traveling with a freely suspended load should avoid slopes unless the load is within the allowable charted capacities of the crane's on-rubber pick and carry load chart.

FIGURE 14.4 A Lampson Manitowoc 1800 crawler crane is shown here while traveling and placing very heavy concrete wall panels at a warehouse in Washington State.

Chapter 14 Review Quiz

1. During pick and carry operations with a rough-terrain crane, you should
   a. position the load over the rear of the crane
   b. position the load at a radius of 15 feet
   c. position the load hoisted 10 feet off the ground
   d. keep the crane boom as short as possible

2. Rough-terrain mobile cranes are designed to travel with a suspended load
   a. on uneven rough-terrain such as mud and loose dirt
   b. on construction sites with any soil or site conditions
   c. only when no rain or other moisture is in the weather forecast
   d. on firm, level ground capable of supporting the crane and the suspended load

3. Crane operators should plan a path of travel during pick and carry operations
   a. only if the path is required to travel on public highways, streets, or bridges
   b. never; all-terrain cranes are specifically designed for these types of operations.
   c. only if hazardous conditions exist on the jobsite
   d. always; a path of travel should always be planned, and the manufacturer's rules and specifications should always be reviewed.

4. When is the crane operator required to de-rate the load chart during pick and carry operations?
   a. Each all-terrain crane is provided with an on-rubber load chart to cover all pick and carry operations, and de-rating is not necessary.
   b. In some cases, if the tires are in poor condition, you may consider de-rating the load chart.
   c. If the crane is of older vintage, you are required to de-rate the load chart 20%.
   d. Some manufacturers may require you to de-rate the load chart, depending on the ground conditions or other special requirements.

5. You are planning a pick and carry operation on your construction project. Who will determine whether the crane is capable of on-rubber operations?
   a. If the crane's tires are properly inflated and in good condition, the crane operator can decide.
   b. If the crane has passed a periodic inspection, the lift director can approve.
   c. It can be approved only if the travel plan has been approved by the supervisor.
   d. It can be approved only if the manufacturer has approved it for pick and carry operations, and under those specifications using the on-rubber load chart.

6. When should crane operators consider the crane's fore-and-aft slope conditions when planning a pick and carry operation?
   a. Never; pick and carry operations are approved for straight-ahead direction.
   b. Never; pick and carry operations are approved for forward and reverse movement only.
   c. Always; crane manufacturers may have important instructions or limitations that are often overlooked.
   d. Fore-and-aft considerations are built into the limitations of the on-rubber charted capacities.

Summary—and Then Some

Accidents involving lifts while on rubber tires during pick and carry operations can cause massive production delays, devastating property damage, and loss of life. A well-developed lift plan put into action by a group of experienced and talented team members is extremely important. The crane operator and all the ground team members must understand that side loading of the crane boom can easily occur during this type of operation. The load chart rating applies only to freely suspended loads and when the load is picked up directly under the boom tip.

If the load is to either side of the boom tip, side loading occurs. Side loading is one of the most common causes of boom failure and happens without warning, often during pick and carry activities. The load radius may increase while traveling with a load, causing an overload condition, and a crane accident and loss of stability can occur. Increasing the load radius reduces the rated capacity of the crane (Figure 14.5). To avoid side loading and increasing the load radius, the crane operator must be aware of boom deflection and plan ahead for the possibility of a radius increase causing a potential loss of stability. The crane must move slowly with the load carried as low as possible to control inertia of the load as much as possible. Pick and carry operations produce shock loading on the crane as well, so every precaution must be made to provide for a slow, smooth operation.

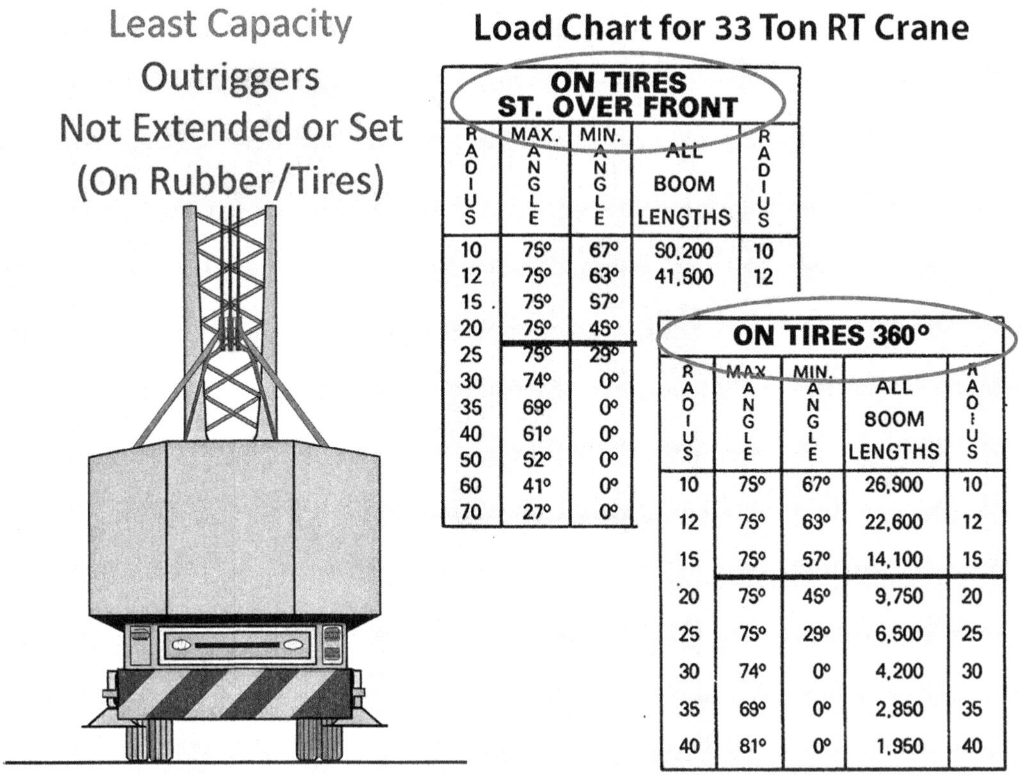

ON TIRES ST. OVER FRONT				
RADIUS	MAX. ANGLE	MIN. ANGLE	ALL BOOM LENGTHS	RADIUS
10	75°	67°	50,200	10
12	75°	63°	41,500	12
15	75°	57°		
20	75°	45°		
25	75°	29°		
30	74°	0°		
35	69°	0°		
40	61°	0°		
50	52°	0°		
60	41°	0°		
70	27°	0°		

ON TIRES 360°				
RADIUS	MAX. ANGLE	MIN. ANGLE	ALL BOOM LENGTHS	RADIUS
10	75°	67°	26,900	10
12	75°	63°	22,600	12
15	75°	57°	14,100	15
20	75°	45°	9,750	20
25	75°	29°	6,500	25
30	74°	0°	4,200	30
35	69°	0°	2,850	35
40	81°	0°	1,950	40

FIGURE 14.5 The least capacity a mobile crane has is on rubber tires.

Load charts do not allow for shock loading. Shock loads are caused by rapid acceleration, sudden stopping, or load release. Shock loading can occur in any situation where the load on the crane suddenly increases. The crane's structural integrity is designed to hoist up heavy loads slowly and steadily. They are not designed to withstand sudden increases or decreases to the actual weight of the load caused by shock loading. Skilled crane operators are a company's first defense against shock loading. Lifts made on rubber as well as pick and carry operations set up the perfect scenario for side loading, shock loading, and radius increase to occur. This truly is the time you need a "smooth operator" who is experienced and understands all the dangerous forces that will take place in these kinds of lifting operations.

Sometimes, the decision to shut down a jobsite and crane operation for the sake of safety does not come as easily as it should in some cases. Other management personnel within a company's structure may be influenced by the cost of shutting down a jobsite, even for a short duration. Let's imagine on our jobsite we have a 50-ton rough terrain (R/T) crane that rents for $200 an hour to $2,000 a day. Next we have the cost for our certified crane operator, riggers, signal persons and a support ground crew of seven that costs an average of $80 an hour. Also, we have the cost of moving our crane onto our jobsite, setup, insurance, fuel, and permits. As you can see this all adds up in a hurry and these costs represent only our crew not the entire project. Our project supervisor is a good person and would never intentionally put any employee in harms way. However, in this case, the supervisor thinks they have seen a few lightning strikes and heard some thunder that's likely miles from the jobsite. Our supervisor directed everyone to continue working and keep an eye on things. Decisions like this are made because of a lack of understanding and training. Today, more than ever before, good training should be provided at all levels of the operation from the owner and supervision, continuing though every employee on the project, including any office personnel who could be affected.

Employers are legally obligated to have a plan for outdoor work that addresses the recognized hazard of lightning. Work must be stopped immediately when a thunderstorm is a possibility. OSHA Standard 29 CFR 1926.1431 specifically addresses weather conditions during personnel hoisting operations, stating that a qualified person must determine if it is safe to lift personnel in dangerous weather conditions. If not, the operation must not begin or must be terminated. OSHA Standard CFR 1926.1417 covers general crane operations and requires that the crane manufacturer's specifications for operating in various weather conditions and how environmental conditions affect the safe operations of the crane must be applied. Employers should implement a site-specific lightning safety plan, which includes all the best practices, and train their employees. Let's take a few moments and dive in deeper about lightning and cranes.

15 Cranes Working around Lightning

Lightning is the occurrence of a natural discharge of electricity of very short duration and high voltage between a cloud and the ground or within a cloud, accompanied by a bright flash and typically also thunder. When a storm is approaching, signs such as dark clouds and an increase in wind speed can give us some warning that lightning may be on its way. The National Lightning Safety Institute tells us that if you see lightning, "flee it." If you can hear thunder, "clear it." The National Oceanic and Atmospheric Administration (NOAA) tells us we must act after hearing thunder or seeing lightning. Too often, we tell ourselves we have time for one more lift and then get caught off guard. Knowing when to suspend crane operations can save equipment and, most importantly, save lives.

Safety Precautions for Lightning

- Stop all operations and cease all crane work as soon as you see lightning or hear thunder. You can be at risk from lightning up to 10 miles away from a storm, so a 30 second delay between lightning and thunder means it's 6 miles closer, putting you and your crew members in a danger zone.

- Power the crane down, turning off all electrical power.
- Secure all equipment and load handling devices by lowering them to the ground.
- Evacuate the crane; all hoisting and rigging support personnel must leave the crane and work area to seek a safe shelter.
- For lattice boom cranes, lower the boom to the ground if possible.
- Secure the crane, lock the doors, and remove the keys.
- Use the 30-30 rule, a lightning safety guideline that recommends seeking shelter if the time between seeing lightning and hearing thunder is 30 seconds or less. After the last thunderclap, you should wait at least 30 minutes before resuming outdoor activities to see to it the storm has completely passed.
- Employers must create and implement a written plan that includes how to notify workers of lightning threats, how to evacuate, and where to find safe shelter.
- Monitor weather continuously. Following the weather forecast and using lightning detection services or apps to track storm activity is the best progressive approach.
- Train employees. Provide site-specific training to employees on lightning hazards, evacuation procedures, and how to identify safe shelters.
- Avoid dangerous areas. Keep employees away from elevated areas, open fields, tall or isolated trees, metal objects, water, and powerlines during a storm.

Unfortunately, OSHA regulations and ASME safety standards do not have a single specific standard exclusively for mobile cranes when lightning is present. Suspending operations will sometimes come to the crane operator's discretion based on being a qualified person. The information provided here is designated to give the owner, supervisor, lift director, crane operator, and all those on the hoisting and rigging crew better information so they are able to make good decisions about shutting down work and when it's safe to start back up. The National Weather Service (NWS) website provides comprehensive information on lightning safety. You can sign up for your local community's emergency warning system to receive alerts about thunderstorms and other hazards. NOAA weather radio is another source for official weather alerts.

FIGURE 15.1 Lattice booms pointing straight up into the sky are perfect lightning rods.

The best shelter is always an enclosed structure. Small outbuildings are not considered safe structures. If a permanent structure is not available, seek hard topped metal vehicles and make sure the windows are rolled up. Stay away from any metal objects, including fences, machinery, and electric equipment. Avoid trees and open fields, and never stand in water.

When a single flash of lightning strikes a taller object near a person, and a portion of the current jumps from the taller object to the person, it is called a *side flash*. A crane boom is definitely a taller object and is nothing more than an extended lightning rod.

A lightning strike travels in stages at different speeds. The initial discharge, the *stepped leader*, moves toward the ground approximately 220,000 to 270,000 miles per hour or about 100,000 m/s to 120,000 m/s. The "return stroke" travels much faster, up to 220,000,000 miles per hour or about one-third the speed of light. The light we see from the strike travels at the speed of light, the actual movement of the electrical current within the lightning channel is slower, however, superfast from a human perspective. From our perspective on the ground, the entire process is instantaneous usually lasting only a few microseconds. The sound of the thunder travels at a much slower speed of sound, about 760 miles per hour. The difference in speed is why you see the flash of lightning before you hear the thunder, a principle used to estimate the distance of a strike by counting the seconds between the flash and the sound.

The "Flash-to-Bang" Rule

The flash-to-bang method is a simple technique used to estimate how far away a lightning strike is during a thunderstorm. This is how to use the method.

1. Watch for the lightning flash.
2. Count the seconds. Immediately start counting the seconds until you hear the corresponding clap of thunder. You can use the second hand on a watch or a stopwatch. The simple way is to count "one thousand one," "one thousand two," "one thousand three," and so on to total the amount of seconds.
3. Divide by 5, take the total number of seconds, and divide it by 5. Example: If you count 10 seconds between the flash and the bang, then 10 seconds ÷ 5 = 2 miles.

The flashes we see because of a lightning strike travel at the speed of light, 670 million mph. An actual lightning strike travels at a rate of 186,000 miles per second, moving about 1 mile every 5 seconds. If you can hear thunder or see lightning, then you are close enough to be struck by lightning. The next strike can be at your jobsite regardless of whether it is raining or not. Lightning is very dangerous and spontaneous; therefore, it is highly important we are prepared. Everyone should stand away from the crane to avoid contact with the machine itself. Lattice boom cranes should be boomed down to the ground whenever room will allow. Telescopic boom cranes should retract all sections and return the boom to the stowed position. Lightning protection systems for the mobile cranes are available for sale. Smartphones offer weather applications that can notify the user of lightning. The best plan is to understand it an be prepared for it. If forecasts show lightning is likely to be near the jobsite, perhaps crane activities should be scheduled for another day. OSHA enforces requirement through the General Duty clause mandating employers to provide a workplace "free from recognized hazards," which include lightning. No job is worth the risks of serious injury or death from a lightning strike.

Do not resume work until at least 30 minutes after the last sound of thunder or flash of lightning, according to NOAA. After 30 minutes or when it's safe to resume crane operations, conduct a thorough inspection, carefully checking for any damage.

- The crane is a metal structure that acts as a conductor, and a strike can electrocute anyone in contact with the crane or rigging.
- Inspect the machine after a lightning strike or a storm has passed. A qualified person should inspect the crane for any damage, especially to electronic systems and wire ropes, before returning it to service.

Lightning Damage to Consider and Check

- The load moment indicator (LMI), anti-two block systems, and other essential safety controls can be immediately put out of service or destroyed. The A2B limit switches, alarms, and cut-off functions must be tested.
- The main computer, electronic controls and wiring may burn out through the entire crane. Check for damages.
- Motors and inverters can be badly damaged or destroyed if not powered off and grounded.
- Radios and communication equipment can be badly damaged or destroyed.
- Structural and mechanical damage can occur anywhere on the crane and boom. Inspect closely and carefully.
- The intense heat and physical force of a lightning strike can compromise the crane's physical components.
- Wire rope can be damaged. The extreme heat can melt steel at point of contact, severely compromising the wire rope's integrity, which requires instant replacement regardless of the amount of broken wires.
- The force of the strike can cause pitting, arcing, and burning on the metal surfaces of the crane. It can also create hidden defects in structural steel that may lead to failure in future operations.
- Inspect the outer surfaces of the crane's boom, jib, and hoisting cable to find evidence of where the electricity entered or exited the crane. Look for signs as arcing, melting, heat damage or discoloration. Structural members must be checked for signs of deformed, cracked, or corroded parts and loose bolts or rivets.
- The intense heat from a strike can damage or melt internal strands or steel rope, even without visible external signs of damage. All wire ropes, pendant lines, straps, and boom hoist lines are links in the electricity "pathway" that must be inspected. Any suspect items must be replaced. Dark scorched areas on the surface of the wire rope often indicate the point where the lightning entered or exited the crane structure. Extreme heat can cause strands, fittings, or adjacent metal parts to melt, deform, or fuse together. The metal may show signs of heat tinting, such as a blue discoloration, which indicates exposure to extreme temperatures.
- Mechanical components must be checked including pins, bearings, shafts, gears, rollers, and all locking or clamping devices for wear, cracks, or distortion.
- Check that the brake system is fully operational, checking for wear on parts, linkage, linings, pawls, and ratchets.

- Check the operator controls by going into the cab, and performing a functional test to see to it all controls are responsive and working properly, including telescoping, swinging, and hoisting mechanisms.
- Inspect the entire electrical apparatus for signs of damage such as pitting, arcing, burning, or melting, including control panels, supply connections, circuit breakers, and wiring harnesses.
- The presence of a white powder residue is typically a sign of surface breakdown and oxidation of the steel surface layers due to the intense heat and electrical energy.
- Look for "barber pole staining," a circular discoloration along any of the wire rope. This can be caused by environmental factors. In conjunction with other indicators like burn marks, it can be evidence of electrical arcing and heat.
- The slew ring and its bearings can be damaged by the electrical current passing through the contact points.
- Electricity can travel through the outriggers into the ground, possibility damaging their parts and components and posing a potential electrocution hazard to ground personnel.
- Hydraulic systems can be damaged. Electrical discharge can disrupt or damage hydraulic controls and components.
- The intense heat can ignite flammable substances, leading to a fire in the superstructure or on board the crane.

FIGURE 15.2 Always refer to the manufacturer's specific procedures for information about what to do following an electrical incident.

After a suspected or confirmed lightning strike, the crane must be removed from service at once. A competent person must perform a thorough inspection following the manufacturer's specifications. The main idea is to identify all damage, specifically in the path the electricity took, to see to it all affected components are addressed before going back into service. Always refer to the crane manufacturer's specific procedures for information about what to do following an electrical incident.

All inspection findings and subsequent repairs must be documented. While frequent, periodic, and annual inspection records have specific retention periods, detailed logs of severe incidents like lightning strikes are often kept for the entire service life of the crane, which is a best practice many companies adopt that are above and beyond OSHA requirements. By following these procedures, crane operators can protect members of the hoisting and rigging crew, and help prevent future incidents and equipment failure.

All wire ropes, pendant, lines, straps, and boom hoist lines are links in the electricity "pathway" and must be inspected after a lightning strike. Lightning kills about 20 to 30 people per year in the United States, with hundreds more injured each year. The number of fatalities has decreased since the mid 2000s, when there were over 40 deaths per year in the same 12 month periods. Survivors of lightning strikes often suffer long-term neurological damage.

Training should occur for each outdoor jobsite so that supervisors can let workers know in advance where a worksite's safe shelters are and the time it takes to reach them. Employers should train supervisors, crane operators, the lift director, and the entire ground crew to provide lightning safety warnings in sufficient time for everyone to reach a worksite's safe shelter and take other appropriate precautions.

Chapter 15 Review Quiz

1. The National Oceanic and Atmospheric Administration tells us that after hearing thunder or seeing lightning, we should go to a safe shelter and remain there for
   a. 10 minutes after hearing the last sound of rainfall
   b. 20 minutes after hearing the last sound of heavy winds
   c. 30 minutes after hearing the last sound of thunder
   d. 40 minutes after seeing signs of dark clouds

2. The best shelters to go to during thunder and lightning storms are
   a. hard-topped metal vehicles
   b. outbuildings
   c. enclosed structures
   d. under large trees or heavy foliage

3. Lightning will travel about one mile every
   a. 5 seconds
   b. 30 seconds
   c. 60 seconds
   d. 90 seconds

4. After a lightning strike near or on a jobsite where your crane is located, it is important to
   a. replace all synthetic slings, shackles, fasteners, and clips
   b. replace the load moment indicator on the crane
   c. replace all rigging hardware and wire rope slings
   d. perform a complete and thorough inspection of the crane

5. Since a lightning strike can produce an extreme form of heat, what must always be replaced?
   a. The load moment indicator
   b. All the hydraulic fluid
   c. All the motor oil
   d. All the wire rope

6. On average, how many people are killed in the United States every year from lightning?
   a. 13 to 20 people
   b. 20 to 33 people
   c. 33 to 43 people
   d. 53 to 65 people

Summary—and Then Some

People who work in the crane hoisting and rigging industry are intense and focused on their work. When signs of lightning start, they may be so focused on their job, the lightning passes right by their thought process. Needless to say, this is a problem!

Another problem is that good, comprehensive training regarding the correct steps to take when lightning is present has been lacking in recent years. This problem starts at the top with the crane owners and supervisors, and continues through the crane operator and lift director, including the entire ground crew. The time for great training about lightning is now. Please share your knowledge with others. This problem is widespread in our industry, so it will take everyone's involvement to improve.

FIGURE 15.3 When struck by lightning the extreme heat can melt steel at point of contact, severely compromising the wire ropes integrity, which requires instant replacement regardless of amount of visible broken wires.

Accidents during assembly and disassembly of lattice boom cranes are one of the major causes of crane-related fatalities. According to OSHA 1926.1404 assembly and disassembly, lattice boom operations must be directed by a person who meets the criteria for both a competent person and a qualified person, or by a competent person who is assisted by one or more qualified persons. This highly trained and knowledgeable person is the assembly and disassembly director (A/D director). The A/D director should be present during the assembly to ensure that each step of the method detailed by the crane manufacturer is respected.

16 Lattice Boom Crane Assembly and Disassembly Responsibilities

The A/D director must know the parts of a lattice boom and understand the applicable assembly and disassembly procedures (Figure 16.1). Employers must comply with all manufacturer procedures or develop their own procedures by a qualified person. Upon completion of assembly of the lattice boom cranes, and before the rig is put to work, the equipment must be inspected by the qualified person.

The assembly/disassembly director supervising the assembly or disassembly operation must address the hazards associated with the operations. These would include:

- Site and ground conditions
  - The site and ground conditions must be adequate for safe assembly and disassembly operations and to support all the equipment.
  - Blocking material

- The size, amount, condition, and method of stacking the blocking must be sufficient to sustain the loads and maintain stability.
  - Proper location of blocking
- When used to support lattice booms or components, blocking must be appropriately placed to protect the structural integrity of the equipment and prevent dangerous movement and collapse.
  - Boom and jib pick points
- The points of attachment of rigging to a boom or sections including jibs must be suitable for preventing structural damage and facilitating safe handling of these components.

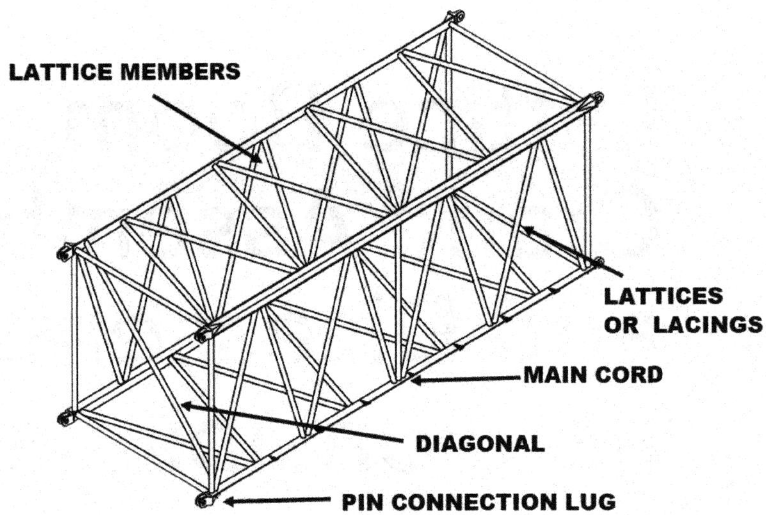

FIGURE 16.1 Parts of a lattice boom. The square frame at each end of the boom insert is commonly referred to as the *picture frame*.

When using an assist crane, the loads that will be handled and hoisted during the assembly or disassembly operation must be verified to be within the charted capacity of the crane and be handled with care (Figure 16.2).

Other potential hazards include:

- Identification of the center of gravity
  - The center of gravity of the load must be identified if that is necessary for the method used for maintaining stability. Measures designed to prevent unintended dangerous movement resulting from inaccurate identification of the center of gravity must be used.
- Snagging
  - Suspension ropes and pendants must not be allowed to catch on the boom or jib connection pins, cotter pins, keepers, and locking pins.
- Stability upon pin removal
  - The boom sections and boom systems, such as gantry A-frames and components, must be rigged or supported to maintain stability upon the removal of the pins.
  - When dismantling a lattice boom, always remove the lower set of pins first.

FIGURE 16.2 Lattice boom sections must be handled with care.

- Struck by counterweights
  - There is the potential for unintended movement from inadequately supported counterweights and from hoisting counterweights.
- Boom hoist brake failure
  - Each time resistance is to be placed on the boom hoist brake to prevent boom movement during assembly or disassembly, the brake must be tested prior to such reliance to determine if it is sufficient to prevent boom movement. If it is not sufficient, a boom hoist pawl or other locking device, or another method of preventing dangerous movement of the boom from a boom hoist failure, must be used, such as an assist crane.
- Loss of backward stability
  - This can occur before swinging the upper works, during travel, or when attaching or removing equipment components.
- Wind speed and weather
  - When installing a lattice boom into a crane, operators must prioritize safety by understanding that wind impacts stability. The main considerations are following the manufacturers specific wind limits, constantly monitoring onsite wind conditions, and understanding how wind affects the large surface area of boom sections.

Here are six more absolutes:

1. Never store jib sections inside boom sections.
2. Only experienced personnel should be involved in the assembly and dismantling procedures.
   - The operator should be in charge whenever possible.
3. Always follow the crane manufacturer's instructions for the correct boom section order layout.
4. Never attempt to "cantilever" any boom lengths longer than specified by the crane's manufacturer. *Cantilever length* means the length of boom suspended beyond the last pendant hookup.

5. Lay sections out in the proper order according to the manufacturer's manual.
   - There are often pendant or jib backstay attachment points that must be located at specific locations; this will require the relocation of different-length sections.
6. Never work or walk under the boom (Figure 16.3).
   - This rule applies whether the boom is being assembled, dismantled, shortened, lengthened, or raised or lowered, whether the assembly or dismantling has been completed or not.
   - When pins or other hardware are being removed, workers must not be under the crane boom or jib sections unless the operation constraints require them to be in such a position. In these cases, the A/D director must provide procedures to minimize the risk of unintended movement during the duration of exposure under the boom.

FIGURE 16.3 The correct way to work when assembling a crane boom. Never work under a crane boom.

Before the assembly or disassembly can start, the A/D director must make sure all the crew members understand these important components:

- When assembling lattice booms, the manufacturer's instructions must be followed.
- What are their duties and tasks.
- What are the hazards associated with those tasks.
- The locations of all hazardous positions they must avoid.

Crew members must inform the crane operator anytime they are going to a location around the operation where they will be out of the operator's sight or view. Whenever the crane operator is aware a crew member is out of sight or in a potentially dangerous location, the crane operator must not move any part of the crane, load, or equipment until the crew member has returned to a safe position.

Often when handling heavy loads as part of the assembly or disassembly operation, we are required to deploy the outriggers on the crane. Whenever possible, the outrigger beams should be fully extended and set. If partial extension work must be done, the crane operator must follow the manufacturer's specifications and any load chart limitations. After assembly has been completed and before raising the boom, the crane's outriggers should be fully deployed and set and should be provided with the correct size and amount of blocking under each float. Proper outrigger deployment will help give the stability necessary to boom up or raise the newly assembled crane booms. In some cases, an assist crane may be needed to help complete this process (Figure 16.5).

FIGURE 16.4 Riggers prepare a lattice boom section for assembly, paying particular attention to the location of the center of gravity.

FIGURE 16.5 Some lattice boom assembly operations require another crane to assist in holding the boom sections.

When dismantling crane booms or jibs, never try to remove any pins holding the boom pendants if they are under tension or are under load. In addition, when dismantling crane booms or jibs, never try to remove any pins on the top or bottom of the boom sections located between the pendant attachment points and the crane house or cab that are under tension or load. When the boom is being supported by the uppermost boom section or supported by blocking, no attempt should be made to remove pins. Most importantly, none of the pins on the boom sections on the cantilevered portion of the boom are to be removed until the cantilevered section is fully supported. Loaded pins under tension spell danger. Never drive out any pins or work on any boom sections unless you are supervised and directed by the A/D director.

FIGURE 16.6 Qualified professionals at specialized crane repair companies, dealerships, or service centers can repair lacings on a mobile crane lattice boom.

Chapter 16 Review Quiz

1. The supervisor in charge of lattice boom crane assembly and disassembly operations is titled the
   a. assembly foreman
   b. assembly and disassembly rigger
   c. disassembly supervisor
   d. assembly and disassembly director

2. When dismantling a lattice boom, which pins should be removed first?
   a. Left side
   b. Right side
   c. Both lower
   d. Upper and lower

3. During the assembly or dismantling of the lattice booms, the manufacturer's reference to "cantilever length" means the
   a. amount of boom you are going to assemble
   b. length of boom suspended beyond the last pendant hookup
   c. length of assembled sections on the blocking
   d. length from the boom foot to the first pendant hookup

4. When assembling lattice booms,
   a. the shorter sections always go closest to the base section
   b. the long and short sections should alternate
   c. the longer sections must always be closest to the tip section
   d. the manufacturer's instructions must be followed

5. When assembling or dismantling lattice booms, working or walking under the boom
   a. is allowed as long as there is tension on the boom pendant lines
   b. is only permitted to remove the bottom pins
   c. is allowed if the sections are being held up with another crane
   d. must never be allowed unless the A/D director makes special provisions

Summary—and Then Some

The employer must make sure the A/D director and all the members of the ground crew are able to read and understand all the details of the crane manufacturer's manual. At the end of the assembly or disassembly operation, you must be able to verify that the crane has been inspected, tested, and maintained in accordance with the manufacturer's recommendations. The need to read and follow the manufacturer's instructions for proper boom layout and assembly is paramount. Always read the fine print, because, among other things, that is where you may be informed that some manufacturers' or government regulations require a load test before the first lift is made. As a crane owner or crane operator, always remember that just because you are told that the crane is assembled properly does not mean that it actually is. It is your job to check and double-check before your crane goes to work.

FIGURE 16.7 Only a qualified professional or specialized company meeting specific criteria should conduct structural repairs on a crane boom. In some cases, the boom can not be repaired.

The most important thing to know about multiple crane lifts is that they require meticulous, detailed planning and precise coordination to ensure load stability has been established and prevent overloading of any single crane. The movement of more than one crane during a lift must be coordinated to happen concurrently. Every attempt must be made to avoid negative influences, such as forces on the crane boom that would cause unintended load distribution. The area where the cranes are to be set up must be properly assessed for good ground conditions suitable for placement of all the cranes that will be involved and also must be assessed for the weights of the loads that will be lifted. Lifting with multiple cranes is always more complex than lifting with one crane, so excellent planning and engineering, with professional crane operators, riggers, signal persons, and ground crew, must be in place to manage the risks associated with multiple-crane lifts.

17 Critical and Multiple-Crane Lifts, the Lift Plan, and the Lift Director

When multiple cranes are involved in one lift, they are typically controlled by the crane operators, so the human factor must be considered. Crane operators must be properly trained, certified, and qualified for these types of operations. Even if the crane operators work together without any issues during a multiple-crane lift, it is very difficult to perfectly integrate these types of operations. One wrong move by any crane operator could lead to a very precarious situation or accident. Any lifting operation involving more than one crane hoisting a common load with shared loading of all cranes meets the criterion of being classified as a *critical lift*. A qualified person must be appointed to act as the lift director and be in attendance, with direct control of all multiple-crane operations. Only the lift director shall give instructions to the crane operators at the controls of the cranes.

A qualified signal person should give any hand signals or radio communications in accordance with the ASME B30.5 mobile crane safety standard during the lift under supervision of the lift director. However, an emergency stop signal may be given by any person observing a hazardous or unsafe condition during the operation.

All the hitch points of the load should be visible from one position. If that's not possible, it may be necessary to place other qualified personnel at various positions around the lift and report their observations to the lift director. The lift director shall review the hoisting operation using a well-developed lift plan (Figure 17.1) with all the crane operators and other members of the lifting team prior to the operation to define all rules and procedures, discuss communications, and put in place a plan in case of emergencies. It is always a great idea to make a trial run or test lift without a load to execute all the movements in order to confirm the load radius, boom angles, boom lengths, and correct configuration of the machines.

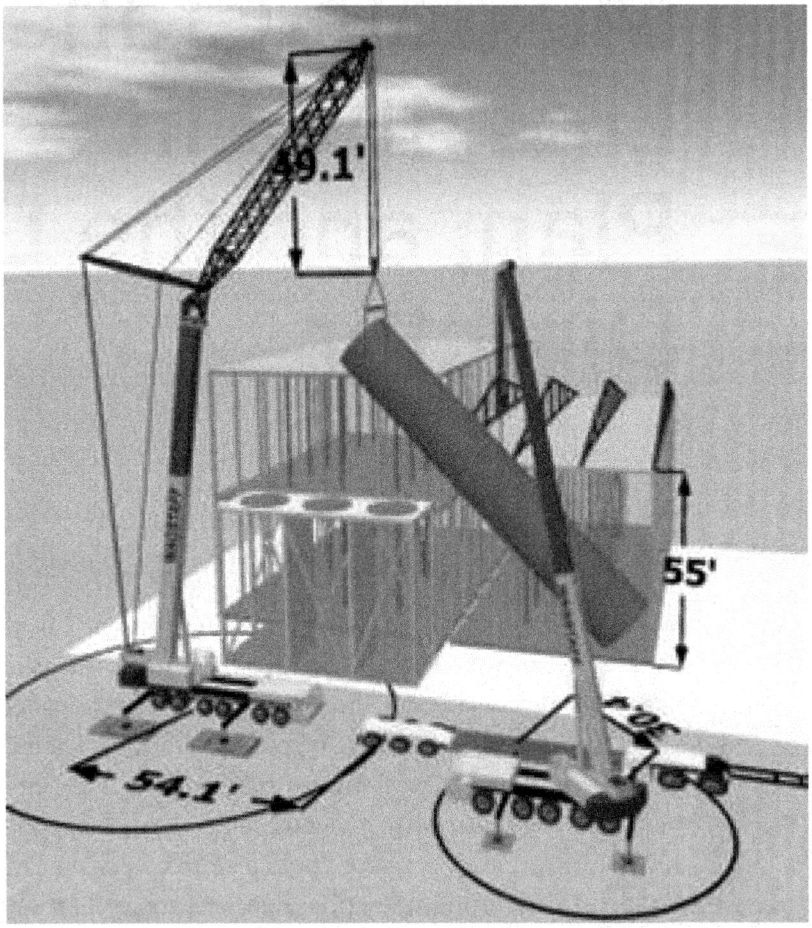

FIGURE 17.1 An image created by a computer software program.
Computer software programs designed for planning multiple crane lifts are very helpful.

The sequences of movements between the rigs should also be reviewed. A multiple-crane lift is a complicated operation, so more precautions are necessary when planning the lift than when planning regular hoisting operations using only one crane. Detailed control measures using experienced, qualified, and certified lifting personnel are the key to success.

When performing multiple crane lifts, a documented engineered lift plan and procedure must be prepared by a competent person, such as an engineer, and closely followed. The lift plan should have a step-by-step approach with intermediate checkpoints during the lift to ensure everything is going as planned. The lift plan must be clearly understood by all those involved in the lift at all levels of the operation.

Understanding the load distribution between the cranes is the first step to a safe operation. Always conduct a "dry run" during pre-lift planning to determine the load share and how it will be managed between the cranes during the operation.

During multiple-crane lifts, even the slightest movement of the load will create extra loading on the other cranes. Crane motions should be coordinated very closely by the lift director. These loadings influence the crane boom, and additional forces will influence the load itself. These forces can occur from misalignment of the cranes. Load shifting can be created by not locating the exact center of gravity of the load prior to lifting (Figure 17.2). To help compensate, all crane motions should be performed at a slower, reduced speed.

FIGURE 17.2 Locating the center of gravity can be very difficult, especially in icy and snowy conditions.

As noted earlier, crane movements must be closely coordinated. Any deviations from the lift plan must lead to an immediate stoppage of the operation. It may be necessary to shift or realign the cranes or make adjustments to the rigging. Key considerations include correct load distribution, making sure cranes are positioned to avoid side loading using synchronized movements to prevent collisions, and having a preplanned communication strategy with standardized signals. A clear understanding of the lift sequence, including the lift-off and set-down order, is decisive for managing the dynamic load shifts between cranes. The correct sequence is the crane that initially carried the lighter share of the load should set its end down first. This allows the load to shift gradually, as its placed in its planned final resting point without overloading either crane at any point during the process. Any changes must be documented and communicated to the entire lift team to make sure everyone totally understands those changes before the lift can start over.

Lifting operations using two or more cranes at the same time should use cranes of similar models and configurations and hoist blocks reeved with the same number of parts of hoist line whenever possible (Figure 17.3). If a lift is planned using mobile cranes along with a different crane type, a conscientious and meticulous assessment must be made by a competent person, such as a registered engineer, before any hoisting occurs and a load is hoisted into the air.

FIGURE 17.3 A row of cranes of the same type. Cranes of the same type and with similar characteristics should be used when planning multiple-crane lifts.

When the loads to be hoisted are equally shared between two or more cranes of the same model and capacity, including comparable hoist line speeds and other similar characteristics, the lift will be refined, the capacities shown on the cranes' load charts will be realistic, and the safety margins will be more believable. Whenever the cranes have different characteristics, such as the hoisting line, swing, and booming speeds, it will be very difficult to harmonize and integrate a smooth lift between them.

An important control measure is to calculate the distribution of the load and how it will be shared between the cranes—most importantly how the load distribution might change at any time during the lift. Only by knowing the exact location of the center of gravity of the load, the load's size and weight, and the location of the rigging hitch points can an accurate calculation be made to determine how much load weight will be shared between each crane. It is important to understand that the closer a crane is to the center of gravity, the more load weight the crane will receive (Figure 17.4).

The cranes should be set up and positioned to minimize side loading on the crane booms. If possible, set up the cranes to minimize swinging of the load during the lift as much as possible. The chance of side loading is very high during the swinging of the crane booms during multiple-crane lifts. In addition, extra precautions, such as avoiding shock and dynamic loading during the lift, are paramount and very important to understand. It is important to determine what the actual load is on each of the cranes throughout the hoisting, moving, and placement process. It is critical for everyone on the lifting team to understand how the load sharing on the cranes can be altered by manipulating the rigging hitch points. Moving a hitch point by only a few inches can have a huge impact on the load sharing between each and every crane involved in the lift. Rigging techniques, such as equalizer beams, can help equalize the load between cranes and balance the percentage of their load chart capacities.

FIGURE 17.4 The closer a crane is to the center of gravity, the more load weight will be influenced on the crane itself.

Another highly important consideration that is often overlooked is weather conditions. Extra attention should be given if the lift you are planning is on a jobsite that is known for windy conditions. An assessment must be made of potential wind sail areas on surfaces of the loads you are planning to handle. Lightning storms are often overlooked, which is a horrible mistake—after all, an erected crane boom in these instances is essentially a lightning rod. Paying attention to weather conditions must be part of every crane operation and lift plan. Monitoring and scrupulously paying attention to weather conditions is part of the lift director's job; however, all those on the crew should keep themselves aware of any wind, fog, rain, snow, hail, extreme heat, bitter cold, or any other adverse weather condition that is expected around the lift area. Any weather that could have unusual consequences on the work and lifting operations must be known in advance, and when necessary, the job should be canceled or postponed until conditions improve. Do not operate cranes in wind speeds that exceed the manufacturer's specifications as gusty winds can severely impact stability and control.

Many conditions may exist or develop on a jobsite where you are conducting or planning a multiple-crane lift operation (Figure 17.5). A reduction of each crane's charted capacities should be considered anytime the crane operator or other members of the lifting crew have a concern. Building a safety factor into a lift because of your own concerns is good business.

FIGURE 17.5 Pre-planning and planning are the keys to success when using more than one crane to make a lift.

Here are some additional steps to consider:

- Conduct a lifting analysis during the planning phase, including load sharing and how it will be managed during hoisting operations.
- Identify any structures, objects, and most importantly, power lines that could restrict crane movement.
- Check the combined weight of the rigging hardware, hook blocks and balls, slings, equalizer beams, spreader bars, and any other attachments that could affect the net capacity of the cranes.
- Seek advice from any crane manufacturers for machines used in the lift.
- Use equalizer lifting beams and spreader bars when they can be properly utilized (Figure 17.6).
- Remove nonessential personnel from the work area. Those and only those involved in the lift should be present. The lift director may define an exclusion zone. This is an area where people can go when dangerous work is being conducted to avoid the risk of being injured by the crane or falling objects.
- Make sure all cranes are properly aligned to avoid side loading of the crane booms.
- Ensure that all cranes used in the lift have load moment indicators and that the indicators are properly programmed and working.
- Provide clear methods of communication between all those involved in the operations, and ensure the methods are in place. When using radio equipment, make sure it is of good quality—only use top-notch equipment. When using radios, have headsets with hands-free capability. Keep charged batteries on hand, including several sets of backup batteries.

Multiple-crane lifts are always unpredictable no matter how well they are planned, so particular care must be exercised whenever they are to be undertaken. Listed below are some important rules (some of which were discussed earlier) that should be observed:

FIGURE 17.6 Use equalizer lifting beams and spreader bars when they can be properly utilized. Whenever possible use spreader bars and equalizer beams like the ones shown here, to lift the load.

- **Rule 1.** Use experienced planners, site supervisors, lift directors, qualified riggers, and certified crane operators.
- **Rule 2.** Keep the number of cranes used to make the lift to a minimum.
- **Rule 3.** Use cranes of the same kind, size, and configuration whenever possible.
- **Rule 4.** Provide a full written lift plan for all multiple-crane lifts. A lift plan, signed by site supervision, project management, crane operator, and lift director, shall be prepared on all critical lifts. The lift plan shall include the items necessary to address the special conditions of each lift.
- **Rule 5.** Make sure that the cranes are properly set up to within 1% of level and on a firm, level supporting surface. Make sure blocking is provided under all floats, with the outrigger beams fully extended and set. When using crawler cranes, timber matting should be considered, with blocking under the tips of the tracks. Crawler cranes must be set up within 1% of level and on a solid lifting surface.
- **Rule 6.** If cranes are to travel with the load, do not permit any other movement at the same time, such as swinging or booming of the load.
- **Rule 7.** Whenever possible, make sure the load to be lifted does not exceed 75% of the crane's charted capacities.
- **Rule 8.** Ensure that the hoist lines of all the cranes involved in the lift remain plumb at all times.
- **Rule 9.** Ensure, whenever possible, that only one person is responsible for directing or signaling the cranes during the lift.
- **Rule 10.** Remember, there is no substitute for a formal pre-lift meeting with all those, and *only* those, involved in the lift.

When planning a multiple-crane lift, and changes or revisions have been made to the original lift plan, make sure participating personnel meet to discuss and review those changes. Multiple-crane lifts can be tricky business. Many believe multiple-crane lifts are much more precarious than a single crane lift, and even more difficult as more cranes are added. This is due to the load interaction between the machines. The most notable accidents in recent years have been those involving multiple-crane lifts. When there are two or more cranes involved in an accident, the influence and consequences of the accident are often compounded.

Critical Lift Plan

Any nonroutine lift made with a crane requiring detailed planning and including additional or unusual safety precautions is considered to be a critical lift. In addition, a lift is considered a critical lift anytime more than one crane (or derrick) is required to make the lift.

A critical lift is a lifting operation that involves increased risk factors. This includes any crane operation that involves the use of nontraditional crane configurations or those that are at risk of encountering special hazards. All crane operations require in-depth planning to go correctly; however, critical lifts take the planning to a higher level.

A good critical lift plan includes all the risk factors involved with the lift. When finished, the plan is a document that virtually becomes a safety plan that must be shared with all those involved in the lift. When properly put into action, it should mitigate all the hazards listed in the lift plan. It will require extra equipment, resources, and preparation to ensure excellent execution. Without a clear understanding of all those requirements, it could fail. Even a small mistake could lead to a catastrophe, with injuries and fatalities.

The decision to develop a critical lift plan should be based on the extra risks discovered in the planning process. Let us start with making a thorough analysis of the net load, including all the rigging and load handling devices of the gross load.

Identify the load's weight, center of gravity, shape, and size. Recognize any parts or components that could shift during the lift, and determine how to secure them. Locate all hitch points, making sure they will handle the weight to be hoisted while in flight and securement when ready to unhook the rigging.

Make certain that the load handling equipment is capable of handling the load you are planning to hoist, including load block and ball hooks, equalizer beams, spreader bars, and all rigging hardware, as well as lifting lugs when applicable (Figure 17.7). Establish that all the load handling equipment meets all the site requirements, including federal, state, and local regulations including industry safety standards.

Ascertain the correct procedures to set up, erect, install, or dismantle any load handling equipment on the jobsite, including any site-specific requirements. Pin down the crane load testing or inspections that need to be performed, including any hoisting equipment involved in the lift.

Make certain the rigging methods that will hold and secure the loads are suitable for load handling activities. Confirm that the rigging process and hardware have the capacity to support the load in each hoisting activity required. Serious attention must be given to the dynamic effects on the rigging during the lift, including any adverse environmental conditions such as high or low temperatures and many other factors. Once the rigging is in place, attention must be given to the position of the center of gravity relative to the hitch points used. Whenever wrapping rigging around a load, attention must be given to the *D*/*d* ratio—the diameter of the load (*D*) in comparison with the diameter of the slings (*d*) selected for the lift. Critical lift plans should be prepared when any of the following conditions is present:

- Any lift that requires the use of multiple cranes
- Any lift that exceeds 75% of the crane's rated capacity within all the crane's configurations
- Any item to be lifted that requires special care in handling because of size, weight, close tolerance installation, high susceptibility to damage, or other unusual factors
- Any item to be lifted that requires exceptional care in handling because it is being hoisted above an important item
- Undetectable damage that would jeopardize the future operations or safety of a facility
- Any loads that are extremely valuable, irreplaceable, or unrepairable
- Lifts that result in the loads leaving the direct view of the crane operator, requiring the operator to work in the blind
- Loads that are potentially unstable in flight
- Loads that will require suspension directly above the rigging personnel

FIGURE 17.7 Be sure the load handling equipment is capable of handling the load you are planning to hoist. A capacity could be more or less than the charted capacity of the crane.

The lift director who is in charge of the lift has the responsibility for the load to be hoisted and has the authority to require it be handled as a critical lift. The lift director shall ensure that a step-by-step procedure is prepared for all critical lifts. Individual procedures can be prepared for one-time lifts, and general procedures may be developed to handle routine recurrent critical lifts.

A critical lift procedure should address these important areas:

- Items to be hoisted and placed
- Special precautions, including cribbing, blocking, or mats for crane tracks or outrigger floats
- Weight of the loads to be lifted, including rigging and all load handling devices
- Location of the center of gravity of all loads to be handled and placed
- A list of equipment that will be involved in the operation, including trucks, trailers, forklifts, rigging hardware, and spreader bars (Figure 17.8)

- All cranes listed by type, manufacturer, and rated capacity

The classification of a lift as being critical is often determined by company policy, but it may also be determined independently by site supervision, project management, or a qualified person. Critical lifts can occur anytime when lifting, preparing for a lift, lattice boom assembly, or disassembly operation.

FIGURE 17.8 The lift director will provide a list of all rigging hardware and spreader bars that will be involved in the operation, including their size and working load limit ratings.

Lift Directors

The ASME B30.5 safety standard for mobile cranes requires a lift director to be on every jobsite during crane and hoisting operations. In addition, OSHA inspection guidelines and enforcement procedures used when inspecting a jobsite where cranes are involved clearly call for a lift director to be onsite. The lift director must be a qualified person. Many companies identify the lift director onsite by issuing a specially colored hard hat or safety vest for the director to wear.

The areas that the lift director is responsible for include:

- Making sure the rigging performed on the jobsite is carried out by a qualified rigger.
- Making sure all signal persons communicating with the cranes are trained and qualified.
- Ensuring that distances of approach when working near power lines are respected along with all other safety standards and regulations pertaining to work around power lines.
- Making sure all members of the operation understand their responsibilities. Members must understand the risks involved. The tasks they are to complete during the lift must be clearly understood.
- Making sure the cranes involved will be sitting on firm and level ground and all safety measures are in place.
- Being present onsite during all phases of the hoisting operation.

Chapter 17 Review Quiz

1. Any lift that involves more than one crane is considered to be a multiple-crane lift as well as a
   a. complex lift
   b. tandem lift
   c. critical lift
   d. structural lift

2. How much of a crane's charted capacity should not be exceeded when making a tandem lift?
   a. 55%
   b. 65%
   c. 75%
   d. 85%

3. According to OSHA, a critical lifting operation is present when
   a. any lifts involving more than one crane are considered critical lifts
   b. the lift exceeds 65% of the rated capacity of the crane
   c. the lift exceeds 70% of the rated capacity of the crane
   d. the lift exceeds 75% of the rated capacity of the crane

4. When two or more cranes are used to lift a load, who is responsible for the operation?
   a. A person who meets the criteria for both a signal person and a qualified rigger, called the *bull rigger*
   b. A person who meets the criteria for both a qualified person and a certified crane operator, called the *lift foreman*
   c. A person who meets the criteria for both a certified rigger and a qualified person, called the *lifting supervisor*
   d. A person who meets the criteria for both a competent person and a qualified person, called the *lift director*

5. According to the ASME B30.5 safety standard, the person responsible for control of a multiple-crane lift is the
   a. certified rigger
   b. qualified signal person
   c. certified crane operator
   d. project lift director

6. What is the exclusion zone used for during mobile crane lifts?
   a. It is a defined area where people are not allowed due to radioactive materials present.
   b. It is a defined area where only supervisors are allowed to go.
   c. It is a defined area where people are not allowed to go due to hazardous waste in the area.
   d. It is a defined area where people are not allowed to go when particular work is being done to avoid the risk of being injured.

7. Which of the following can be considered a critical crane lift?

a. Any lift being hoisted in the rain

b. Loads above 5 tons

c. Lifting of personnel

d. Loads being hoisted in the afternoon

8. Particularly strict, no-nonsense attention must be given to all the cranes set up and to the alignment being used for multiple-crane lifts to avoid what type of pressure on the cranes' booms?

a. Dynamic loading

b. Shock loading

c. Static loading

d. Side loading

9. According to OSHA regulations and ASME safety standards, a critical lift means

a. any lifting operation made in the morning hours

b. any lifting operation made during the evening hours

c. any lifting operation that requires heavy bull rigging

d. any lifting operation that involves increased risk factors

Summary—and Then Some

Lift plans and the responsibilities of the lift director are complex, especially when a multiple-crane lift is involved. Many companies, when faced with a multiple-crane lift or a heavy, complicated lift, will decline and subcontract to a mobile crane rental company or crane consulting firm. The risk is high, and a large amount of liability is clearly on the line, so this is a viable alternative for many employers.

After completion of a successful critical lift, there is still work to be done. The lift director should review the development, planning, and execution of the operation with the members of the lifting team. Together they should discuss areas of the operations that could improve future critical lift activities. Multiple-crane lifts are among the most technically complex lifts to plan and execute. Crane and rigging accidents are caused 90% of the time by human error. Hire only the most experienced, highly trained, certified, and qualified craftworkers for all hoisting and rigging operations.

FIGURE 17.9 Two Sickelsteel cranes work in unison to provide this very heavy lift. The two crane operators make a precision smooth lift with constant situational awareness since the two operators are out of each other's view. Clear communication protocols are an essential part of this lift plan.

FIGURE 17.10 Two giant Lampson Transi-Lift crawler cranes make this very heavy and risky lift over the Columbia River in Washington State. Good engineering and exceptional planning was the key to success for this lift.

If you are planning a lift with a mobile crane, it is highly important you get an accurate load weight. The lifting equipment used to lift the load must support not only the load weight, but its volume, height, center of gravity, and any other important aspects of the load. The load weight will determine the specifications of the equipment used to make the lift.

Equipment or fabricated components delivered to your worksite may have the load weight marked on the load. Always look for permanently stamped load weight indicators. Oftentimes, the bill of lading or shipping papers will have that valuable information about the component, including its size and weight. A manufactured product such as a piece of machinery may have its specifications in a catalog or brochure. Engineering prints or design plans are another excellent source to find an assembled weight of a product. Sometimes smaller, lighter loads can be weighed with industrial floor scales. In some cases, if no load weight information is provided, you will need to calculate it yourself using available load weight tables.

18 Determining Load Weights

Let's imagine a large 40-foot shipping container arrives, sitting on the back of a truck trailer to your jobsite. You are asked to unload the unit with the crane you are operating and place it in the jobsite yard for future storage. You notice an identification plate attached to a door section indicating it weighs 9,260 pounds. As you look over the load, you see lifting lugs positioned on the top four corners designed for hoisting purposes. You are thinking this appears to be a straightforward lift. About then, one of your riggers opens the doors on the front of the container to discover a C15 Caterpillar engine mounted to the rear floor of the container. After a few minutes of discovery, a plate is found to say the engine weight is 3,239 pounds. In addition, a 25-gallon fuel tank is found full of diesel fuel. You calculate the diesel fuel to weigh 7 pounds per gallon, so you add another 175 pounds to the lift. What was thought to be an easy-as-pie lift has now become a perplexing job. Your ground crew, through wise discovery, has determined the original container weight of 9,260 pounds, along with all the extra components found, including the rigging, adds up to a much heavier lift.

9,260 pounds (shipping container) + 3,239 pounds (C15 Caterpillar engine) + 175 pounds (diesel fuel) + 225 pounds (rigging) = 12,674 pounds

Careful inspection of the container and discovery of the extra components found inside changes this lift to be much heavier than first thought. Finding the Caterpillar engine mounted at the rear of the container has changed the location of the center of gravity (COG). Hitch points for rigging the lift must be relocated. The crane operator and the lifting team in this story have done a good job adjusting to several factors found in the field. Never judge a load by its first appearance. Discovery and careful inspection of all components of this lift saved the day.

Dynamometers or load cell devices can be involved as part of the rigging with many lifts. This technology can give the crane operator a report of what a load weighs when it is lifted a few inches off the ground. These units are hitched in line with the crane's hoist line, hook, rigging hardware, and slings. The load cell calculates the weight of the lift by measuring the force applied to unit using a strain gauge, or pneumatic pressure inside the load cell. Cranes equipped with a load moment indicator (LMI) can be utilized in the same way today's technologies provide the crane operator and lifting team with more good tools for planning hoisting operations than ever before. Load cells can come in the form of a rated lifting shackle with integrated electronics and microprocessors to provide the weight of a lift once it's hoisted off the ground. Valuable information, including the load weight, is sent to a hand-held remote device that will warn the crane operator, lift director, and other members of the hoisting operation of an overload condition. Exceeding the charted capacity of the crane can be stopped, avoiding possible structural failure or loss of stability of the machine, putting members of the lifting team at risk, including innocent bystanders. The first and most important step in any rigging operation is to determine the weight of the load to be hoisted. If this information cannot be obtained from shipping papers, design plans, catalog data, or other dependable sources, it may be necessary to calculate it. On erection plans, the size of the steel beams is usually supplied together with their length and weight per foot, so it is easy to compute the weight of the beams you plan to lift.

For other materials, the weights are normally based on their weight per cubic foot, so you have to determine how many cubic feet of material, or volume, you are hoisting in order to estimate the weight. Table 18.1 lists the weights of frequently used materials.

TABLE 18.1 Calculating Load Weights

Weights of Frequently Used Materials (Based on Volume)			
Material	**Weight per Cu/ft**	**Material**	**Weight per Cu/ft**
Metals		**Earth**	
Aluminum	165	Earth, dry	75
Brass	535	Earth, wet	100
Bronze	500	Sand and gravel, dry	105
Copper	560	Sand and gravel, wet	120
Iron	480	**Timber, Air Dried**	
Lead	710	Cedar	22
Steel	490	Fir, Douglas, seasoned	34
Tin	460	Fir, Douglas, unseasoned	40
Masonry		Fir, Douglas, wet	50
Brick masonry, soft	110	Fir, Douglas, plywood	36
Brick masonry, common	125	Hemlock	30
Brick masonry, pressed	140	Pine	30
Concrete, cinder, haydite	100–110	Popular	30
Concrete, stone	144	Spruce	28
Concrete, stone, reinforced	150	**Liquids**	
Building Materials		Alcohol, pure	49
Asphalt	80	Gasoline	42
Cement, portland, loose	94	Oils	58
Crushed rock	90–110	Water	62
Glass	160	Ice	56

The weights of angles can be measured with close enough results for safe job use. An angle is a structural shape that can be considered as a bent plate with some additional metal at the center for strength. If an angle is flattened out, a large square plate is the result.

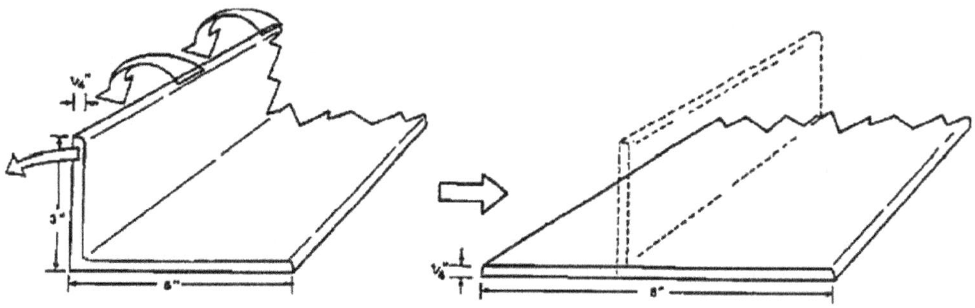

FIGURE 18.1 Imagine an angle flattened out to form a plate.

The weights of any structural shape, not just angles, can be computed by separating the parts and flattening them into rectangles, which, in turn, become parts or multiples of a square foot of steel an inch thick. Plates are often rolled into tanks or other shapes we need to hoist. It is necessary to determine the square foot area of all the parts before the weight can be calculated. This requires learning the two simple formulas for computing the circumference and the area of a circle.

The first formula is for calculating the circumference or distance around the outer edge of a circle. It is found by multiplying the diameter, which is the distance from one edge of the circle to the other, by 3.14 (pi).

As an example, a pipe 6 feet in diameter would have a circumference of 6 feet × 3.14, or 18.84 feet. To compute the weight of this pipe, which is 30 feet long and made of ⅜-inch plate, mentally unroll it and flatten it out, as seen in Figure 18.2. This represents a plate that is 18.84 feet wide by 30 feet long by ⅜-inch thick.

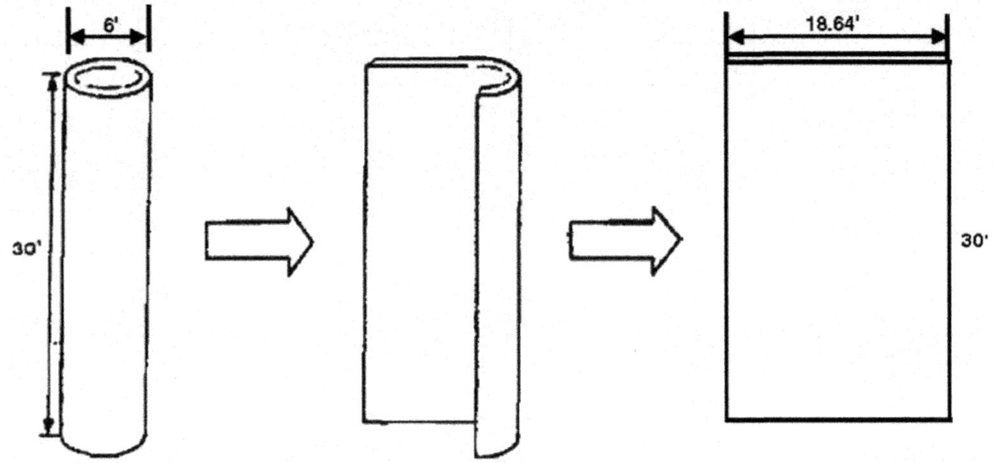

FIGURE 18.2 Roll out the pipe to form a flat plate before making calculations.

The second formula you will need gives the area of circular objects. Area equals 3.14 × diameter divided by 4. The pipe has an end cap ⅜ inch thick and is 6-feet in diameter, the end cap would have a surface area of 3.14 × 6 feet × 6 feet divided by 4 = 28.26 square feet.

Chapter 18 Review Quiz

1. The basic weight of 1 square foot of steel 1 inch thick is approximately 40 pounds. How much will 2 plates of steel each measuring 1½ inches × 3 feet × 6 feet weigh?

   2 (plates) × 1.5 inches (thick) × 3 feet × 6 feet × 40 pounds (per square foot) = ___ pounds

FIGURE 18.3 Typical steel plates in storage.

2. You have a 5-inch × 3-inch × ¼-inch angle flattened out to approximately an 8-inch × ¼-inch plate. The weight per square foot is 40 pounds. How many pounds would the flattened out plate weigh?

   40 pounds (per square foot) × 8 inches × 12 inches (1 foot) × 0.25 (¼) inch = ____ pounds per square foot

3. You are planning to lift a unit of spruce lumber that is 12 feet long, 3 feet high, and 4 feet wide as illustrated in Figure 18.4. The weight per cubic foot of dry spruce is approximately 28 pounds. How much does this unit of lumber weigh?

   12 feet × 3 feet × 4 feet × 28 pounds (per cubic foot) = ____ pounds

FIGURE 18.4 Unit of spruce lumber you are planning to lift using the load weight tables in Figure 18.1.

4. A pipe 6 feet in diameter has a circumference of 6 feet × 3.14, or 18.84 feet. To compute the weight of this pipe, which is 30 feet long and made of ⅜-inch plate, mentally unroll it and flatten it out. This represents a plate that is 18.84 feet wide by 30 feet long by ⅜ inch thick. The weight per square foot is 40 pounds.

FIGURE 18.5 Large pipe laying in a construction storage area.

a. How much would this 30 foot long pipe weigh?

30 feet × 0.375 (⅜) inch × 40 pounds (per square foot) = ___ pounds

b. The pipe has an end cap ⅜ inch thick and 6 feet in diameter (Figure 18.6). It would have a surface area of 3.14 × 6 feet × 6 feet divided by 4 = 28.26 square feet (we'll round it off to 28.3). How much would this 6-foot diameter pipe end cap weigh?

28.3 square feet× 0.375 (⅜) inch × 40 pounds (per square foot) = ___ pounds

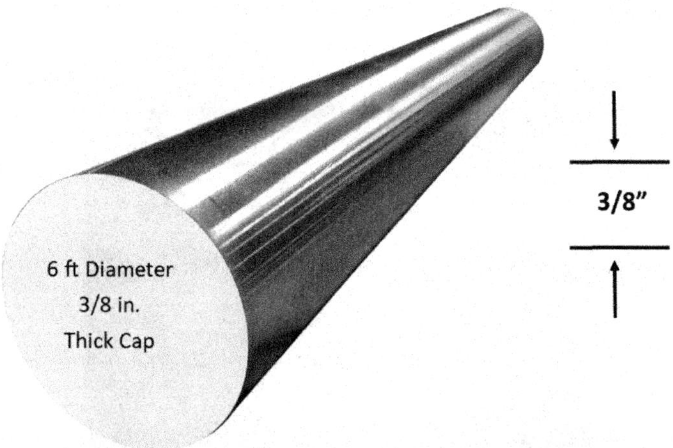

FIGURE 18.6 A 6-foot-diameter pipe with end cap.

c. What would be the combined weight of the pipe and the end cap?

28.3 square feet × 0.375 (⅜) inch × 40 pounds (per square foot) = ____ pounds

Summary and Then Some

Always evaluate the load you want to lift. Inspect the load for any identification or marks that indicate the weight. Make sure the weight represents the entire load, not just a single component. Check supporting documentation for load weights. Look over any available drawings or diagrams that are available. If necessary, estimate the load's weight using the load weight tables as shown in Table 18.1. Always consider adding your own safety factor of additional weight for extra protection.

FIGURE 18.7 Using load cells provides benefits in crane and rigging operations. They enhance safety by giving the crane operator real-time information to make decisions and avoid overloading risks that can cause accidents or damage equipment.

There are many considerations that need to be made before a lift is ever executed. Crane operators, supervisors, lift directors and riggers must know the capacity and restrictions of all lifting devices used to hoist the load. The weight of the load, center of gravity, rigging equipment, and the size and capacity of the crane being used are all equally important to make certain the lift will be safe. Riggers in charge of hoisting and rigging operations must be accountable for knowing that the weight of the load, location of the center of gravity, approved lifting devices, and rigging equipment have been chosen in compliance with all OSHA regulations and ASME industry safety standards. In addition, all rigging equipment must be adequate for the load to be hoisted based on the number of sling legs, lifting lugs, hitch point arrangements, and effects of sling angles. Knowledge about sling angles when making a lift is critical. Understanding how sling angles can affect the working load limit (WLL) of a sling is highly important when rigging a load and helps create a greater level of safety during hoisting and rigging operations on general industry and construction jobsites.

As sling angles decrease, the sling stress increases. To calculate lifting capacities, first we must know the weight of the load to be lifted. The rated lifting capacity of the sling must be greater than the load to be lifted. The angle of pull on the leg of a sling varies with the sling leg angle. Forces exerted on a sling can be very high. The sling angle will affect the tension and stress put on each sling leg. Company representatives will appoint someone competent and qualified for jobsite rigging operations. This person's responsibility is to make sure the proper sling angles are achieved and the load is rigged safely before every lift. Safety is the most important part of all crane and rigging operations.

19 The Effect of Sling Angles

The rated capacities of multi-leg slings will depend on their size, the number of legs, and the angles formed by the slings to the horizontal line of the load. For example, two slings that are used to lift a 1,000-pound load will have a 500-pound load on each of the legs, provided both slings remain at 90 degrees to the load. As the angles of the slings to the load decrease, the load on the slings will increase. At a 30-degree sling angle, the load on each sling will be equal to the weight of the load. Lifting loads with sling angles 30 degrees or lower is considered dangerous and should be avoided if possible. If not, special steps must be taken.

When the sling angles that are horizontal to the load fall below 30 degrees, the stress on the sling legs increases dramatically (Figure 19.1). At a 20-degree angle, the load on each sling lifting the same 1,000-pound load will be increased to 1,462 pounds. At 10 degrees, the stress on each sling will be approximately three times the actual weight of the load being lifted. At 5 degrees, the stress will be almost six times the weight of the load.

A load will remain level provided the slings are correctly applied and the load's center of gravity stays directly beneath the hoist block of the overhaul ball and hook of the crane. If the load is allowed to tilt because of incorrect rigging, the stress on one leg of a two-leg sling will increase while the other decreases. A load tilted about 12 degrees will see the stress on one of the sling legs increase from 70 to 80%.

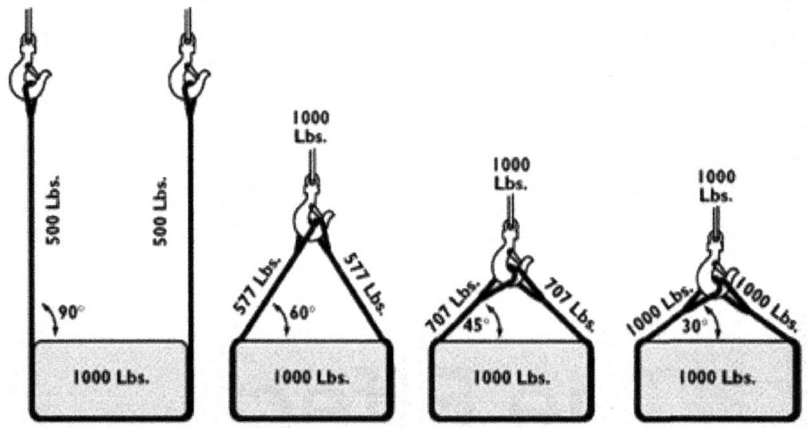

FIGURE 19.1 Sling angles of 30 degrees or less should be avoided because the stress from the load on the slings is very high.

Sling Angles and Working Load Limit

Sling angles are crucial in determining working load limits for sling configurations. Sling angle table capacities are given for bridle and basket hitches at angles of 60, 45, and 30 degrees. Measuring these angles can be difficult, since the measuring tools required are not always available.

Two of these angles can be easily determined before consulting the tables for a bridle hitch. A 60-degree sling angle will be created when the distance between the attachment points is equal to the length of one of the slings (Figure 19.2). For a basket hitch, that angle will be created when the distance between the points at which the sling first contacts the load is equal to the length of one inclined leg of the sling. A 45-degree sling angle is created when the angle formed by the two legs of a bridle hitch attached to the crane hook or link is 90 degrees.

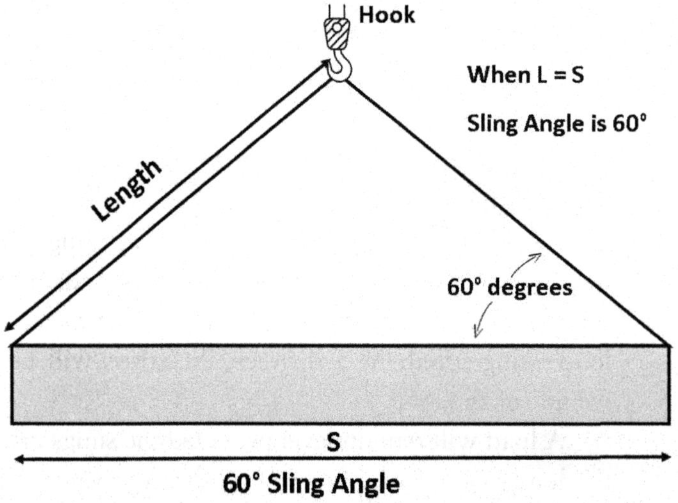

FIGURE 19.2 A 60-degree angle will be created when the distances between the attachment points are equal to the length (L) of the slings (S).

Calculating Safe Working Loads on Multi-Leg Slings

Prior to lifting any loads, it is important to ensure that the slings selected have the required capacity to make the lift. To accomplish this, the length and working load limit of the single vertical hitch of the selected slings must be known. Consult available sling tables, or approximate their WLL by multiplying their diameter in eighths squared times 275.

Example: $\frac{3}{4}$-inch-diameter rope = $\frac{6}{8}$-inch-diameter rope

6-foot × 6-foot × 275 equals a WLL of approximately 9,900 pounds.

Try the math yourself by completing the calculation posed in Figure 19.3.

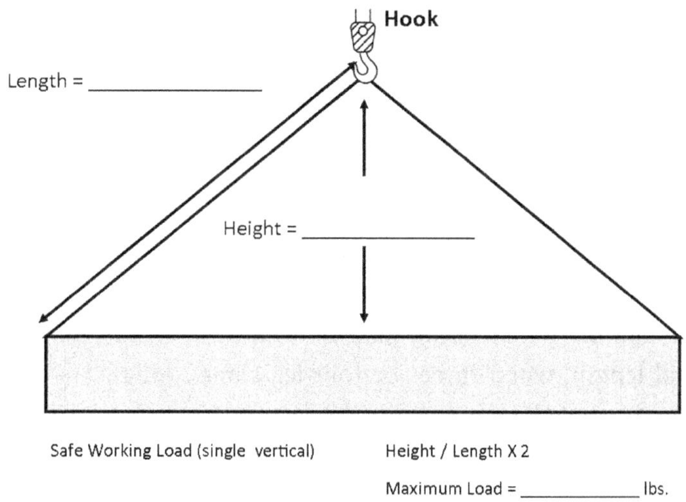

FIGURE 19.3 Calculate the maximum load that can be lifted by two $\frac{5}{8}$-inch-diameter slings.

Rule of Thumb for Working Load Limits

The rule of thumb for estimating the safe working loads for common sling configurations is based on the WLL of a single vertical hitch of a given size and material and the ratio *H*/*L*, where *H* is the height measurement from the saddle of the hook to the top of the load and *L* is the length of the sling.

If you cannot measure the entire length of the sling, then measure to a convenient point, and call this distance *L*. From this point, measure down to the load, and call this distance *h*. The ratio *h*/*L* will be the same as the ratio *H*/*L*.

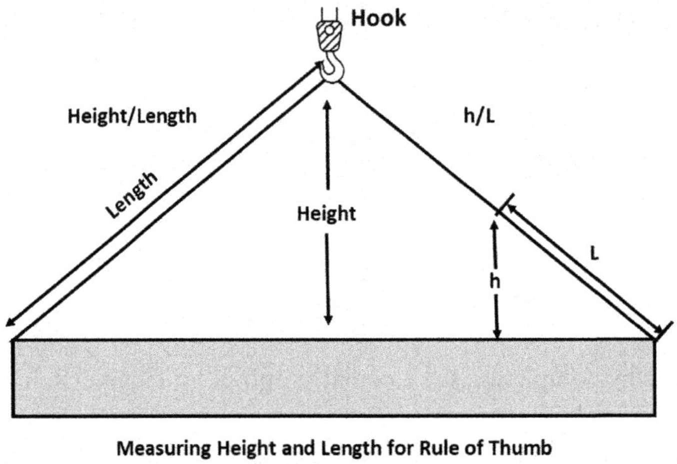

FIGURE 19.4 Measuring height and length for the rule of thumb for working load limits.

Calculating the Actual Load on Each of the Slings

With the addition of components such as eye bolts, shackles, or turnbuckles to the sling configuration, it is also important that their rated capacities not be exceeded. To ensure this it may be necessary to calculate the actual load being placed on each of the slings.

The formulas for calculating the working load limit and the actual load on slings are based on using two slings of equal length, when three- or four-leg sling configurations are used or when the lengths of the slings are not equal (Figures 19.5 to 19.7).

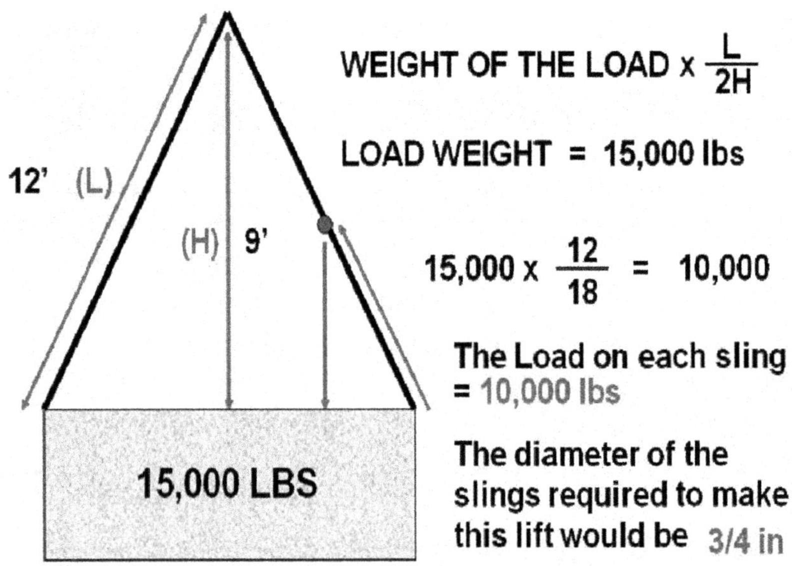

FIGURE 19.5 Calculating the actual load on each sling.

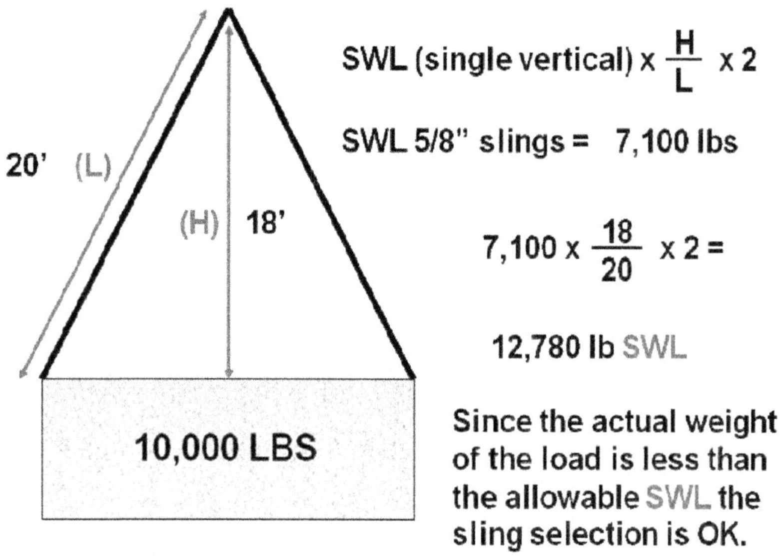

FIGURE 19.6 Determining the safe working load of preselected slings.

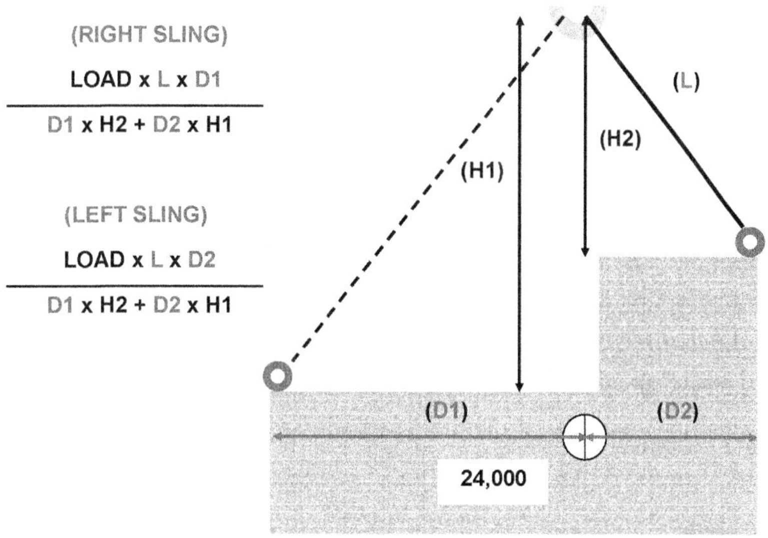

FIGURE 19.7 Calculating the load on unequal slings, where *L* = length, *D* = horizontal distance, and *H* = height.

Hoisting Loads with Sling Angles Less than 30 Degrees

The ASME B30.9 Safety Standard calls for the development of a critical lift plan for loads rigged with sling angles less than 30 degrees. Compliance with this standard means certain proper safeguards are taken to ensure all members of the hoisting operation are protected. A critical lift plan will help mitigate the risks of hoisting loads with 30 degree sling angles or less. A qualified person or the lift director will approve the lift plan and help supervise the lift. Lifting with small sling angles is risky business that requires precise calculations and professional execution. A spreader bar lifting attachment should be considered to help distribute the load weight more adequately and helps produce better sling angles.

Chapter 19 Review Quiz

1. When lifting with two slings with a 60 degree angle to the load, what would the length of each sling be if the distance between the lift points is 12 feet?
   a. 6 feet
   b. 12 feet
   c. 18 feet
   d. 24 feet

2. When using two slings to lift a load, at what angle to the load will the load on each sling equal the weight of the load?
   a. 15 degrees
   b. 30 degrees
   c. 45 degrees
   d. 60 degrees

3. According to the ASME B30.9 safety standard for slings, a critical lift plan must be developed whenever sling angles are less than
   a. 20 degrees
   b. 30 degrees
   c. 40 degrees
   d. 50 degrees

4. Who must approve a critical lift plan when low sling angles are used?
   a. A competent person
   b. A qualified rigger
   c. A qualified person
   d. A certified crane operator

5. As a sling angle decreases, the sling
   a. will stretch
   b. stress decreases
   c. stress increases
   d. reaches its safety factor

6. A load tilted 12 degrees will see the stress on one of the sling legs increase
   a. 40 to 50%
   b. 50 to 60%
   c. 60 to 70%
   d. 70 to 80%

Summary—and Then Some

Wrap Up Summary and Then Some

Sling manufacturers provide capacity charts for their equipment that include sling angles configurations in the tables. They are very helpful and should be available to everyone on a lifting team. Slings are designed with a 5-to-1 safety factor. The safety factor is created to allow for wear and tear on them, unbalanced loads, dynamic forces, and shock loading. Never exceed the charted capacity ratings of any sling. Sling angles do matter. If not respected, rigging gear can be overloaded to a point of failure, leading to more problems. It is crucial to have a good understanding of sling angles. Poor sling angles can be detrimental to the working load limit (WLL) of the sling, putting everyone involved with the lift at risk of a serious accident. Be on the lookout, and create a culture of safety within your hoisting operations. Be proactive, making sling angles a top priority in your rigging practices so everyone goes home safe.

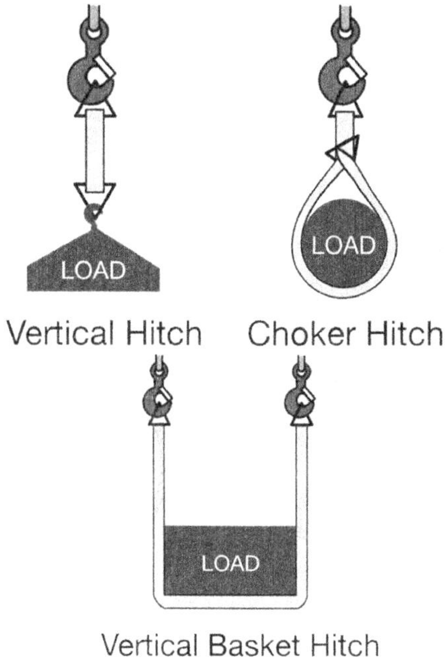

FIGURE 19.8 Three basic hitches.

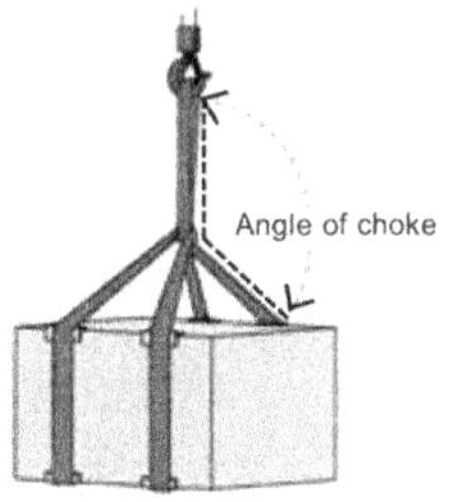

CHOKE HITCH REDUCTION CHART	
Angle of Choke (Degrees)	Loss Factor
120 - 180	1.00
90 - 119	.87
60 - 89	.74
30 - 59	.62
0 - 29	.49

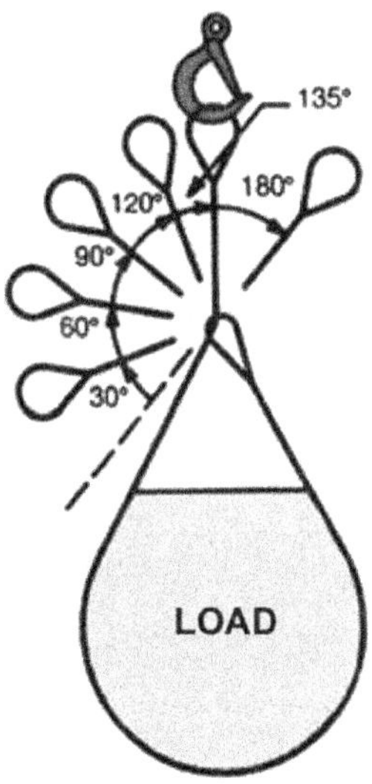

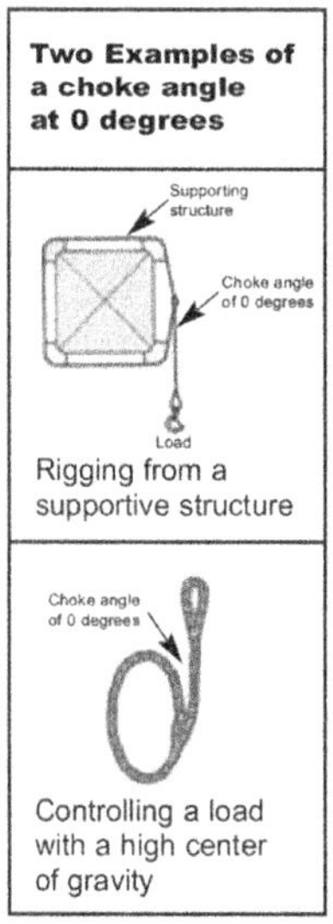

FIGURE 19.9 Choker reduction chart.

To properly make heavy lifts safely, it is highly important that the crane operator, including each member of the lifting team, have a comprehensive knowledge of the many types of rigging equipment, slings, and hardware and how to properly use them. When all members of the hoisting and lifting team are familiar with all the different rigging components' functions, you will have established a foundation for a safe and structured operation, always in compliance with safety standards and regulations. When taking part in hoisting and rigging operations, all workers are required to pay close attention to all safety procedures and protocols because of the risk of endangering all those on the hoisting and rigging crew. According to the U.S. Bureau of Labor Statistics, over half of crane-related deaths between 2011 and 2017 involved an employee being struck by an object or piece of equipment. Falling objects are to blame for 91 deaths during this period. Understanding safety standards and regulations and keeping a detailed eye during all lifting operations are critical in this industry. By doing this, as well as knowing why it is important to comply with all safety standards and regulations during your hoisting operations, you will help ensure efficient craning activities, minimize risks, and prevent accidents, while boosting productivity on the jobsite. Investing in training certification programs ensures that workers are highly skilled, equipped with competent knowledge, and confident while performing their duties in this dangerously fast-paced industry.

20 Specialized Rigging, Slings, and Hardware Used in Craning

The Importance of Training

It is important to understand what preparation is required in hoisting and rigging before hoisting the load, as well as what to do when lifting while the center of attention is on raising or moving the load. Both areas are equally important to understand when establishing safe hoisting and rigging operations, and both require skills that involve working closely with the crane operator. The most comprehensive training for hoisting and rigging personnel will require learning a wide range of technical skills involving a large variety of rigging equipment and hardware while knowing the different applications they've been designed for. This requires training with the specific purpose of developing skills in hoisting, rigging, crane operation, load assessment, load handling, prioritizing, procedural guidelines, rigging configuration, hazard identification, safety protocols, and other specialized skills. Focused, comprehensive training in these fundamental concepts, and knowing the differences among them, will begin the transformation of personnel into professionals in their craft—experienced experts in preparing and securing loads for safe hoisting operations. Each will have the valuable knowledge required to establish flawless communication with their colleagues preparing a load or executing lifts. The result—true teamwork among a well-rounded, highly trained group of riggers and lifting personnel.

Understanding the Difference Between Lifting and Rigging

Lifting, or *hoisting*, refers to the task of raising a load. *Rigging* involves the entire procedure of preparing, securing, and controlling the load for safe craning operations.

Lifting Operations

- When hoisting or lowering heavy objects on the job, cranes, hoists, and forklifts are some of the most common ways of hoisting loads.
- A hoisting operation is the physical movement of the load from one location to another.
- Lifting requires the use of specialized equipment, such as hooks, slings, rigging hardware, and lifting lugs, to name a few.

Rigging Operations

- Rigging incorporates the process of preparing the load for hoisting by ensuring the load is properly hitched, positioned, and controlled.
- Rigging operations include properly selecting the correct rigging gear, calculating load weights, and developing lift plans
- Riggers oversee managing load control systems, assessing load stability during the lift, and ensuring all is in compliance with safety standards and regulations.

The definitions of *lifting hardware* and *rigging hardware* can be used comparably. It is highly important to remember that lifting hardware can be referred to as rigging; however, not all rigging can be used for hoisting, or lifting. These terms can be used differently by the construction industry and general industry or even by different manufacturers. Know the difference between the terms, and use them correctly when discussing equipment—especially when planning lifts. Not all rigging equipment that is used to secure, or lash down, loads is approved for overhead lifts, which require equipment with higher ratings and working load limits.

Sling Protection

The proper application of the slings can help maintain the safety of the lifting operation and increase the efficiency and security of the slings.

- Never use kinked or damaged slings or hoist ropes. To provide maximum operating efficiency and safety, all slings and fittings should be given thorough periodic inspections as well as daily inspections for signs of wear and abrasion, broken wires, worn or cracked fittings, kinking, crushing, flattening, and excessive corrosion. Special care should be taken in inspecting the areas around thimbles and U-bolt fasteners.

- While applying a sling to a load, be sure the arc of the sling, where the sling makes contact with the load, is equal to 1 rope lay in length. This is the distance it takes 1 strand to rotate 360 degrees around the rope's core and multiple stops. This approximate distance is equal to the wire rope's diameter multiplied by 3. When using wire rope slings, if the bend is too sharp, the strands on the outer side of the rope will be forced to stretch. This stretching can become permanent, leaving a sharp kink in the sling.
- Keep all slings away from oxygen, acetylene cutting, and electric or arc welding operations.
- Avoid any sling contact with solvents and chemicals that could cause corrosion and weakening of the sling, shown in Figure 20.1.

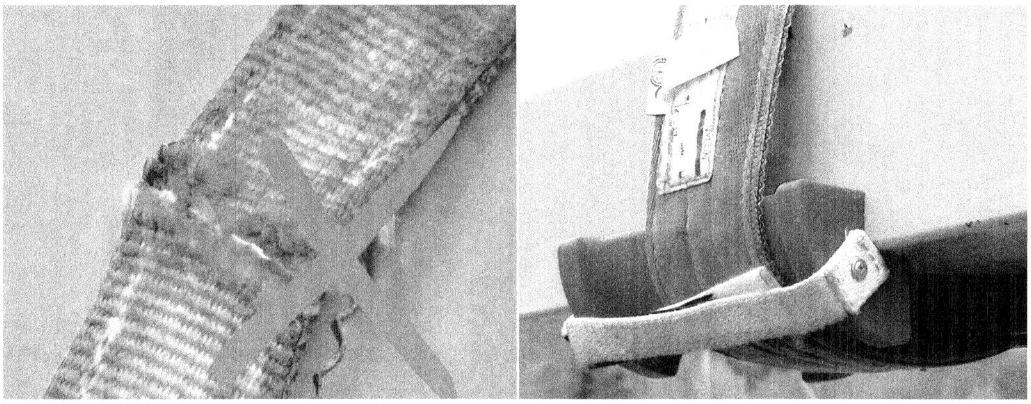

FIGURE 20.1 Always use sling protectors or softeners to avoid dangerous cuts to synthetic slings.

All slings, especially synthetic slings, can be damaged when lifting a load if they are not properly protected. Edge protection with sling use is critical in preventing sling failure and is a requirement in current sling safety standards.

- Nylon protectors attach with strong magnets to the steel corners and keep the sling from contacting the load. Made of solid nylon, these magnetic sling protectors allow for quick and easy attachment and removal, depicted in Figure.20.2.
- Cut and damaged slings are the leading cause of most rigging-related accidents.

FIGURE 20.2 Sling protectors are also known as "softeners."

Slings in contact with edges, corners, or protrusions should be protected with a material of sufficient strength, thickness, and construction to prevent damage to the sling, as stated in ASME B30.9 and depicted in Figure 20.3.

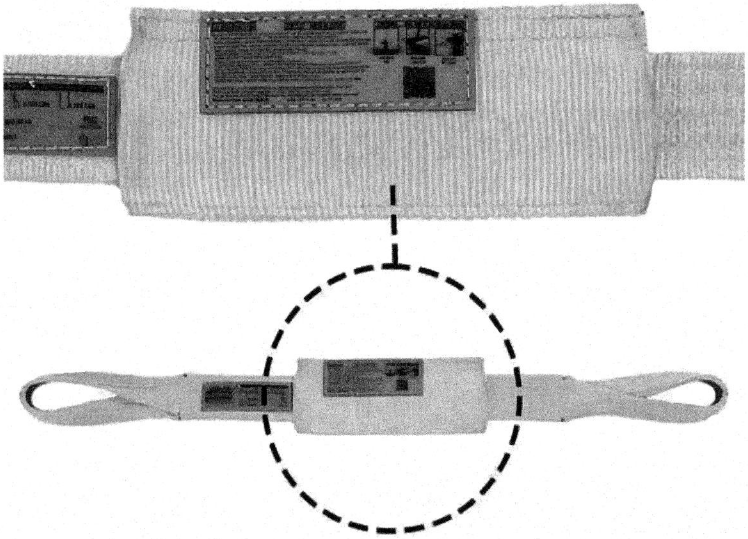

FIGURE 20.3 Wear sleeves are available to protect your wire rope slings.

Figure 20.4 shows a variety of softeners that are available from hosting and rigging dealers.

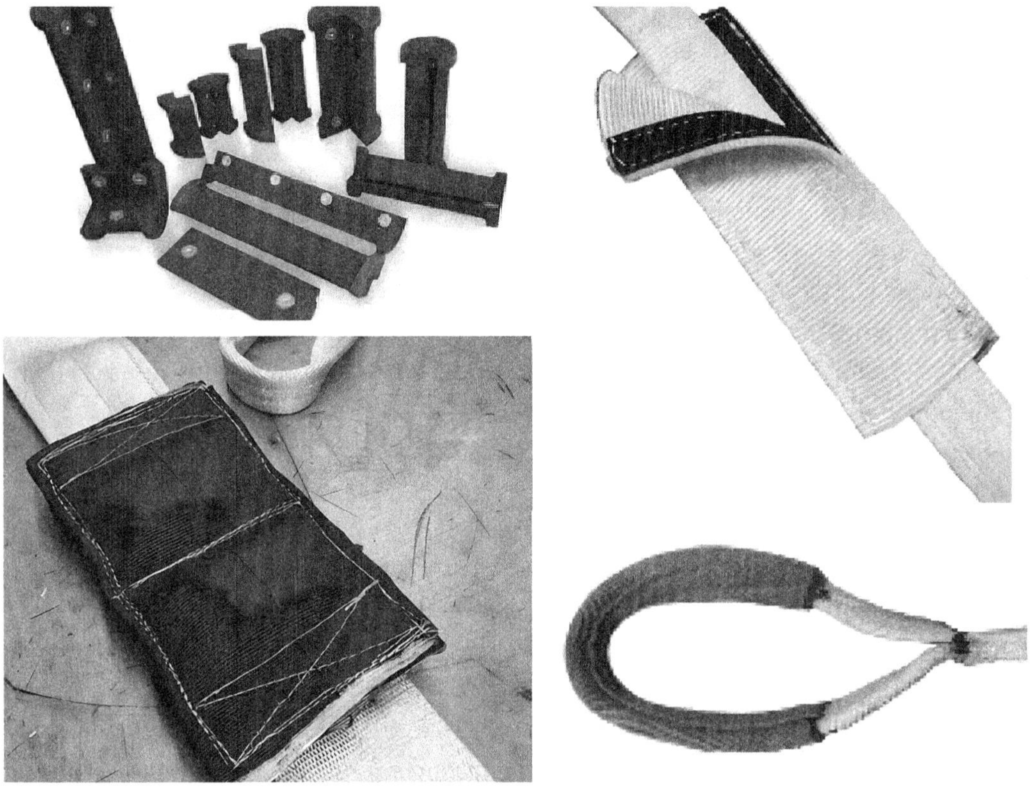

FIGURE 20.4 With so many choices available to protect your slings, there is no excuse not to.

How Rigging Equipment and Hardware Are Rated

All equipment used to hoist a load, most importantly starting with the crane, must be rated. After all, the crane and the equipment used to rig the load must support the load's weight once it has been lifted into the air. Each piece of equipment and hardware, including any slings, shackles, eye bolts, or rings, must be individually selected according to each of their ratings to minimize the chances of failing in all hoisting and rigging operations. Understanding the importance of this step eliminates risking the possibility of your fellow workers, or worse, the general public, being injured by falling loads.

Rigging Hardware Rating for a Component's Weight Limits

- **Rated capacity (RC).** This term, referred to as the *working load limit* or the *safe working load*, indicates the maximum load that a piece of rigging equipment can handle under ideal conditions.
- **Minimum breaking load (MBL).** The minimum breaking load, or force, specifies that the force that must be reached before lifting equipment has the probability of breaking or undergoing change. The MBL is also the amount of force required when testing for tear.
- **Safe working load (SWL).** The safe working load is the maximum weight that a piece of rigging equipment can hold under specific conditions. The SWL is determined by the manufacturer and takes into account the material, construction, and specific application for which the rigging hardware is designed.

- **Working load limit (WLL).** The working load limit is the maximum load that a piece of equipment can safely handle under normal operating conditions. This is a very important safety measure that ensures that the equipment is not pushed to its breaking point during regular use, preventing accidents and equipment failure.

Calculating WLL

The WLL is calculated by dividing the breaking strength of a component by the safety factor. In this example, we will use 6:1 as the safety factor and 60,000 pounds as the breaking strength: WLL = 60,000 pounds divided by 6 = 10,000 pounds.

The breaking strength is always a higher number than the working load limit because the WLL includes the safety factor in order to account for potential stress and variations in use. The breaking strength is the upper limit. The WLL is the recommended safe operating load limit.

Special important note: *SWL* is an older and better-known term that has been used in hoisting and rigging operations for many years.

The Difference Between SWL and WLL

The term *WLL* is more specific and standardized than SWL. The term *SWL* is more of a generalized term and can be understood in different ways. Today, SWL has been replaced by WLL because of the potential legal complications regarding the use of SWL as a result of it suggesting guaranteed safety—a guarantee that isn't possible in any hoisting and rigging operations, due to human error and unexpected circumstances. Using WLL is more precise and spells out specific working conditions and the importance of staying within the equipment's limits, rather than suggesting absolute safety.

Dynamic Amplification Factor

The dynamic amplification factor (DAF) is a variable factor that represents the dynamic force that the working load is exposed to during hoisting or lowering the load and lowering the load when placing it on the ground. The DAF is used to make calculations below the crane hook and to determine the lifting capacity of a crane.

When gearing up for a hoisting and rigging operation, it is most important to ensure that the rated capacities of the rigging hardware and equipment are correct for the application they're being used for. Remember, the equipment's WLL may be affected by the materials used in its manufacturing and in the rigging equipment's design. For example, working around chemicals or high temperatures may influence the material that the rigging hardware or equipment selected is made of. Always consider which forces the hardware might experience during a lift, such as shock loading from sudden load movement and rapid speeds during hoisting or suddenly stopping the load.

It's always a great idea to select rigging hardware and equipment with a higher WLL than needed for the load you plan to lift. Having this extra safety factor built into the lift by the rigging team, which exceeds the WLL of the equipment being used, is always good practice. Safety is always the most important thing to consider in any hoisting and rigging operation. Crane and rigging accidents put workers, and in many cases the public, in danger. The cost of equipment and property damage adds to this problem. As noted at the beginning of the chapter, statistics from the U.S. Bureau of Labor Statistics show that from 2011 to 2017, falling objects and rigging equipment were involved in more than half of fatal crane injuries in the hoisting and rigging industry. Rigging hardware equipped with a WLL increases safety by providing a limited margin for error; however, that margin must never be mistaken as extra operating strength gained to make a heavier lift.

Keep in mind the following important safety tips to maintain the highest level of safety in any hoisting and rigging operation:

Before the Lift

- Make sure all personnel are properly trained, qualified, and certified for the work they'll be performing.
- Understand all the manufacturers' specifications of all the equipment and rigging hardware used during the lift.
- Identify any power lines before lifting, and contact the local power provider for the proper protocols before any lift is made.
- Know the WLL limits of all the rigging hardware and lift equipment, and when possible, allow yourself a margin for error by selecting gear rated higher than necessary for the safe execution of the hoisting operation.
- Inspect all equipment before the lift. Ensure it is ready to go by ensuring it is in good condition and by eliminating any components found to be defective and removing them from the jobsite.
- Select the correct gear based on size, weight, diameter, length, and width of the load being handled.
- Keep the lift director thoroughly informed of all the details about the equipment.
- Develop pre-lift plans, and hold meetings with all members of the lifting team.

During the Lift

- Keep all personnel away from the lifting area except those appointed by the lift director.
- Use only highly trained, certified lift directors, crane operators, riggers, and signal persons.
- Never allow anyone under a suspended load.
- Operate the crane you're using for the lift within its charted capacities while never exceeding the WLLs of any of the hardware or rigging equipment used for the lift.
- Closely follow all rigging load capacity charts and safety requirements.
- Keep personnel alert and aware of all safety precautions identified in the lift plan.
- Communicate with the crane operator using only approved ASME B30.5 hand signals and radio protocols.

Following the Lift

- Review the completed lift to ensure all safety procedures were followed. Identify any problem areas for future improvement as well as discuss successful areas, and thank the lifting team for a job well done.
- Inspect all rigging equipment, and place hardware in a safe, organized storing area.

Rigging Equipment and Hardware Inspections

Let's face it, crane and rigging operations take a heavy toll on the equipment we use when making heavy lifts. Prolonged exposure to extreme temperatures and harsh environments is hard on hoisting and rigging equipment. Lifting heavy loads that are subjected to side and shock loading is also going to have an adverse effect on the structural integrity of the various components. And of course, there is the chance the equipment has been used incorrectly or has been pushed past its limits a few times.

Unfortunately, there is a chance a component has been overloaded, taken past its WLL on occasion. So do we really want to use any of our gear before we take a very close look? OSHA clearly states that equipment should be inspected prior to each use, before each shift, and as necessary during the shift to ensure it is being used safely. If the equipment is damaged in any way, it should be withdrawn from service immediately. We should always refer to the manufacturers' recommendations for correct inspection procedures and replacement criteria. Maintain good records of all inspections, including the date, the inspector's name, and any problems located. All members of the hoisting and rigging crew should be highly trained on inspection procedures.

Inspections must be carried out by a qualified person. Annual inspections are the minimum; however, many companies require more frequent inspections and keep records on a monthly or quarterly basis, especially when working in harsh, severe conditions or when the equipment is frequently used to lift heavy loads. Let's say your team has just made a very heavy lift—say, 85% of the charted capacity of the crane and running close to the WLLs of some of the rigging equipment or hardware. Wouldn't it be a good idea to inspect all the gear, including the crane, before the next lift? Yes, regular, thorough inspections are critical for ensuring safety. Initial, frequent, and periodic inspections must be conducted per the manufacturers' instructions and specifications. By conducting these inspections and adhering to OSHA regulations and ASME safety standards, we can reduce the risks of accidents, have a safe lifting environment, and have everyone go home to their families after every shift.

Shackles

Shackles are critical for different types of construction, industrial lifting, and rigging applications. They provide you with safety and efficiency when handling heavy loads. It is, however, necessary to choose shackles designed for the application they're being used for, because not all shackles are created equal. Not doing so could cause damage to equipment and nearby property and result in serious injury or fatalities. Knowing about different types of shackles and their uses will help you choose the correct one. A shackle is usually a U-shaped metal link that comes with an opening secured by a clevis or a bolt pin. It can also be a hinged metal loop with a locking pin mechanism.

A typical shackle consists of the following four parts, shown in Figure 20.5.

- **Crown or bow.** The curve-shaped body of a shackle.
- **Ears or lugs.** The threaded part of the shackle where the pin is attached or inserted.

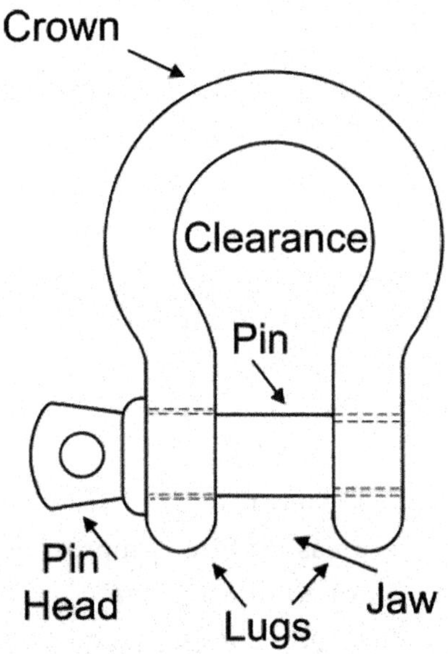

FIGURE 20.5 Learn the parts of a typical shackle.

- **Screw, bolt type, or round pin.** A steel bolt that runs across the two ears. This is what perfectly secures the shackle to a load.
- **Shoulder.** The part of the pin that comes into contact with the ear when the pin is fully threaded.

Choosing the correct shackle, as well as knowing which design is best suited to the application it will be applied to, isn't always easy. Picking an appropriately designed shackle depends on many deciding factors, such as the size of the load, the shackle's working load limit, and many other safety factors. The different body designs of a shackle constitute one of the most critical factors that determine how it should be used; the most popular body shapes include D- and bow-shaped variants. Carefully taking into consideration all these factors will assist you in choosing the shackle design needed for the lifting and rigging operations it will be applied to (see Figure 20.6).

FIGURE 20.6 A professional rigger carefully selects rigging equipment.

Before selecting shackles, keep the following in mind:

- Always factor in the working load limit of a shackle before use.
- Consider the angle of any slings attached to hoist the load. The working load limit will change, depending on this angle.
- Make sure to thoroughly inspect the shackle and all other rigging equipment before each use.
- Discard shackles with any apparent kinking, corrosion, bending, warping, stretching, or other obvious signs of physical damage.
- Conduct an official inspection at a minimum of once per year or as often as directed by the manufacturer's recommendations and as frequently as needed if harsh or heavy usage is apparent.

Shackles include the following types.

Chain Shackles

Also called "D shackles," chain shackles have a D-shaped body. This type of shackle is narrower than the anchor or bow shackle and features a threaded pin, referred to as a *clevis-type* pin. Chain shackles are made out of a variety of steel alloys consisting of stainless steel, zinc-plated steel, and galvanized steel (Figure 20.7). These shackles are designed to lift moderate to heavy loads in manufacturing, construction, and shipping applications that are vertical, straight line pulls. They are not suitable for side loading. If side-loaded, significant stress is placed on the pin and bolt side, twisting the shackles and potentially causing critical failure.

FIGURE 20.7 Shown here is a bolt type D shackle.

Anchor Shackles

People often use the terms *anchor shackles* and *bow shackles* interchangeably, but they are not the same. Anchor shackles have a large O-shaped body (Figure 20.8), and their design makes them suitable for various industrial applications, including automotive, manufacturing, and warehousing, and for use in steel mills. Usually made from galvanized steel, these shackles can be used for side pulling while allowing you to fasten chains and straps to the loads safely and efficiently. When using these shackles for multiple-leg industrial lifting, you have to factor in a relevant reduction in their working load limit.

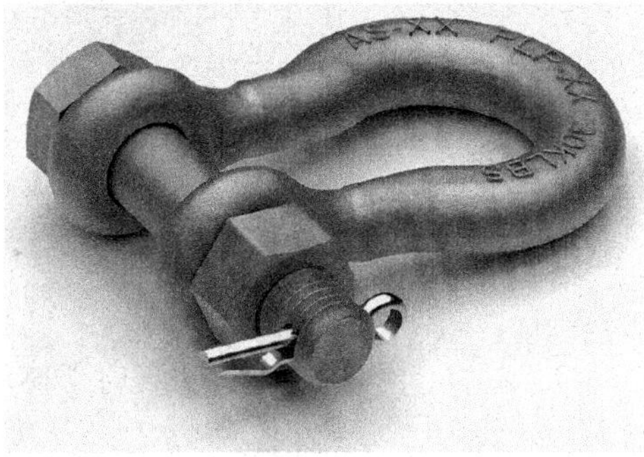

FIGURE 20.8 The anchor shackle is approved for side pulling.

Bow Shackles

Although they look similar to anchor shackles, the O-shaped body of bow shackles is much larger. Stainless steel and titanium are the most popular materials for manufacturing these types of shackles (Figure 20.9). Bow shackles are more suitable for hoisting, as they can handle multiple loads from different angles. Various industries like manufacturing, construction, shipping, oil and gas, and heavy engineering use these shackles for lifting heavy loads.

FIGURE 20.9 Bow shackles are very popular in the construction industry.

Twist Shackles

Twist shackles are one of the special types of lifting devices. They are very similar to D-shaped shackles, but their body has a 90-degree twist. They are made from stainless steel to ensure higher strength (Figure 20.10). Due to their shape, the attachment points remain perpendicular to each other. As there is no torque, twist shackles make rotating loads safer and more efficient. They are ideal for reducing kinks when lifting or rigging heavy loads (Figure 20.11). You can find them on assembly lines or in shipping yards, where rotating the load is necessary.

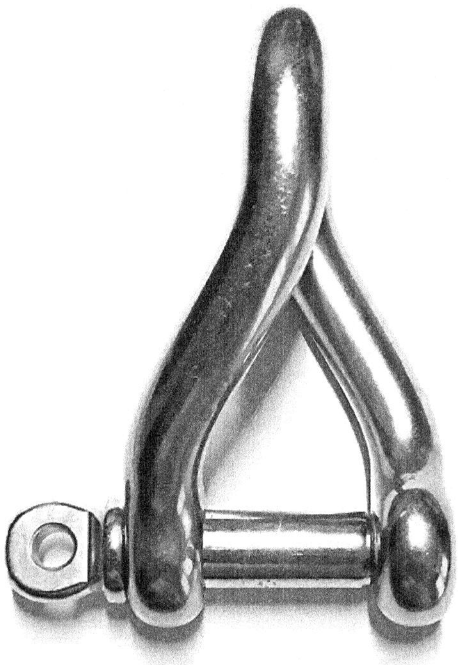

FIGURE 20.10 Stainless steel twist shackle.

FIGURE 20.11 Because of the perpendicular shape of the twist shackle, the attachment points remain across from each other.

Wide-Body Shackles

Wide-body shackles are one of the most popular shackles. They are specifically designed to attach synthetic slings, web slings, and round slings (Figure 20.12). The wide-body shape helps improve the service life of synthetic slings and prevents kinking. Usually made from galvanized steel, they are suitable for industrial applications involving tie-downs and rigging (Figure 20.13). Shackles are available in many different shapes and sizes.

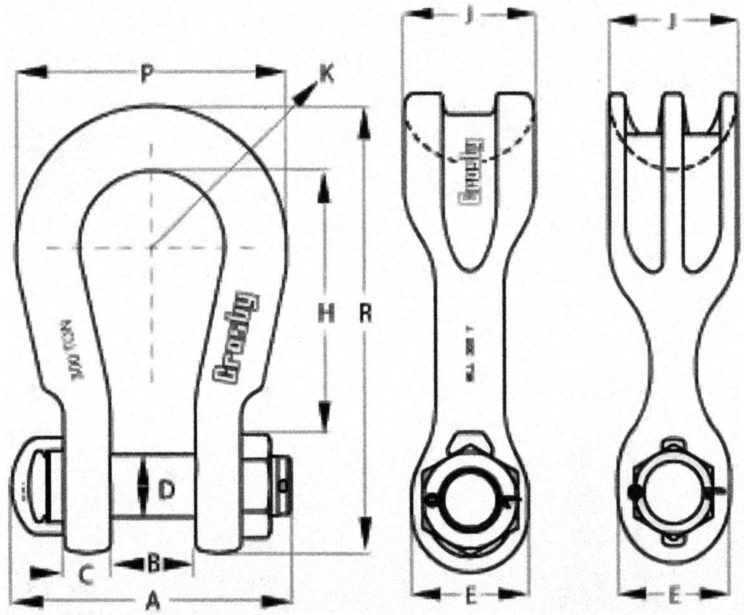

FIGURE 20.12 A wide-body shackle.

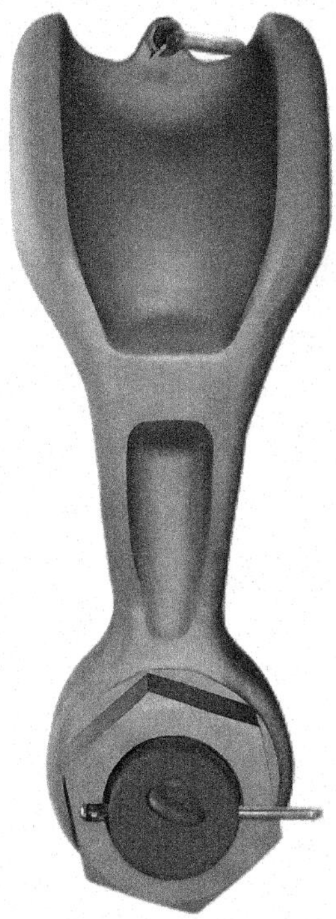

FIGURE 20.13 Be sure to choose the correct shackle for the appropriate job.

Bolt-Type Shackles

Made from steel alloys, bolt-type shackles come with a split pin that holds a nut and bolt together. They are not easy to set up, as the bolt-type pin comes with three to four components. For this reason, they are not designed for temporary connections. However, they offer highly secure connections, making these shackles more suitable for lifting applications involving permanent attachments. It is also better to use these for industrial lifting requiring load movement. Using screw-type shackles for hoisting is dangerous. The pin might come unscrewed, resulting in damage or injuries (Figure 20.14).

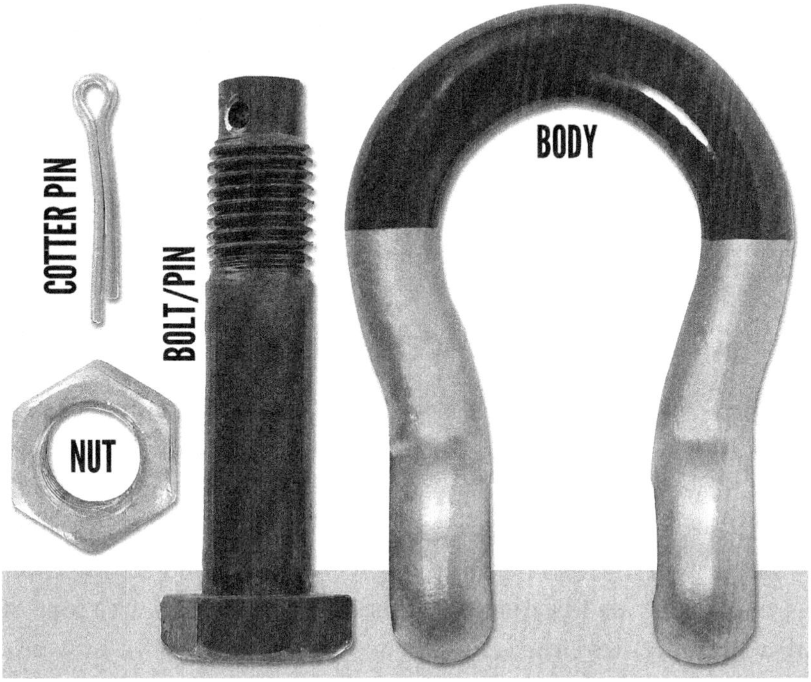

FIGURE 20.14 Bolt-type shackles work best when making long-term attachments.

Pin-Type Shackles

Pin-type shackles, also referred to as *screw-pin shackles*, usually consist of a bolt and cotter pin. They are made from different types of steel alloys, including galvanized steel and stainless steel. You can easily screw the pin into the ears of the body of a shackle. There are only two components involved, and so setting up screw-pin shackles is easy. As a result, they are often more suited for industrial applications requiring quick assembly and disassembly (Figure 20.15).

To sum up, shackles are available in many different shapes and sizes (Figure 20.16), so be sure to choose the correct shackle for the job.

FIGURE 20.15 Pin-type shackles.

FIGURE 20.16 Shackles have two main parts: the bow or body and its pin.

Wire Rope Clips

Wire rope is an extremely versatile mechanical device that can be used to help support and move an object or load. Whether for use on cranes or for other lifting applications, it's important to have a solid understanding of the rigging components that are being used to attach to and lift a load. As a rigger, and often user of wire rope, it's necessary to understand the types of wire rope end termination that can be used at the ends of a length of wire rope, the most common being wire rope clips. Wire rope clips can be used to form a load-bearing eye at the end of a cable or wire rope or to connect two cables together with a lap splice. Wire rope clips are popular because they can be installed in the field and provide 80–90% efficiency of the wire rope's breaking strength, depending on the diameter of the wire rope (Figure 20.17). Always remember, they are not to be used for making slings. As the ASME B30.9 slings standard states, "Mechanical wire rope terminations . . . that require periodic adjustment to maintain efficiency shall not be used to fabricate slings." There are two main types of wire rope clips, U-bolt wire rope clips and double-saddle clips. Both types are discussed a bit later in the chapter. U-bolt wire rope clips (Figure 20.18) are the most common and may be made of forged or malleable metal. Never use malleable metal clips for overhead lifting.

FIGURE 20.17 Wire rope clips are available in many different sizes.

FIGURE 20.18 A 3 ¼" drop forged cable clip, preventer, or clamp is pictured here.

Dropped Forged Wire Rope Clips

This type of wire rope clip is essentially a U-bolt, two nuts, and a metal base saddle that can be made from forged steel or cast iron. Careful consideration and attention must be given to how the U-bolt wire rope clips are installed. The base of the wire rope clip is made from forged steel. Forged clips are heated and hammered into the desired shape, resulting in a consistent grain structure in the steel. Forged wire rope clips are used for critical, heavy-duty, overhead loads such as winch lines, crane hoist lines, support lines, guy lines, towing lines, tie-downs, scaffolds, and more (Figure 20.19).

FIGURE 20.19 Shown here is a $\frac{9}{16}$–$\frac{5}{8}$" dropped forge wire rope clip.

Malleable Wire Rope Clips

Malleable wire rope clips are used for making eye-termination assemblies only with right-regular-lay wire rope and only for light-duty uses with small applied loads, such as handrails, fencing and guard rails. The base of the wire rope clip is made from malleable cast iron, as shown in (Figure 20.20). Malleable cast iron clips may fracture under heavy use and does not have the same desirable metallic properties as steel or the beneficial grain structure that a forged base has. Malleable clips can be found in most neighborhood hardware stores. They often get confused with quality-rated forged U-bolts. Do not make that mistake when planning overhead lifting operations.

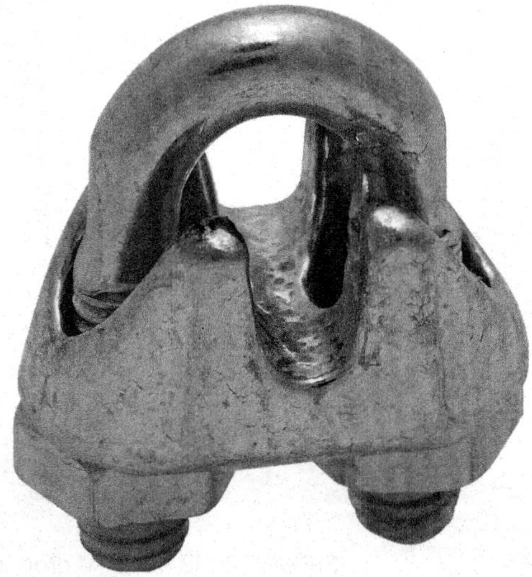

FIGURE 20.20 A ¼" malleable wire rope clip.

Double-Saddle Wire Rope Clips

Double-saddle wire rope clips consist of two saddles, each with a leg, and two nuts—one used on the top and one on the bottom (Figure 20.21). Double-saddle wire rope clips can be used in either direction, so they take the guesswork out of installation during the application of the live end and dead end of a piece of wire rope.

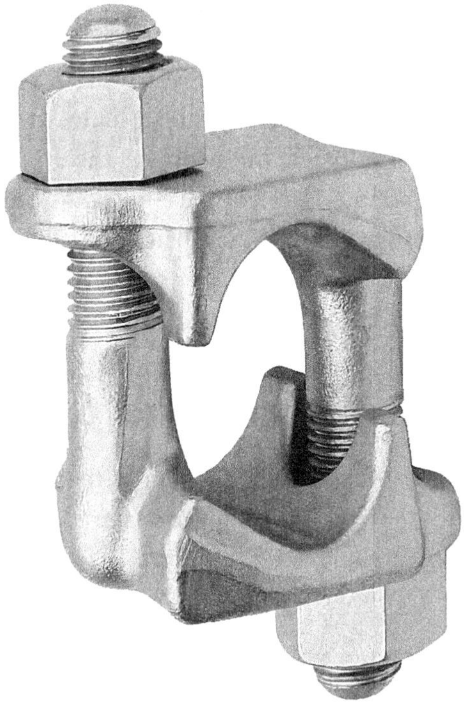

FIGURE 20.21 Double-saddle wire rope clips are currently very popular in the craning industry.

U-Bolt Wire Rope Clips

There are two sides of a U-bolt-style wire rope clip, the saddle and the U-bolt. When securing a wire rope eye, it is important to place the clip on the correct end of the rope. The saddle of the clip rests on the live end of the rope, while the U-bolt always rests on the dead end of the rope. A saying commonly used in hoisting and rigging to help you remember this is, "Never saddle a dead horse!" In other words, never put the saddle on the dead end of the rope. This old saying has been around for generations, so memorize it. And to help you remember to never saddle a dead horse, carefully consider the following (Figure 20.22).

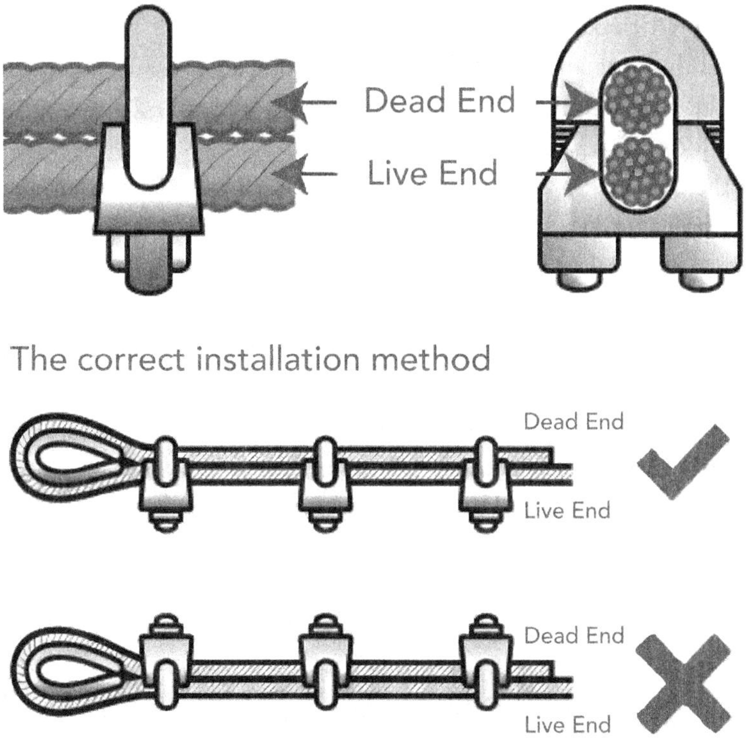

FIGURE 20.22 Improper installation of a clip can cause serious accidents.

The Importance of Using a Torque Wrench

In order to function properly, wire rope clips require the use of a torque wrench. If the nuts on the clips are too tight, or, on the other hand, too loose, it can cause the clip to malfunction and fail. If the clip is over torqued, it could damage the threads of the wire rope. But if the clip isn't torqued tightly enough, the holding power of the clip is diminished, and the wire rope could slip through. A properly installed wire rope clip has an efficiency of 80 to 90% of the wire rope's breaking strength, depending on the diameter of the rope. For smaller diameters, such as $\frac{1}{8}$ inch to $\frac{7}{8}$ inch, the efficiency is typically 80%. Larger diameters; 1 to 3.5 inches, can reach up to 90% efficiency when installed correctly.

Incorrect Amount of Turnback When Forming an Eye

The turnback is the portion of the wire rope eye that runs from the end of the bearing eye to the live end. Having less than the suggested amount of turnback will decrease the efficiency of the wire rope eye and could lead to failure.

- **Using the wrong type of clip for the application.** It is important to use the correct wire rope clip, forged or malleable, for the lift or any other work you're planning. Malleable clips can only be used for noncritical uses, such as being used as tension rope to form a perimeter around a parking lot; they are never used for overhead lifting.
- **Always use correct spacing between clips.** Depending on the number and size of the wire rope clips, there is a proportional amount of space required with the placement of each clip on the rope.
- **Always use the recommended number of clips.** There are a minimum number of clips required for use related to the wire rope diameter. Using less than the number of specified clips could result in decreased efficiency and possible failure.
- **Regularly inspect the rope and retighten all clips.** After installing the clips, it is necessary to regularly cycle the rope and retighten the clips. Monitoring the torque on the nuts is important, as they will loosen over repeated use.
- **Always match the clip size to the wire rope diameter.** The clip size used, whether it be a $\frac{1}{8}$-inch clip, a 3-inch clip, or any other size, must match the diameter of the wire rope. If it doesn't, the wire rope could slip out of the clip (see Figure 20.23).

Steps for installing a wire rope clip include:

1. Wrap the wire rope around the thimble or eye, and turn back the correct amount of rope, specified by the manufacturer. (OSHA and the U.S. Army Corps of Engineers specifically address the use of thimbles in wire rope eye splices and other rigging operations. Thimbles prolong the life and enhance the strength of wire rope assemblies.)
2. Apply the first wire rope at the end of the dead end, with one base width of space, and use a torque wrench to tighten the nuts on the wire rope clip. The ASME B. 30 series of safety standards provide guidelines for the selection, construction, installation, and inspection of the wire ropes' lifting components, including those with thimbles.
3. When applying the second clip, if required, place it as close to the eye loop, or thimble, as possible. Once more, properly tighten the nuts of the clip with a torque wrench. During

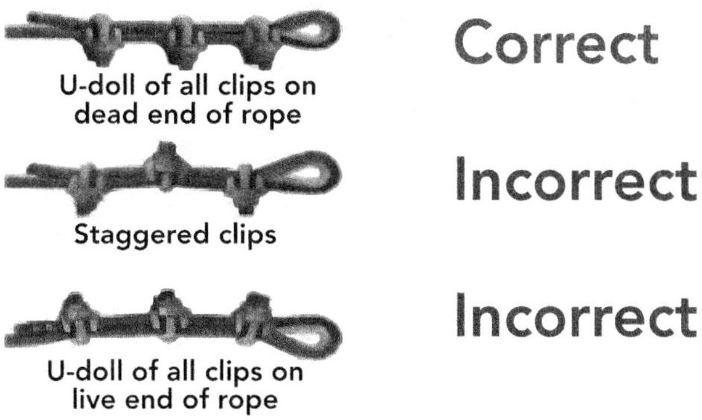

Rope (in)	Minimum Number of Clips	Amount of Rope Turnback from Thimble (inches)	Torque in Foot-Pounds for Unlubricated Bolts
5/16	2	5 1/2	30
3/8	2	6 1/2	45
7/16	2	7	65
1/2	3	11 1/2	65
9/16	3	12	95
5/8	3	12	95
3/4	4	18	130
7/8	4	19	225

FIGURE 20.23 Correctly spaced and torqued wire rope clips to form a wire rope thimble assembly.

instillation, it is crucial to use the correct size, and the correct type, of thimble needed for the wire rope used. The thimble must be an exact match, with strands being kept in tight contact with the thimble during instillation to prevent potential misalignment or damage, commonly known as *high stranding*.

4. If more than two clips are required, make sure that the clips are evenly spaced between the two that were placed first, while ensuring that the thimble is securely fastened within the wire rope, using wire rope clips or crimping sleeves. Be sure to always do this in accordance with manufacturer's instructions.

Two more things to keep in mind:

- Safety standards and regulations are available for more guidance, including ASME B30.9 and OSHA 1910.184.
- Protection of the eye or the looped end of a wire rope sling prevents the rope from fraying, bending, or becoming damaged during the lift. This helps maintain the original shape of the loop and reduces localized pressure, minimizing the chance of the rope flattening or breaking.

Wire rope clips are a common and necessary piece of rigging hardware when it comes to using wire rope and forming end terminations. Wire rope clips are used to form an eye or to connect two cables together. It's important to understand how to correctly install a wire rope clip, as incorrect installation will lead to the decreased efficiency of the wire rope assembly. Wire rope remains a valuable component in hoisting and rigging operations, contributing heavily to the lifespan of a wire rope assembly when installed correctly.

Eye Bolts

Used as a lifting or securement point, eye bolts are one of the most popular types of rigging hardware. Like slings, sling hooks, and shackles, eye bolts come in a number of different designs and configurations. Eye bolts can be used as a connection point for rigging, anchoring, pulling, or pushing and also are used in hoisting applications. Although eye bolts are commonly used in construction and general industry operations, they are often misinterpreted and used incorrectly. It can be confusing to understand what type of eye bolt to use for your operation in conjunction with your other rigging gear.

Eye bolts come in different design configurations based on their operation and their intended use. Here are some terms you should familiarize yourself with to help you understand different types and designs of lifting eye bolts (Figure 20.24).

- ➤ **Eye.** The loop formed at the top that can be either bent, welded, or forged
- ➤ **Shoulder.** The "skirt" where the eye and shank come together, designed to resist bending
- ➤ **Shank.** The threaded shaft that attaches to the eye

FIGURE 20.24 Knowing which eye bolts are acceptable for overhead lifts is critical in crane operations.

Always ensure that the eye bolt that you select meets or exceeds the working load limit for the lift you have planned and is suitably higher than the WLL of the slings and other rigging hardware you have selected.

Eye Bolt Markings/Identification

According to the ASME B30.26 Safety standards for rigging hardware; Each eye bolt must be labeled to identify the following:

- **Name of manufacturer or trademark.** This allows hardware to be easily traced to its maker
- **Size and rated load capacity.** This is critical to ensure the eye bolt is rated for the intended lift
- **Grade for alloy eye bolts.** This designation is required for alloy eye bolts in order to distinguish between them and other types of eye bolts

These markings are required to be present throughout the lifetime of the hardware, and it is the responsibility of the user that they remain legible.

When selecting the proper eye bolt for your application, one of the most important considerations is whether you need a shouldered or nonshouldered (plain pattern) eye bolt. Eye bolts should be used for vertical, straight-line pull. Nonshouldered eye bolts should only be used for straight line pull or vertical lifts and never used for angular lifts.

- **Shouldered eye bolts.** Shouldered eye bolts are also commonly known as *shoulder pattern* eye bolts. These eye bolts are designed with a shoulder at the point where the eye and the shank come together. This shoulder design reduces bending stresses on the shank and allows the eye bolt to be used for angular lifting if the shoulder is properly seated in the load. Both shouldered and nonshouldered eye bolts are available (Figure 20.25).

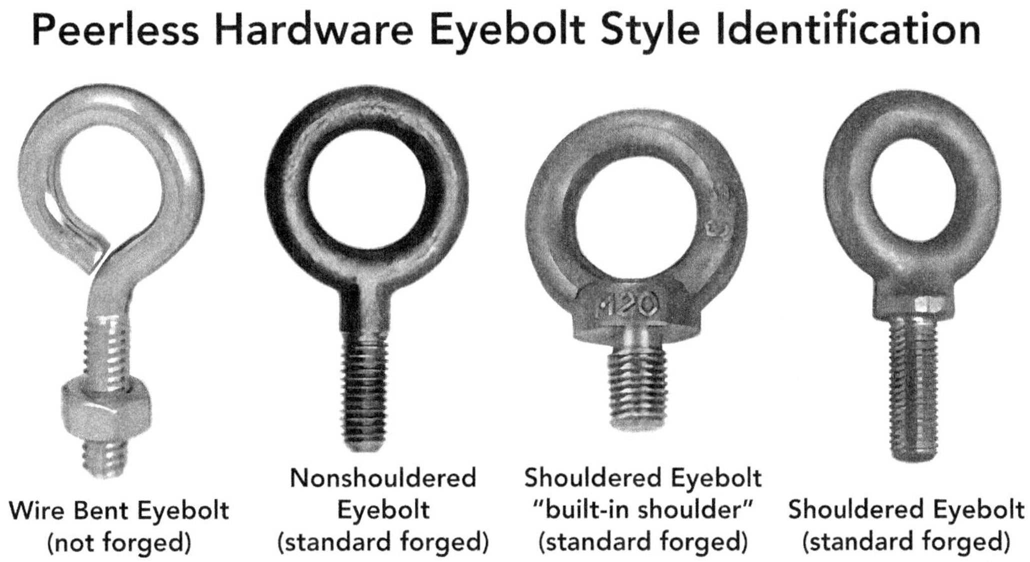

FIGURE 20.25 A shouldered and non-shouldered eye bolt.

When used for side loading or angular loading, you must ensure that the shoulder is completely flush to function properly. Always follow the manufacturer's specifications and reduction in capacity based on the different angle of hoisting the load.

- **Nonshouldered eye bolts.** Nonshouldered eye bolts (Figure 20.26) are also commonly known as *plain pattern* eye bolts. Designed without a shoulder, they can only be used for vertical or inline lifts. Nonshouldered eye bolts are not designed to be used for any type of side loading or angular hoisting.

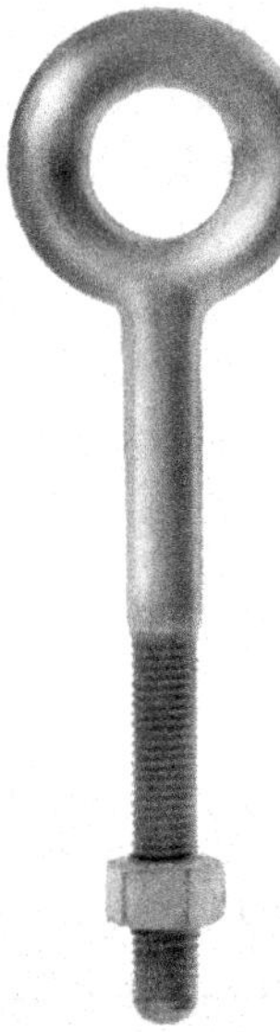

FIGURE 20.26 Nonshouldered eye bolts have a simple, smooth transition from the threaded rod to the looped eye.

- **Forged eye bolts.** Forged eye bolts (Figure 20.27) are hammered or pressed into form. This changes the grain structure of the metal to create a stronger, denser, and more durable product. Forged eye bolts are useful for hoisting and rigging lifts because they are much stronger than other bolts and can resist load weight up to the tensile strength of the material they're made from. They are available in a wide selection of small and large sizes. When equipped with a shoulder, forged eye bolts can be used for straight vertical pull lifting of the load.

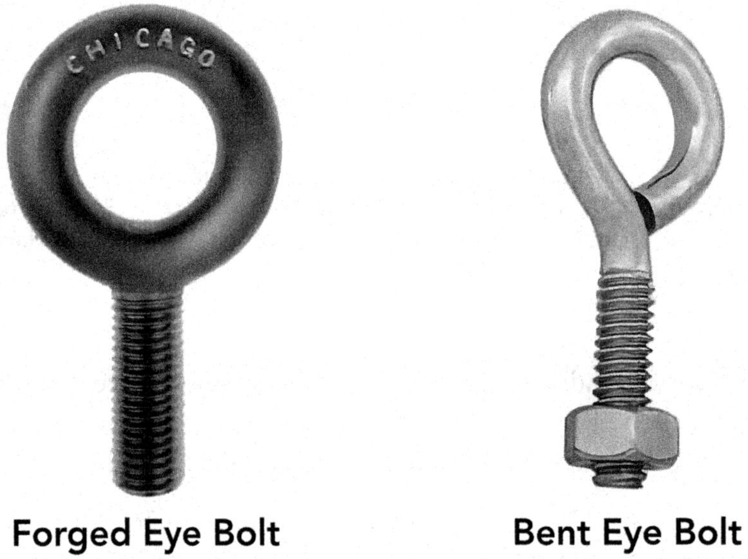

FIGURE 20.27 Used as a lifting or securement point, eye bolts are one of the most popular types of rigging hardware.

- **Bent eye bolts.** Also referred to as *turned eye* or *wire eye bolts*, these eye bolts are made from a single part that is shaped to form the eye. Bent eye bolts are used primarily for lighter load hoisting and should never be used to attempt angular hoisting. The bolt may bend or break and cause a serious accident or even death.

- **Nut eye bolts.** In nut eye bolts, the bottom of the shank protrudes through the bottom of the load and is secured in place with a nut. Nut eye bolts can be shouldered or non-shouldered and can be fabricated through forged, welded, or bent eye procedures.
- **Machinery eye bolts.** Machinery eye bolts have a shank that is threaded and designed to be installed into tapped hitch points that have been engineered specifically as hitch points. When equipped with a shoulder, they can be used with angular loads up to 45 degrees. When installing machinery eye bolts, the minimum tap depth can be determined using this equation:

Minimum tap depth = basic shank depth + half the nominal eye bolt diameter

- **Screw eye bolts.** Screw eye bolts are simply a screw with an eye. They are commonly referred to as a *lag eye*. A screw eye bolt has an eye formed on one end and a threaded shank that tapers to an eyepoint. Screw eye bolts are primarily used for light applications and are screwed into a wood or lag anchor. They do not have a working load limit and are not recommended for use in overhead lifts or hoisting heavy loads. Nut eye, machinery eye, and screw eye bolts are shown in Figure 20.28.

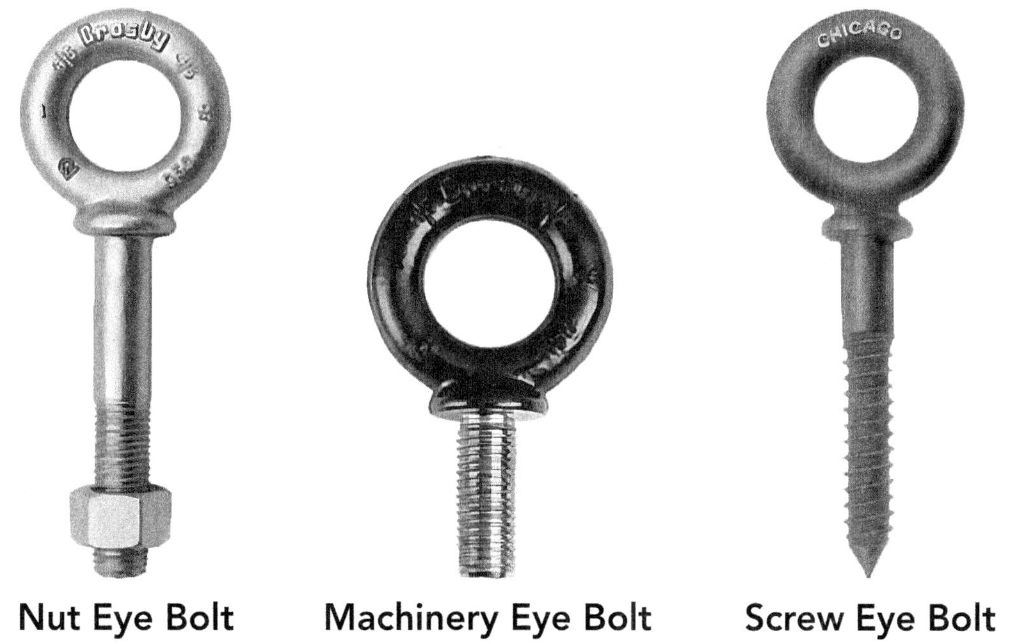

FIGURE 20.28 Eye bolts are commonly used in industrial applications. They are also commonly used incorrectly. Be aware!

- **Galvanized eye bolts.** The galvanization process adds a thin zinc coating to protect the eye bolt material from corrosion and oxidation. Galvanized eye bolts can be used in industrial applications where moisture isn't a primary concern, but where the eye bolts still need to be protected against elements that may cause them to corrode or deteriorate.
- **Stainless steel eye bolts.** Stainless steel will always maintain corrosion resistance if scratched or damaged and is a perfect choice for rigging operations, tie-downs, or any other demanding

applications. There are many different grades of stainless steel, 304 being the most common. Grade 316 is often used in saltwater applications because of its resistance to ocean spray.

Working load limits for eye bolts are based on a straight vertical line pull in a slow and careful lifting motion. Angular lifts will significantly lower working load limits and should be avoided whenever possible. If an angular lift is required, a properly seated shoulder pattern eye bolt must be used. Loads should always be applied to eye bolts in line with the eye, not at an angle to the plane. Angular lifts must never be more than a 45 degree pull. Always refer to shoulder eye bolts' WLL reduction to vertical angle charts provided by the manufacturer (Figure 20.29). To better understand the correct type of eye bolt you need for your operations, ask yourself these important safety questions:

- Why is the load being hoisted?
- How much does the load weigh?
- Where is the lift going to be made?
- Where will the load be placed?
- What material is the load made of?
- Where will the eye bolt be applied?
- Where will the hitch points be located?
- Is the lift going to be a straight vertical line pull or an angular lift? What type of slings will be used for the lift, and what will the horizontal sling angle be for hoisting?
- Will there be any other rigging hardware used to make the lift? If so, determine which components will be used and the WLLs for each item.

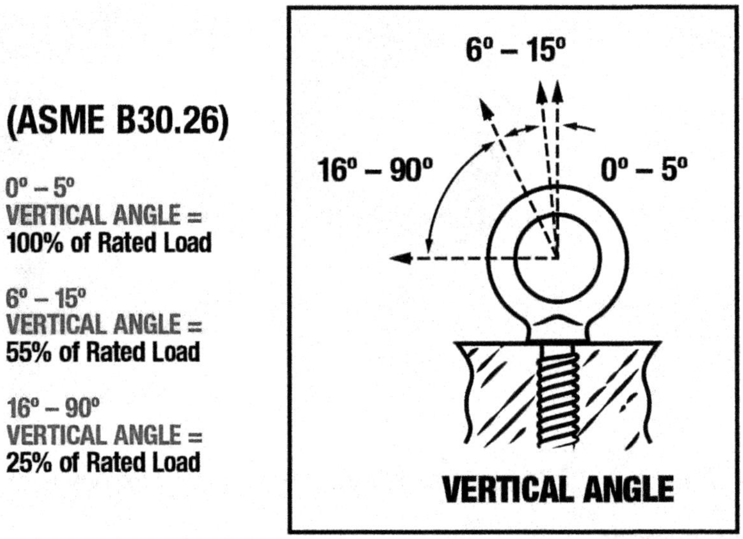

FIGURE 20.29 Eye bolts come in different designs and configurations based on the application.

Work with your local rigging dealer to purchase the best type of eye bolt for your operation, and make sure all eye bolts you choose have a WLL that is appropriate for the jobs you will be doing. For a better understanding of the parts that make up an eye bolt, see Figure 20.30.

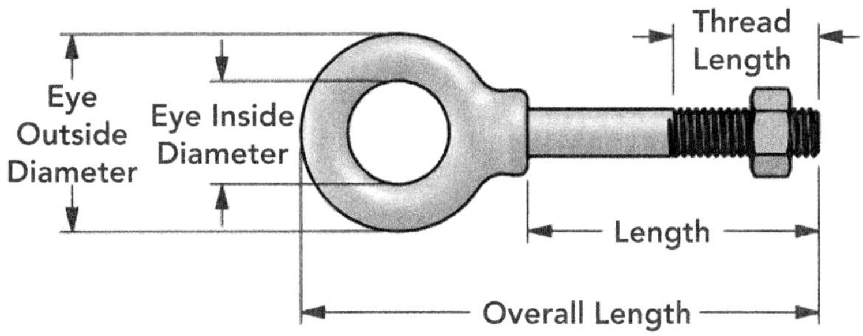

FIGURE 20.30 A drop forged eye bolt does not have any welded point.

Links and Rings

Links and rings are a basic type of rigging hardware, consisting of a single metal loop. Those who are new to the crane and rigging industry may find it challenging to see what makes these pieces of hardware so valuable when rigging an overhead lift. Links and rings are used in different hoisting and rigging equipment as connection points between different hoisting gear and can be used to connect any two separate pieces of rigging assembly. For example, they could be used to attach two synthetic or wire rope slings together. They are also used to make connection points between rigging and sling assemblies and are commonly used as the connection point in multi-leg sling assemblies, typically manufactured as chain or wire rope. A link may be the connection point for one, two, three, or four single-leg configurations. Master links and rings, oblong master links, master rings, and pear-shaped master links may all also be referred to as *collector rings*, or *links*, as they collect multiple sling legs into a single link. In addition to being used in sling assemblies, links and rings can also be used as a connection point between virtually any two parts of a rigging assembly. For example, you may use a link or ring to connect a load to a crane hook, a sling to a hook, or a shackle to a crane hook (Figure 20.31).

FIGURE 20.31 This is a good example of using a link to connect a crane hook.

Types of Links and Rings

There are several different types of links and rings that can be used in an assembly. The most commonly used types of links and rings are:

- Oblong master links
- Master link subassemblies
- Pear-shaped links
- Master rings
- Coupling links

FIGURE 20.32 A great view of using oblong master links.

Oblong Master Links

Oblong master links (Figure 20.32) are oblong-shaped, permanently closed loops that are often found at the top of a multiple-leg chain assembly or wire rope bridle. In this design, the oblong master link is the connection point that collects the legs that make up the sling assembly. Also known simply as *oblongs*, these links are used to gather several sling legs to connect to a crane hook.

While commonly used as connection points in multiple-leg slings, oblong master links are also designed to be used as connection points between rigging and hardware, and their oblong shape makes these links ideal when attaching into the bowl of a crane hook, also known as the *saddle*.

Master Link Subassemblies

If there are more than two sling legs in an assembly, a master link subassembly (Figure 20.33) may be used in place of a single master link. While it is possible to have three to four legs attached to a single master link, this often requires extremely heavy, thick master links that are difficult to manage.

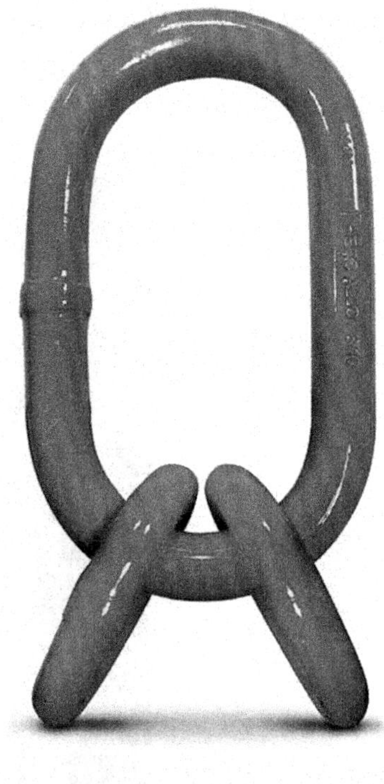

FIGURE 20.33 A sub-assembly ¾" chain link.

Pear-Shaped Links

Pear-shaped links (Figure 20.34), like oblong master links, are also used for multiple-leg chain slings, wire rope bridles, and various rigging connection points. However, pear-shaped links are limited to accommodating smaller sling assemblies with two legs or less. The pear shape of these links makes them ideal for attaching smaller lifting hooks. In some cases, a pear-shaped link will provide a tighter fit than an oblong master link does, which will prevent side-to-side movement.

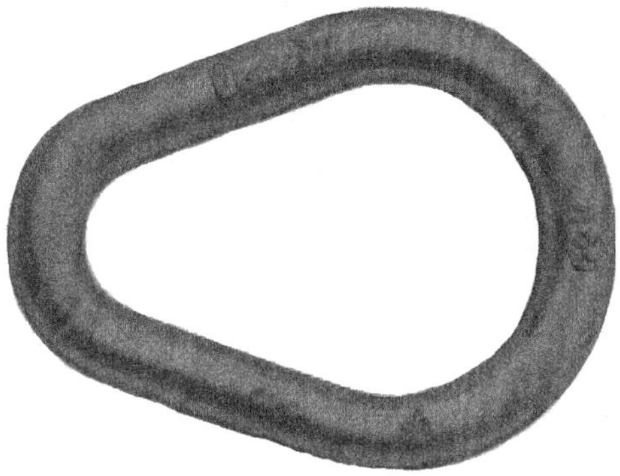

FIGURE 20.34 A $^3/_8$" pear-shaped link with a working load limit (WLL) of 20,500 lbs.

Master Rings

Master rings are circular, permanently closed rings. Like master links, they can be used with wire rope bridles, chain sling assemblies, and other rigging connection points. While master rings can be used to accommodate multiple-leg assemblies, the round shape of a master ring makes it less ideal than an oblong master link for connecting to large crane hooks, as shown in Figure 20.35. Master rings are most often used in fabrication or small machine shops and are otherwise rarely used in hoisting and rigging operations.

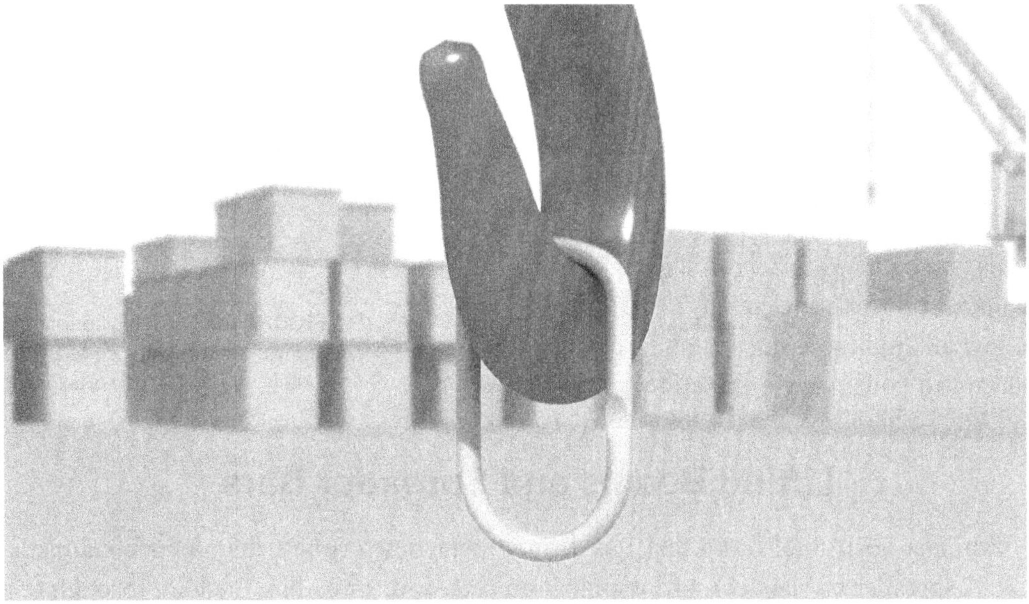

FIGURE 20.35 A master ring is used to hook up to a crane hook.

Coupling Links

Coupling links are primarily used to connect a portion of chain to a master link or to a fitting. They may also be used to create a connection between master links, hooks, or other pieces of rigging hardware. Coupling links may be welded or mechanical:

- **Welded coupling links.** Welded coupling links (Figure 20.36) are connected to the master link or end fitting and welded shut to form a connection. These are very versatile and can be used to bridge setups to connect multiple slings to a hitch point. They are classified as master links.

- **Mechanical coupling links.** Mechanical coupling links (Figure 20. 37) consist of multiple parts that may include a bushing, bolt, and spring at the center. These mechanical coupling links act as attachment points that hinge at the center. Mechanical coupling links are available in Grades 80, 100, and 120. Connecting links are often made from high-strength alloy steel. When working in marine environments or corrosive environments, stainless steel or quenched and tempered steel may be used.

FIGURE 20.36 A $\frac{7}{8}$" coupling link with a WLL of 42,700 lbs.

Links and rings are important hardware that are needed to achieve safe, reliable connections when multiple-legged slings are used in crane operations. Be sure to choose the correct design, made out of the right material, for the job you are performing, and know your safety standards before you go to work while keeping these concepts in mind:

- Do not exceed the WLL for any hardware component.
- Make sure the WLL of a link is equal to or greater than any slings being used for a lift.
- Inspect all slings, additional hardware, and equipment carefully before each use.
- Be aware of any angular loading of the link or ring being used that will affect its WLL.
- Never use damaged or deformed slings or rigging equipment. Promptly discard them.
- Follow all applicable safety standards and regulations relevant to your rigging operation.

FIGURE 20.37 This coupling link is designed for use with grade 80/100 chain slings. It has a 4 to 1 design factor.

Lifting Beams and Spreader Bars

A lifting beam is a horizontal beam that has a single attachment point that connects directly to the crane hook. A spreader bar has two lifting points on each end, spreading the load through tension in the slings that are attached to the crane hook. Both a lifting beam and a spreader bar are below-the-hook devices used to distribute the weight of a load during hoisting operations. The design of each of these below-the-hook devices determines how much force is disseminated and how it is utilized. A lifting beam requires less overhead clearance and uses less rigging. A spreader bar holds two slings apart and distributes the weight lifted over two or more hitch points. The spreader bar is less costly to use than the lifting beam. The spreader bar is typically a lighter unit and is often used for wide loads.

Lifting Beams

Lifting beams (Figure 20.38) are reliable units used with cranes to stabilize and distribute the load weight during a lift. They are best utilized for lighter and shorter span lifts. Because of the single lifting lug on top of the beam that is hooked to the crane hook, the beam will use less headroom. Lifting beams keep lifting slings at a 90 degree angle to the load, providing a straight- line vertical lift. Lifting beams provide multiple hoisting points under the beam and can use multiple hoisting points. They allow for more precision lifting and accurate placement of the load. In addition, lifting beams absorb a lot of the stress generated from the load. Lifting beams work well with flexible loads requiring support along their length. Lifting beams are manufactured to meet requirements of the ASME safety standard B30.20. They must be regularly inspected and must be rated for a specific load. Tag lines may be helpful to help control the load when using a lifting beam during a lift. When a load is lifted with a lifting beam, the load induces bending forces on the beam. The beam will absorb much of this stress, minimizing the stress on both the crane and rigging being used for the lift. Only properly trained and qualified crane operators and members of the hoisting and rigging team should use lifting beams.

FIGURE 20.38 This lifting beam has multiple lift point positions and a 2-ton WLL.

Lifting beams are great lifting devices that improve efficiency and safety when hoisted by a crane. They can handle a large variety of loads and are excellent in areas with low headroom.

Spreader Bars

Spreader bars (Figure 20.39) serve as a helpful piece of below-the-hook lifting equipment in the craning industry. A spreader bar allows the lifting team to attach the load at two hitch points, evenly spreading the weight of the load throughout the bar and any slings used. Spreader bars are highly durable and can handle very heavy weights at two different lifting points. Spreader bars change lifting loads into compressive forces in the bar and tensile forces in the slings used. Spreader bars can help to control the load and reduce the chances of crushing or causing damage to the load. They also reduce the stress on other hoisting equipment by distributing forces across multiple points. Spreader bars are especially useful for lifting objects that are not balanced and where the center of gravity is not easily located because they can counteract uneven weight distribution. Telescopic spreader bars are available with extendable sections for maximum flexibility and the ability to adjust to multiple load sizes. For heavy lifting, spreader bars are very valuable in construction, manufacturing, shipping, and other industries. Types of spreader bars available for use are:

- Container spreader bars
- Forklift spreader bars
- Rotating spreader bars
- Modular spreader bars
- Basket-style spreader bars
- Adjustable spreader bars
- Fixed spreader bars

Spreader bars are valuable tools for safe and efficient material handling. They are useful load-handling equipment used across many industries. They can distribute weight, improve stability, and help prevent load damage. They are very valuable when conducting hoisting operations with a crane.

Safe practices for lifting beams or spreader bars include the following:

- Never remove any safety decals, labels, or markings that were placed by the manufacturer.
- Be sure that only trained, designated individuals use below-the-hook lifting equipment.
- Never try to change, modify, or alter these types of lifting devices.
- Be sure that crane operators are trained and knowledgeable on how to use these types of hoisting equipment.
- Never allow special lifting equipment to contact any obstruction. Especially be aware of any overhead power lines in the vicinity of using this type of hoisting equipment.
- Never use any lifting equipment that is damaged or tagged "Out of service" for hoisting operations.
- Never use spreader bars or lifting beams for side pulls or to be side-loaded.
- Remove out-of-service tags on bars or beams only with the permission of an authorized person.
- Never subject spreader bars and lifting beams to shock loading.
- Avoid sudden acceleration or sudden deceleration and stops when using this type of hoisting equipment.
- Thoroughly inspect spreader bars or lifting beams before every use and then after. Always follow the manufacturer's specifications when using this type of equipment.
- Ensure that all ground personnel other than those appointed by the lift director are kept clear when raising, lowering, or using this type of hoisting equipment.
- Always make sure any slings or rigging hardware are rated to match or exceed the ratings of the spreader bar or lifting beams used for a lift.
- Use nonconductive tag lines whenever necessary to control bars and beams as well as the load, especially near power lines.
- Never use these types of lifting equipment to hoist personnel.

- Never allow personnel to be under a suspended load.
- Never leave a suspended load on a spreader bar or lifting beam for an extended period of time or over night.
- Never exceed the rated WLL of any spreader bar or lifting beam

Both spreader bars and lifting beams play an important part in safe hoisting operations. Each has its own distinct advantages, depending on the requirements of the lift. Understanding these differences is very important so that the correct tool is chosen, providing a safe hoisting operation. For more information about spreader bars and lifting beams used by cranes for hoisting loads, refer to and carefully read the ASME safety standard B30.20 on below-the-hook lifting devices.

Wire Rope Slings

Wire rope slings for lifting materials are the most common choice and provides several advantages over other sling types. They are a great choice because of their strength, durability, and ability to conform to the shape of a load. They are commonly found in the construction and manufacturing industries, as well as in the oil and gas industries. They can be used in corrosive environments and for handling hot materials. Wire rope slings can withstand lots of weight and resist abrasion, cutting, and crushing, making them popular for heavy-duty hoisting and rigging operations. Wire rope slings can be very versatile and are available in several configurations including:

- Single-part slings
- Multi-part slings
- Braided designed slings

Wire Rope Sling Eye Splices

Fold back eyes used in some wire rope slings are formed by bending the rope to the desired eye dimensions and then securing the dead end to the live end by means of a hydraulically pressed-on sleeve or sleeves. Although this type of splice is still commonly used, it is strongly recommended that it not be used for overhead hoisting. Improper swaging or deterioration of the wires within the splice could result in complete failure without warning. The best and most secure method of creating eye splices recommended for all types of rigging and hoisting is the Flemish eye fitted with a pressed metal sleeve. When properly made, this type of splice is one of the most dependable and preserves approximately 95% of the catalog breaking strength of the rope. The strength of this splice is not wholly dependent on the pressed sleeve but has the added friction of the pre-formed interlocking strands that create the eye itself.

The Flemish eye is made by separating the wire rope's individual strands into two parts. The two strands are then looped in opposite directions and laid back together, creating an eye shape at the end of the rope. This creates a strong and durable termination for the wire rope often used in crane hoisting and rigging operations. The ends of the strands are laid back into the body of the rope and secured with a swaged metal sleeve called a *ferrule*. This ferrule is pressed onto the rope, compressing the strands and creating a strong permanent connection. If the sleeve were to break, the spliced construction of the wire rope's Flemish eye would potentially retain most of the wire rope's strength, providing a safety margin.

A mechanical splice refers to a hydraulic machine called a *swager*, which is used to set the sleeve in place and reinforce the eye (Figure 20.39). Flemish eye slings are known to be tough and to withstand abuse (Figure 20.40).

Here is some other valuable information regarding wire rope slings:

- A warning generally occurs before failure by the visible breaking of outer wires. Some reserve strength is provided in the inner wires of the strands and core; both are protected by the outer wires and can possess sufficient strength to carry the load if a reasonable factor of safety is allowed for the sling.
- Properly fabricated wire rope slings are the safest type available for general use.
- Wire rope slings do not wear as rapidly as fiber rope slings, nor are they susceptible to the weak-link ailment of chain slings.
- Wire rope slings show, by inspection, their true condition, and the appearance of broken wires clearly indicates the extent of fatigue, wear, abrasion, and damage due to corrosion.
- All wire rope slings should be made of improved plow steel and should have an independent wire rope core to reduce the possibility of the rope being crushed in service.
- It is recommended that all eyes in wire rope slings be equipped with thimbles and that they be formed with the Flemish splice and secured by swaged or pressed sleeves or fittings. This method will produce an eye that is as strong as the rope itself and provides reserve strength should the pressed sleeve or fitting fail.

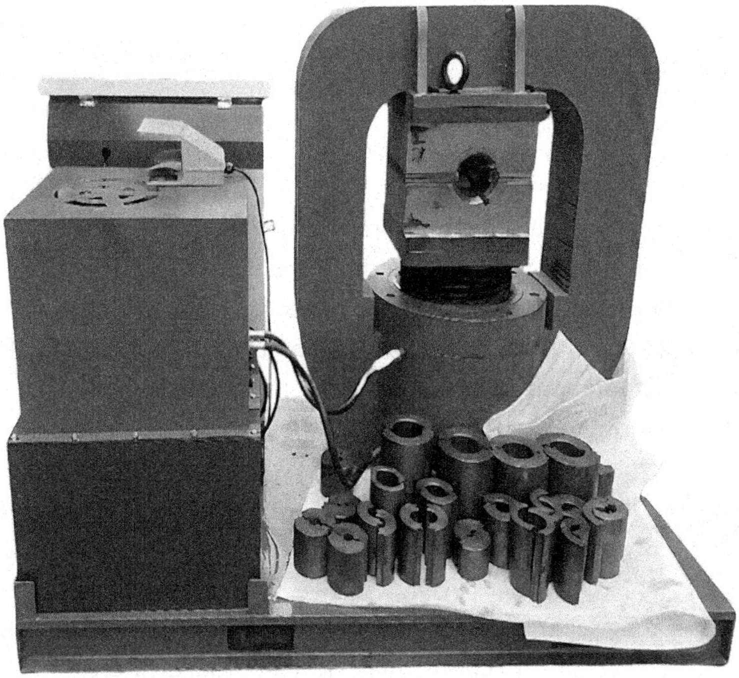

FIGURE 20.39 A sling swager press machine.

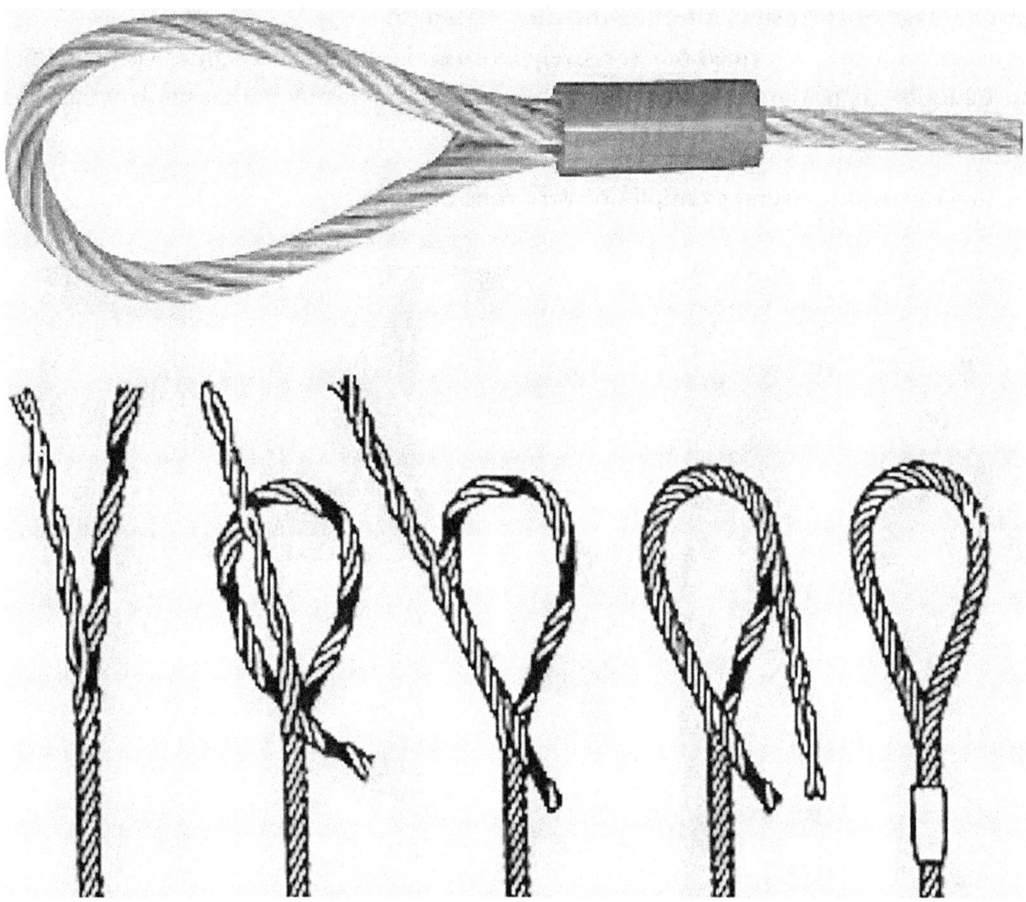

FIGURE 20.40 A view of a Flemish eye sling.

A visual check for major damage or deterioration should be conducted by the person who is planning to use the slings before every use. More detailed inspections should be performed by a designated person with extensive knowledge of wire rope. The inspection should be made at least annually and include a record of the inspection and its condition. The frequency of inspections should be based on common sense. If the sling has been used frequently to make heavy lifts, it should be looked at more frequently. If the crane just made a very heavy lift, and the sling was used close to its WLL, then it makes a lot of sense to carefully look over the slings before and after every lift. Being proactive about slings inspection may prevent an accident from happening or save someone's life. According to the ASME safety standard B30.9, a wire rope sling must be removed from service if any of the following conditions is present:

- Missing or illegible identification tag or markings.
- Broken wires. The number of broken wires allowed varies depending on the type of sling.
- Strand laid slings, including single part slings. Ten randomly distributed broken wires in one rope lay, or five broken wires in one strand in one rope lay, are cause for removal of the slings.
- Severe localized abrasion. A reduction in the nominal diameter of the wire rope of 5% or greater, due to abrasion or scraping.
- Kinking, crushing, bird-caging, or any other damage that distorts the rope structure.

- Heat damage or corrosion, affecting the sling's strength.
- Damaged, cracked, deformed or excessively worn end attachments. Other visible damage that causes doubt about the continued use of the sling, or any problem that renders the sling unfit for use.

Figure 20.41 provides several examples of wire rope damage.

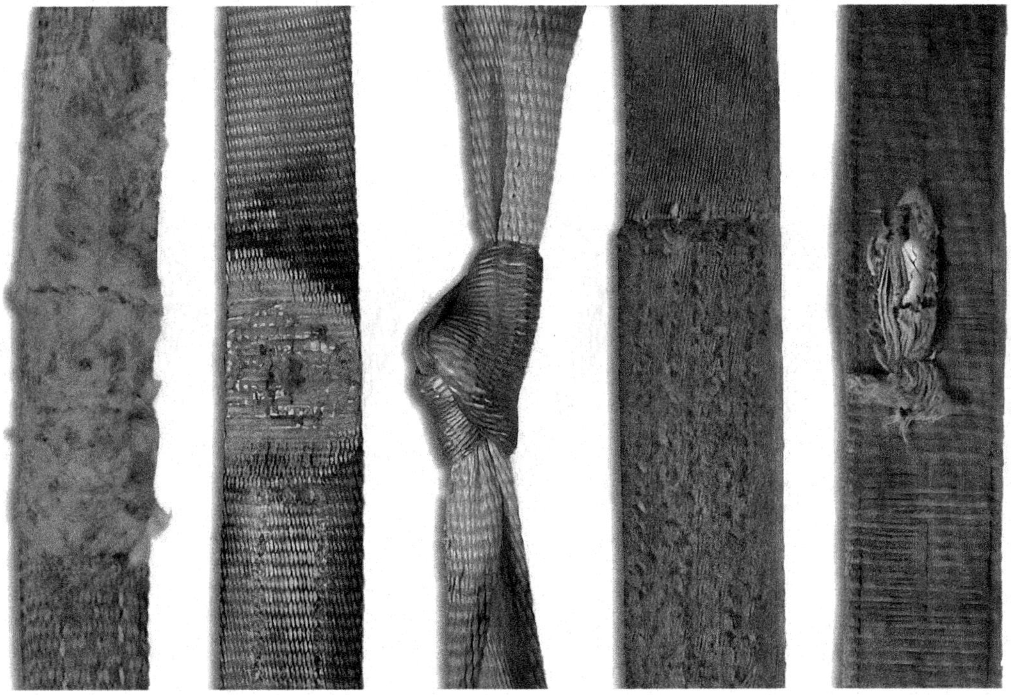

FIGURE 20.41 Safe operating practices OSHA 1926.251 states that slings that are damaged shall not be used.

Inspection Frequency Requirements

Inspections should be done according to the following:

- **Initial inspection.** New or repaired slings require a complete inspection before use.
- **Frequent inspection.** A visual inspection needs to be done by a competent person before each use.
- **Periodic inspection.** Periodic inspections are required at least every 12 months or more often in severe conditions. They need to be done by a qualified person.

Additional Web Sling Information

If you have repairs done on your wire rope slings, the company making those repairs is required to attach its own name and sling tags on the slings it repairs and also include the following information:

- Rated load of at least one hitch type configuration
- Number of legs on the sling if more than one

Not all slings have serial numbers, and certain slings do not provide multiple hitch capacities; you will need to know the configuration and classification of your wire rope slings. For example, you will need to know sling sizes, such as 6 × 19, 6 × 25, or 6 × 37. It is also important to know the correct eye size for the shackle or crane hook you will be making lifts from to make sure you have an acceptable D/d ratio.

In rigging, the D/d ratio is the difference between the diameter of the object around which the sling is bent (D) and the diameter of the sling itself (d). The ratio is very important in determining the sling's capacity and preventing premature wear and damage. The D/d ratio is calculated by dividing the diameter of the bend by the diameter of the sling. A small D/d ratio means a tighter bend that can seriously reduce the sling's lifting capacity. A diagram showing D/d ratios is shown in Figure 20.42. A common D/d ratio for wire rope is 4:1.

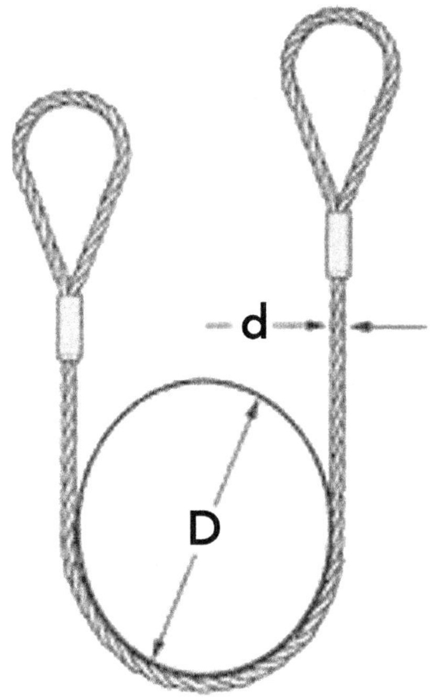

FIGURE 20.42 D/d ratio; D = diameter of bend and d = diameter of wire rope.

Be sure never to shock-load a sling. There is no way to calculate the force applied to a sling by shock loading. The WLL of a wire rope sling can be exceeded by a sudden application of force, and damage can occur to the sling as well as the sudden release of a load. Always remember, wire rope slings are designed for a straight line vertical pull. Anything other than straight line vertical lifts will increase the loading stress on the sling. Pay particularly close attention to the sling angle's charted capacity and other important information in the ASME safety standard B30.9.

Alloy Steel Chain Slings

Although suddenly losing control of the load can be caused by the failure of even one of a sling's chain links, which may be more difficult to predict than the same happening with a wire rope sling, there are certain jobs for which chain slings are more appropriately designed than wire rope. Chains are more robust, are easier to store, have great flexibility, have excellent grip, are resistant to abrasions as well as corrosion, and are particularly well designed for hoisting sharp or jagged loads that would easily damage wire rope or synthetic slings.

Alloy steel chains are tough and are a versatile option for heavy hoisting and rigging operations. They are strong, durable, and resistant to abrasion and heat. Chain slings can conform to the shape of the loads they are lifting and are able to hoist hot materials. Chains come in all sorts of lengths, sizes, and attractive colors at your local hardware stores. *Be careful!* Those usually found at your neighborhood hardware are not rated for overhead lifting operations. Anything less than Grade 80 chain is made for transport or general use, but never for overhead lifting. Grade 70 chain is a transportation chain used to tie down loads. There is a stainless Grade 63 chain available; however, it is used in the food industry and may be used around some chemical applications. Chain slings are made from different grades of alloy. The most common grades in use are 80 and 100. Chains are manufactured in accordance with the American Society for Testing and Materials (ASTM). In the United States, the only three grades of chain acceptable for overhead lifts are:

- Grade 80
- Grade 100
- Grade 120

To properly measure a chain, measure from the top bearing point of one of the ends of the single-leg chain with a hook on each end. It will then be measured from the bearing points of each one of the hooks. Types of alloy chain sling configurations include:

- Single-leg chain slings
- Double-leg chain slings
- Triple-leg chain slings
- Quad-leg chain slings
- Chain slings with adjusters

Using the correct grade and type of alloy steel chain is highly important and must be based on the specific lifting requirements of your hoisting and rigging operations. Always pay special attention to proper load handling procedures and WLLs. Some alloy steel chains will stretch 15 to 20% before breaking, providing a visual warning they are about to fail when rigging around corners or edges. The D/d ratio or the diameter of the object that the chain links around needs to be reviewed to prevent damage to the chain and maintain its WLL.

Regular cleaning, proper lubrication, and correct storage are important for maintenance and will improve the chain's lifespan. Chains showing excess wear as listed in the ASME B30.9 safety standard must be removed from service and destroyed immediately—and destroyed in a way that they can never be used again. Always follow the chain manufacturer's recommendations, and always follow compliance with industry safety standards set by organizations such as OSHA, ASME, and ASTM.

Inspection of Alloy Steel Chain Slings

ASME B30.9 requires periodic inspections of alloy steel chain slings, depending on usage, severity, and environment. Visual inspections should be conducted before each use, and a complete, thorough documented inspection should be performed at least annually. Here is a complete breakdown of inspections:

- **Visual inspections.** ASME B30.9 requires visual inspections of all slings, including alloy steel chain slings, before each use.
- **Periodic inspections.** Additional in-depth, documented inspections are required at least annually for normal service conditions. These inspections must be made by a qualified person.
- **Frequent inspections.** The frequency of specific inspections should be performed based on the severity of use and at least once a day.

See the various examples of chain damage shown in Figure 20.43.

FIGURE 20.43 You should clean the chain prior to inspection to more easily see damage or defects.

Inspection Criteria

ASME B30.9 outlines specific criteria for removing slings from service. They include:

- Severe abrasion or corrosion
- Distortion or deformation of end fittings
- A missing or illegible chain sling identification tag
- Broken welds or signs of brazed joints
- Broken links or signs of excessive wear

If any of the inspection criteria is not met, the chain sling must be removed from service immediately and not returned to use until the chain has been inspected and approved for service by a qualified person. Alloy chain slings have a design safety factor of 5:1. This means that 5 times as much force as the WLL has to be applied to the sling before it would likely fail. Remember, employers must use qualified riggers during hoisting and rigging activities.

Rated Loads

Review sling angles and capacities rated in tons, based on chain size and grade for single-leg, endless, two-leg, three-leg, and four-leg configurations in Chart 20.1.

Review additional information regarding chains' WLL in Chart 20.2.

Alloy Grade(T)80 Chain Slings Manufactured to AS3775-2014

Chain	Single Leg Slings				Slings of 2, 3 or 4 legs				2 Leg Slings
Chain Size mm	Straight or adjustable sling (without derating)	Adjustable sling with derating	Reeved Sling	Basket Sling Max 60°	Straight Sling 60°	Straight Sling 90°	Straight Sling 120°	Reeved Sling Max 60°	Basket Sling Max 60°
6.0	1.1	0.8	0.8	1.5	1.9	1.6	1.1	1.5	2.5
7.0	1.5	1.1	1.1	2.0	2.6	2.1	1.5	2.0	3.4
8.0	2.0	1.5	1.5	2.6	3.5	2.8	2.0	2.6	4.5
10.0	3.2	2.4	2.4	4.1	5.5	4.5	3.2	4.1	7.2
13.0	5.3	4.0	4.0	6.9	9.2	7.5	5.3	6.9	11.9
16.0	8.0	6.0	6.0	10.4	13.8	11.3	8.0	10.4	18.0
20.0	12.5	9.4	9.4	16.3	21.6	17.6	12.5	16.3	28.1
22.0	15.0	11.3	11.3	19.5	26.0	21.2	15.0	19.5	33.8
26.0	21.2	15.9	15.9	27.6	36.7	29.9	21.2	27.6	47.7
32.0	31.5	23.6	23.6	41.0	54.5	44.4	31.5	41.0	70.9

CHART 20.1 The chart is based on chain slings Grade 80.

SINGLE LEG SLINGS				2, 3, OR 4 LEG SLINGS				BASKET SLINGS	
CHAIN SIZE (mm)	STRAIGHT SLING	*ADJUSTABLE SLING	REEVED SLING	STRAIGHT SLING 60°	STRAIGHT SLING 90°	STRAIGHT SLING 120°	REEVED SLING Max 60°	1 LEG Max 60°	2 LEG Max 60°
6	1.1	1.1	0.8	1.9	1.6	1.1	1.4	1.4	2.5
7	1.5	1.5	1.1	2.6	2.1	1.5	2.0	2.0	3.4
8	2.0	2.0	1.5	3.5	2.8	2.0	2.6	2.6	4.5
10	3.2	3.2	2.4	5.5	4.5	3.2	4.2	4.2	7.2
13	5.3	5.3	4.0	9.2	7.5	5.3	6.9	6.9	11.9
16	8.0	8.0	6.0	13.8	11.3	8.0	10.4	10.4	18.0
20	12.5	12.5	9.4	21.6	17.6	12.5	16.3	16.3	28.1
22	15.0	15.0	11.3	26.0	21.2	15.0	19.5	19.5	33.8
26	21.2	21.2	15.9	36.7	29.9	21.2	27.6	27.6	47.7
32	31.5	31.5	23.6	54.5	44.4	31.5	41.0	41.0	70.9

CHART 20.2 More information on chain slings.

Rated loads are based on:

- Material strength, used to manufacture the chain.
- Design factor or safety factor.
- Type of hitch being used to lift the load.
- Angle of loading, or chain sling, or the D/d ratio.
- Diameter of curvature over which the sling is used. Do not use horizontal angles less than 30 degrees unless recommended by the sling manufacturer or a qualified person.

Effects of Wear

Do not use chains if the size at any point of a link is less than that recommended by the chain's manufacturer.

Sling Selection

For multiple-leg slings used with nonsymmetrical loads, ensure that an analysis by a qualified person is performed to prevent overloading of any leg. Ensure that multiple-leg slings are selected according to sling rating tables when used at the specific angles given in the tables. Ensure that operations at other angles are limited to the rated load of the next lower angle given in the tables or calculated by a qualified person. Do not use a component unless it is the proper shape and size to ensure that it is properly seated in the hook or lifting device.

Attachments

Only use attachments, such as hooks, rings, oblong links, pear-shaped links, and welded or mechanical coupling links, that have a rated capacity at least equal to that of the alloy steel chain with which they are used. If attachments with rated capacities lower than the chain are used, ensure that the sling is rated to the weakest component used on the sling.

Effects of Environment

Store slings in an area where they will not be subjected to mechanical damage, corrosive action, moisture, extreme temperatures, or kinking. When slings are exposed to extreme temperatures, follow the guidance provided by the sling manufacturer or qualified person. Do not use alloy steel chains that have been heated above 1,000 degrees Fahrenheit (538 degrees Celsius). Remove them from service.

Alloy chain slings exposed to temperatures above 400 degrees Fahrenheit (205 degrees Celsius) have reduced load ratings. Reductions in rated load for Grade 80 and Grade 100 chain slings used at and after exposure to elevated temperatures are given in Chart 20.3. Cold temperature limits are in place if slings are used at temperatures below –40 degrees Fahrenheit (–40 degrees Celsius). The manufacturer must be consulted. The strength of alloy steel chain slings can be affected by chemically active environments. Consult the manufacturer before the sling is to be used in chemically active environments.

Temperature		Grade of Chain			
		Grade 80		Grade 100	
°F	°C	Temporary Reduction of Rated Load While at Temperature	Permanent Reduction of Rated Load After Exposure to Temperature	Temporary Reduction of Rated Load While at Temperature	Permanent Reduction of Rate Load After Exposure to Temperature
Below 400	Below 204	None	None	None	None
400	204	10%	None	15%	None
500	260	15%	None	25%	5%
600	316	20%	5%	30%	15%
700	371	30%	10%	40%	20%
800	427	40%	15%	50%	25%
900	482	50%	20%	60%	30%
1,000	538	60%	25%	70%	35%
Over 1,000	Over 538	Remove from service	Remove from service	Remove from service	Remove from service

CHART 20.3 The effects of elevated temperatures on rated loads of alloy chain.

Proof Testing

Before the initial use of a sling, ensure that every component of a new, repaired, or reconditioned alloy steel chain sling has been proof-tested by the sling manufacturer or a qualified person and meets the requirements of the ASTM specification. Retain the certificates of proof testing, and make them available for examination.

Bad Applications

What *not to do*:

- Do not shorten or lengthen a sling by knotting or twisting.
- Do not rest loads on the sling.
- Do not pull a sling from under a load when the load is resting on the sling; always provide lots of blocking to sit the load on.
- Do not drag slings on the floor or over abrasive surfaces.
- Do not allow shock loading.
- Avoid twisting and kinking.

Cautions to Personnel

What *to do*:

- Ensure that all portions of the human body are kept away from the areas between the sling and the load and between the sling and the crane or hoist hook, including the hook line.
- Ensure that personnel never stand in line with or next to the legs of a sling that is under tension.

- Ensure that personnel do not stand or pass under a suspended load.
- Ensure that personnel do not ride the sling or the load unless the load is specifically designed and tested for carrying personnel.

Rigging Practices

What *to do* in regard to rigging practices:

- Ensure that slings are hitched in a manner providing control of the load.
- Ensure that sharp edges in contact with slings are padded with material of sufficient strength to protect the sling, such as softeners.
- Ensure that slings are shortened or adjusted only by methods approved by the sling manufacturer or a qualified person.
- Ensure that during lifting with or without a load, personnel are alert for possible snagging.
- Ensure that in a basket hitch, the load is balanced to prevent slippage.
- When using a basket hitch, ensure that the legs of the sling contain or support the load from the sides, above the center of gravity, so that the load remains under control.
- Ensure that in a choker hitch, the choke point is only on the sling body, never on a fitting.
- Ensure that in a choker hitch, an angle of choke less than 120 degrees is not used without reducing the rated load.
- Ensure that slings are not constricted, bunched, or pinched by the load, hook, or any fitting.
- Ensure that the load applied to the hook is centered in the base (bowl) of the hook to prevent point loading on the hook—unless the hook is designed for point loading.

A rigger is the person who chooses the proper chain and support equipment to support and manipulate the load. First you must know the weight of the load to be lifted. It must be determined if one single-leg chain can be used from the crane hook down to the load, or if you need a multiple-leg alloy steel chain configuration to get the job done. If that is the case, how many lifting points do you have available to make the lift? Are those points existing lifting lugs, or will other rigging hardware be needed to establish those lifting points?

It is your most crucial responsibility to consider all risks prior to using steel alloy chain or any other rigging equipment or hardware used in combination with the chain to make a lift. Alloy chain users shall be trained in the selection, inspection, dangers to personnel, effects on the environment, and rigging practices in accordance with OSHA regulations and all relevant ASME B30 series safety standards. Use by untrained persons is hazardous. Do not use any sling, rigging equipment, or hardware unless you are positively certain of what you are doing. It is most important that everyone on the hoisting and rigging team remember that when it comes to using hoisting and rigging gear (including chain slings and hardware), lack of training, knowledge, and skill can lead to severe injury or death to you or members of your crew, as well as those just passing by, the innocent general public. We must, by all means available, ensure that everyone goes home safe after every shift on the job.

Synthetic Web Slings

Synthetic web slings are a very—and the most—popular sling. They are a popular choice for hoisting and rigging operations because they are lightweight and flexible. Synthetic web slings offer several advantages for rigging purposes. They are made of nylon or polyester-type yarns. Each synthetic material has its own special properties, but both offer strength, noise protection, convenience, and affordability.

Synthetic slings (Figure 20.44) are available in web and round types. When using a web sling, besides always knowing the weight of the load to be lifted, you will need to know additional information, including:

- The material it's made of—polyester or nylon
- The number of plies in the body
- The type of eye or fitting needed on both ends of the sling
- The length of the sling
- The width of the sling
- Whether additional cut, wear, or abrasion protection is needed, such as softeners

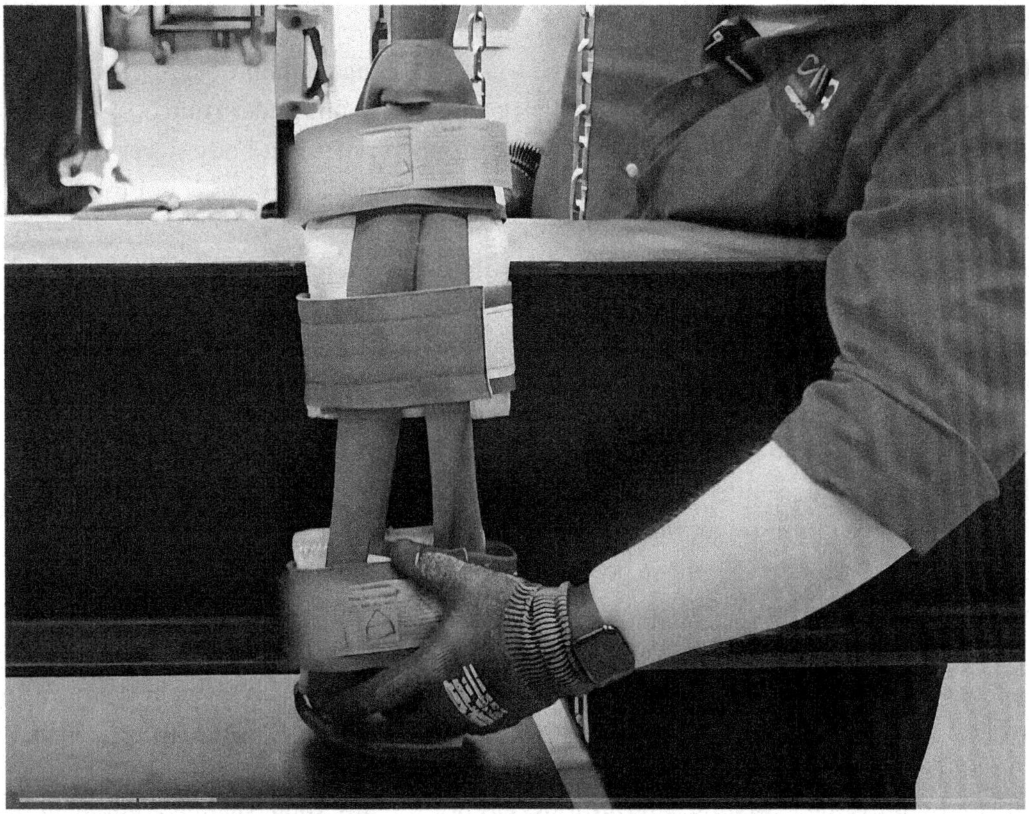

FIGURE 20.44 A qualified person must select materials and methods that protect slings from damage.

Synthetic web slings are susceptible to damage from abrasions, cuts, and certain chemicals. Polyester slings are unaffected by common acids and bleaching agents; however, they should not be used when chemicals rated sulfuric acid or alkaline are nearby. Polyester web slings may be used in temperatures between –40 to 194 degrees Fahrenheit, or –40 to 90 degrees Celsius. Using web slings outdoors in these temperatures can cause damage to the sling, reducing its strength, and wet or frozen slings will suffer a reduced load capacity. Sunlight and prolonged exposure to UV radiation may also affect the strength of a synthetic web sling.

The stretch of a polyester web sling when at its rated capacity will be about 3%. Synthetic round slings are shown in Figure 20.45.

FIGURE 20.45 A collection of polyester round slings.

Nylon is not affected by grease and oil and resists chemicals like aldehydes, ethers, and strong alkalis. Nylon slings should not be used with acids and bleaching agents. The stretch of a nylon sling, when working at its rated capacity, is about 8 to 10%. The choice of using nylon or polyester slings depends on the nature of your hoisting and rigging operation. New slings are marked by the manufacturer to provide this important and valuable information:

- The name of the manufacturer
- The manufacturer's stock number
- The rated load for each type of hitch configuration
- The type of synthetic material used to make the slings

Long-term exposure to sunlight or ultraviolet radiation can affect the strength of synthetic webbing slings. They should be stored when not in use in a cool, dry place. Synthetic web slings are manufactured in several different configurations, including single- or multi-leg bridle assemblies, eyes formed on each end of the sling, and an endless or continuous loop configuration with fittings on both ends.

When determining the length of a sling, you should measure from the top load bearing point to the bottom load bearing point. For example, an eye-to-eye configured sling would be measured from tip to tip of the eyes. Multi-leg sling assemblies are measured from the load bearing point of any attachment, such as rings, fittings, or hooks. Synthetic web slings can be ordered in any length you wish and specifically designed for a lift. Most web slings are standardized and sold from 2-foot to 20-foot lengths. The ply number of a sling is the number of layers of webbing that construct the body of the sling. Web slings can be ordered in 1-, 2-, 3-, or 4-ply configurations. More plies added to a sling add strength to its body. Slings with a wider body can distribute the load over a larger sling surface area. Wider-width slings can provide better protection to delicate, fine surfaces. Sling widths are provided from 1-inch to 12-inch or wider widths when special-ordered.

It's important to protect your synthetic web slings by way of softeners. Softeners can be materials you collect on the jobsites or other places you visit in your travels. Old concrete hose, fire hose, and slotted pieces of pipe or chunks of wood all work well. Synthetic web sling manufacturers offer sling wear pads that can be sewn into the webbing material at certain wear points—these can be ordered. A special coating can be added to the webbing material during manufacturing that protects against moisture and dirt penetration and helps reduce abrasion. Wide-width special shackles made especially for use with synthetic web slings are very helpful, providing an increased area for loading and offering 100% WLL effectiveness.

Specific types of synthetic slings available include the following.

Web Slings

Web slings (Figure 20.46) are made from nylon or polyester webbing. They are used for normal hoisting and lifting activities.

FIGURE 20.46 A polyester web sling.

Round Slings

Round slings (Figure 20.47) consist of a continuous loop of yarn, usually covered by a protective sleeve. Round slings offer good load distribution.

FIGURE 20.47 A polyester round sling is a versatile, convenient, and cost-effective way to lift a variety of materials.

FIGURE 20.48 A plasma sling is made from ultra-high molecular synthetic fiber.

Plasma Slings

Plasma slings (Figure 20.48) are made from ultra-high molecular weight polyethylene (UHMWPE). Plasma slings are very strong and lightweight and are resistant to water and abrasion.

Inspection of all synthetic slings used for overhead lifting must be done prior to any lift, and they should not be used if any of the following faults are found according to the ASME 30.9 safety standard:

- Damage to any attached fittings
- Excessive pitting, corrosion, cracks, bends, twists, gouges, or breaks
- Missing or illegible sling identification
- Damage to the webbing material
- Acid or caustic burns
- Melting or charring of the sling's surface
- Holes, tears, cuts, or snags in the sling
- Broken or worn stitching in the load bearing splices of a sling
- Excessive abrasive wear
- Knots in any part of the sling
- Discoloration and brittle or stiff areas

In addition:

- For hooks and rigging hardware, consult ASME B30.10 and ASME B30.26.
- Other conditions or other visible damage that causes doubt about the continued safe use of the sling warrants removal from service.

Note: Synthetic web slings should be inspected frequently—daily or before each shift—by a qualified person. A qualified person should also conduct inspections at the very least annually. Written records of the most recent periodic inspection should be maintained. Unsafe slings should be removed from service immediately and tagged "Do not use" and properly disposed of or destroyed. The qualified person should conduct additional periodic inspections where conditions justify, such as the severity of the service conditions, the nature of the lifts being made, and the frequency of sling use.

Repairing/Reconditioning

Do not use worn or damaged slings or attachments. Discard or repair them. Use damaged slings only after they are repaired, reconditioned, and proof-tested by the sling manufacturer or a qualified person using the following criteria:

- Ensure that the manufacturer or a qualified person performs repairs.
- Ensure that repairs of hooks and fittings meet ASME B30.10 and B30.26 standards.
- Do not repair cracked, broken, melted, or damaged webbing material of the slings.
- Do not repair load bearing splices.
- Do not make any temporary repairs of synthetic webbings or fittings.
- Mark repaired slings to identify who made the repairs.
- Retain the certificates of proof test, and make them available for examination.

General Information Regarding Synthetic Web Slings

Much of the information in the lists that follow can be applied not only to synthetic web slings, but to the other sling types as well.

Cautions to Personnel

- Ensure that all portions of the human body are kept away from the areas between the sling and the load and between the sling and the crane or hoist hook.
- Ensure that personnel never stand in line with or next to the legs of a sling that is under tension.
- Ensure that personnel do not stand or pass under a suspended load.
- Ensure that personnel do not ride the sling or the load unless the load is specifically designed and tested for carrying personnel.
- Do not use synthetic webbing slings as bridles on suspended personnel platforms.

Effects from the Environment

- Store slings in an area where they will not be subjected to mechanical, chemical, or ultraviolet damage or to extreme temperatures.
- When slings are exposed to extreme temperatures, follow the guidance provided by the sling manufacturer or a qualified person.

- Consult the sling manufacturer for recommended inspection procedures when nylon or polyester webbing slings are extensively exposed to sunlight or ultraviolet light.

Rigging Practices

- Ensure that slings are hitched in a manner providing control of the load.
- Ensure that sharp edges in contact with slings are padded with material of sufficient strength to protect the sling.
- Ensure that slings are shortened or adjusted only by methods approved by the sling manufacturer or a qualified person.
- Ensure that during lifting with or without a load, personnel are alert for possible snagging.
- Ensure that in a basket hitch, the load is balanced to prevent slippage.
- When using a basket hitch, ensure that the legs of the sling contain or support the load from the sides, above the center of gravity, so that the load remains under control.
- Do not drag slings on the floor or over abrasive surfaces.
- Ensure that in a choker hitch, the choke point is only on the sling body, never on a splice or fitting.
- Ensure that in a choker hitch, an angle of choke less than 120 degrees is not used without reducing the rated load.
- Ensure that slings are not constricted, bunched, or pinched by the load and hook, or any fitting.
- Ensure that the load applied to the hook is centered in the base (bowl) of the hook to prevent point loading on the hook.
- Ensure that an object in the eye of a sling is not wider than ⅓ the length of the eye.
- Do not shorten or lengthen a sling by knotting or twisting.
- Do not rest loads on the sling.
- Do not pull a sling from under a load when the load is resting on the sling.
- Do not allow shock loading.
- Avoid twisting and kinking.

Do not use synthetic web slings with loads in excess of the rated load capacities described in the appropriate tables. Ensure that synthetic web slings have suitable characteristics for the type of load, hitch, and environment in which they will be used and that they are not used with loads that exceed the rated load capacities described in the manufacturer's product specification catalog. Consult the sling manufacturer or a qualified person for synthetic web slings without a catalog. Follow other safe operating practices, including the following practices for sling selection:

- For multiple-leg slings used to hoist unbalanced loads, ensure that an evaluation is performed by a qualified person to prevent the overloading of any leg.
- Ensure that multiple-leg slings are selected according to their capacity charts and used at the specific angles given in the manufacturer's specifications catalog. Ensure that operations at other angles are limited to the rated loads of the next lower angle given in the catalog or calculated by a qualified person utilizing the synthetic sling's rating tables, as shown in Figure 20.49.
- Ensure that the fitting is the proper shape and size so that it is seated properly in the crane hook or lifting device.

FIGURE 20.49 Flexibility, weight, strength, and non-marring qualities make the slings popular.

Understanding Bearing Stress

Bearing stress is the contact between two separate bodies that can be high enough to cause damage or even destroy a synthetic sling. Bearing stress is a quick and simple calculation that riggers and other ground personnel can make in the field to ensure your lift does not have too much bearing stress. Bearing stress is often not understood or is overlooked by most members of the hoisting and rigging team. Bearing stress is increased when the sling is unevenly loaded by a curved surface. Anytime you are hooking synthetic slings using rigging hardware with a curved surface, such as hoisting rings, eyebolts, hooks, or shackles, you will have bearing stress.

The Web Sling and Tie Down Association (WSTDA) addresses bearing stress in its standards for the use of synthetic slings. WSTDA standards related to bearing stress in synthetic round slings guide users in selecting the correct rigging hardware and calculating the bearing stress to help prevent sling damage during hoisting and rigging operations. WSTDA says that flat surfaces such as those found on specialized rigging hooks minimize bearing stress and eliminate the need to calculate an effective width. For curved surfaces like those found on anchor shackles, a curvature adjustment is required using bearing stress calculations. WSTDA recommends a bearing stress value of 7,000 psi during sling loading. Pressure over 7,000 psi could have damaging effects on your slings such as rips, punctures, and tears in the sling cover, exposing the core yarns. These damages would require the sling be removed from service.

To calculate bearing stress, divide the applied force or load by the load bearing area, as demonstrated here:

$$\text{Bearing stress } (\sigma) = \text{applied force } (F)/\text{load bearing area } (a)$$

As an example, if the applied force is 6,000 pounds, the effective contact width is 1.5 inches, and the hardware diameter is 0.62 inch, then the load bearing area would be 0.93 square inch. The bearing stress would then be 6,000 pounds/0.93 square inch, or 6,451 psi, which is within the WSTDA's recommended limits.

Other interesting statistics you must keep in mind:

1. Synthetic slings have a design factor of 5:1.
2. Nylon straps will stretch 20 to 30% at the breaking point.
3. A girth hitch can reduce sling strength by as much as 50%.
4. A nylon synthetic sling can stretch up to 8 to 10% at its rated capacity.
5. Tying a knot in a synthetic sling can decrease its strength up to 50%.

Metal Mesh Slings

Metal mesh slings are widely used in demanding and high-heat environments like metalworking facilities, steel mills, or steel processing facilities where the loads may be abrasive and hot. The steel construction of these slings resists abrasion, cutting, and heat damage that synthetic slings would be less resistant to. Also known as *wire mesh slings*, metal mesh slings can be thought of as a perfect marriage between the various types of hoisting slings, such as alloy chain slings and synthetic slings made from polyester or nylon. Using a metal mesh sling offers a combination of strength, durability, flexibility, excellent load grip, and securement of the load.

Metal mesh slings are often preferred in rough environments because of their ability to be repaired and recertified. They also have low stretch and a wide-bearing surface area to firmly grip and secure a load, thereby reducing load damage. They're extremely resistant to abrasion and cutting; however, if there is evidence of even one broken wire in the sling, the entire sling needs to be removed from service. Often the wire mesh material is made from galvanized high-carbon steel, alloy, or stainless steel to resist corrosion. The fittings can be plated or made from an alloy material for corrosion resistance and longer service life. Uncoated metal mesh slings may be used in temperatures ranging from –20 to 550 degrees Fahrenheit, or –29 to 288 degrees Celsius, without a reduction to their WWL. For metal mesh slings that have been coated with elastomers, such as polyvinyl chloride or neoprene, the safe operating temperature range is lower; typically, 0 to 200 degrees Fahrenheit or –18 to 200 degrees Celsius. Beyond carbon steel, they can be made from alloy steel to achieve high WLL ratings or from stainless steel for work around corrosive environments. Some wire mesh slings can be repaired for longer service life and cost savings. Chemically active environments can affect sling strength, requiring careful material selection and consultation with the sling manufacturer. Figure 20.50 depicts a Type I and Type II mesh sling meant to be used with a basket hitch configuration.

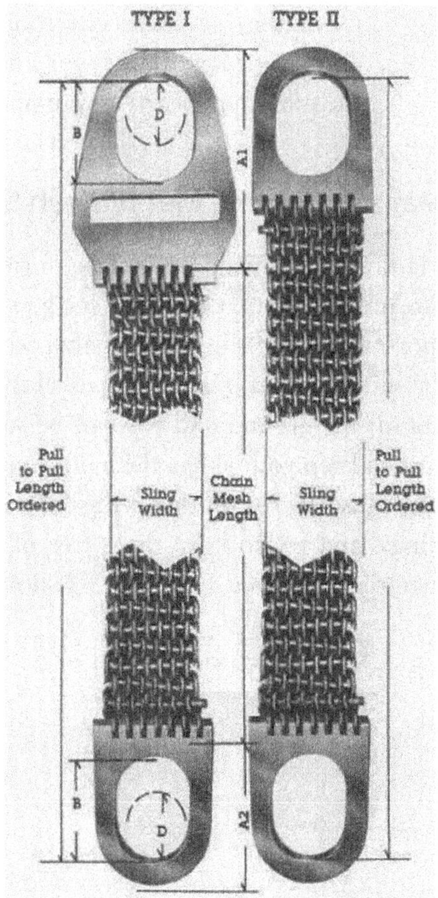

FIGURE 20.50 Metal mesh slings are available in two types. Type I and Type II.

Inspection of Metal Mesh Slings

All inspections shall be performed by a designated person with any deficiencies further examined by a qualified person to identify hazards and determine what additional steps need to be taken to address the hazard. Two industry standards provide the end user with guidelines for inspection and with criteria that warrants removal from service: OSHA 1910.184 and ASME B30.9. Best practice is to inspect metal mesh slings upon receiving them from the manufacturer. Double-check the sling identification to make sure that it's what you ordered and that the rated capacity meets all your project specifications and lifting requirements.

- **Frequent inspections.** Designate a competent person to perform a daily visual inspection of slings and all fastenings and attachments for damage, defects, or deformities. The inspector should also make sure that the metal mesh sling that was selected meets the specific job requirements it's being used for. However, users can't rely on a once-a-day inspection if the sling is used multiple times throughout the day. Shock loads, severe angles, edges, and excessive heat can quickly cause damage to the material, so best practice is to perform a visual inspection prior to each use.
- **Periodic inspections.** A periodic inspection is performed by either a professional inspection provider or a qualified person every 12 months at a minimum, or every month to every quarter in more severe service conditions. Depending on the severity of the operating environment and frequency of use, your business may decide to inspect metal mesh slings more often than the minimum yearly requirement.

Maintenance of Metal Mesh Slings

Maintaining a metal mesh sling during and in between uses is the best way to help extend the sling's life and ensure that the sling stays in service. Inspections are easier to perform, as well as being more thorough, when slings are easily accessible, organized, kept off the ground, and stored in a cool and dry environment. Hang up your slings, or keep them in a designated locker or rigging box where they are off the ground and will not be subjected to mechanical damage, corrosion, or moisture. Also, be sure to keep your slings clean. Prolonged exposure to dust, dirt, or moisture can degrade the materials and cause corrosion that will shorten the life expectancy of the product. Wipe grease or oil off your slings, and try to keep them free of dirt, chemicals, or other particulates that can break down the material over time. Figure 20.51 shows a metal mesh sling being used in a choker hitch configuration.

FIGURE 20.51 Shown here is a metal mesh sling in both a choker and basket hitch configuration.

Chemically Active Environments

The strength of metal mesh slings may be degraded by chemically active environments. This includes exposure to chemicals in the form of solids, liquids, gases, vapors, or fumes. The sling manufacturer or a qualified person should be consulted before slings are used in chemically active environments.

Temperature

When slings are used at temperatures above 550 degrees Fahrenheit (228 degrees Celsius) or below –20 degrees Fahrenheit (–29 degrees Celsius), the sling manufacturer should be consulted. If the sling contains any coatings that change the temperature range of the sling, the manufacturer shall provide a revised temperature range chart.

Metal mesh slings must be removed from service if any of the defects listed here are found, and any other condition that causes any doubt regarding the continued use of the sling (Figure 20.52).

FIGURE 20.52 Mesh slings are specified for applications where the loads are abrasive and hot, or will cut other types of slings.

Criteria for Removing from Service

- Missing or illegible sling identification
- Broken weld along the sling edge
- Broken wire in any part of the mesh
- A reduction in wire diameter of 25% due to abrasion or 15% from corrosion
- Lack of flexibility due to mesh distortion
- Distortion of the choker fitting so the depth of the slot is increased by more than 10%
- Distortion of either fitting so the width of the eye opening is decreased by more than 10%
- A 15% reduction of the original cross-sectional area of any point around the hook opening
- Visible distortion of either fitting out of its plane
- Cracked end fittings of the sling
- Slings with locked spirals or without free articulation or movement
- Fittings that are pitted, corroded, cracked, bent, twisted, gouged, or broken
- Other conditions, including visible damage, that cause doubt as to the continued use of the sling

Generally, metal mesh slings have a higher initial cost compared with other sling types but do have their advantages—like being able to handle hot, abrasive, or sharp loads that would damage other sling materials. Their strength and durability make them a good choice for demanding construction and industrial jobs. Their flexibility and wide bearing surface allow them to mold around and securely grip loads with unusual shapes. New, repaired, or reconditioned slings should be proof-tested by the manufacturer or a qualified person before initial use. Metal mesh slings are valuable in hoisting and rigging operations. Their strength, heat resistance, and ability to support abrasive or sharp loads make them a wise choice for hoisting and rigging operations. Figure 20.53 shows a metal mesh sling and all its parts.

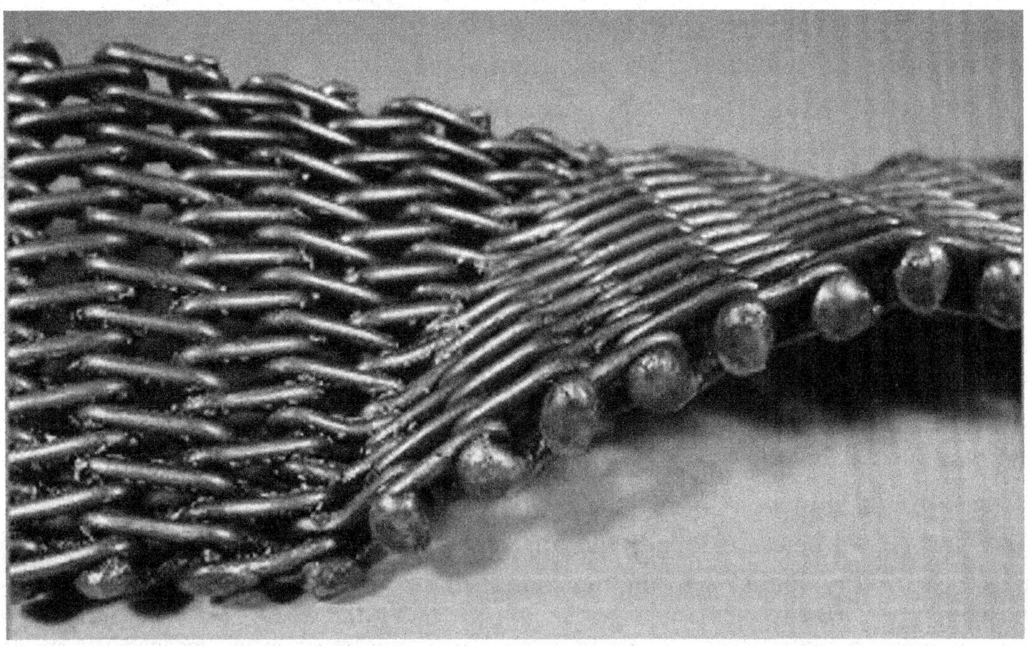

FIGURE 20.53 Do not load slings in excess of rated capacity. Load edges must be straight and flat and in contact of the full width of the mesh at bearing points.

Chapter 20 Review Quiz

1. Lifting and rigging are two separate operations. Lifting involves which of the following?
   a. The process of preparing the load for hoisting by ensuring the load is properly hitched, positioned, and controlled
   b. Managing load control systems and assessing load stability
   c. The physical movement of the load from one location to another
   d. Calculating load weights and development of lift plans

2. What is the most current and effective way to rate rigging hardware for weight limits?
   a. Safe working load (SWL)
   b. Minimum breaking load (MBL)
   c. Dynamic amplification load (DAL)
   d. Working load limit (WLL)

3. Which system used for rating hardware limits has potential legal complications and has been replaced?
   a. Rated capacity (RC)
   b. Minimum breaking load (MBL)
   c. Safe working load (SWL)
   d. Working load limit (WLL)

4. Which system used for rating hardware limits is the most precise in specifying working conditions and the importance of staying within the equipment's capacity limits?
   a. Working load limit (WLL)
   b. Dynamic amplification factor (DAF)
   c. Minimum working load (MWL)
   d. Safe working load (SWL)

5. When the signal person, rigger, or other personnel need to communicate with the crane operator, that person must use approved hand signals in accordance with which safety standard?
   a. OSHA CFR 1926.1400
   b. ASME B30.5
   c. OSHA CFR 1910.184
   d. ASME B30.26

6. Chain shackles are designed to be used for
   a. very heavy loads in construction
   b. straight line vertical pull lifts only
   c. straight or angular load lifting
   d. moderate lifting in general industry only

7. Twist shackles are one of the special types of lifting devices because their body has a
   a. 60-degree twist
   b. 70-degree twist
   c. 80-degree twist
   d. 90-degree twist

8. Wire rope clips used for overhead lifting must be made out of
   a. malleable iron
   b. galvanized iron
   c. drop forged steel
   d. wrought iron

9. Wire rope clips are popular because they can be installed in the field. Depending on the diameter of the wire rope, they provide how much efficiency of the rope's breaking strength?
   a. 60 to 70% efficiency
   b. 70 to 80% efficiency
   c. 80 to 90% efficiency
   d. 100% efficiency

10. What regulation or safety standard refers to the correct use of wire rope clips?
    a. ASME B30.9
    b. ASME B30.10
    c. OSHA CFR 1926.1404
    d. ASME B30.12

11. According to which safety standard do eye bolt markings and the identification requirement come from?
    a. ASME B30.5
    b. ASME B30.9
    c. ASME B30.12
    d. ASME B30.26

12. Which type of eye bolt should never be used for overhead lifts?
    a. Bent eye bolts
    b. Machinery eye bolts
    c. Forged eye bolts
    d. Non-shouldered eye bolts

13. Which type of link is used to group as many as 4 slings together?
    a. Pear-shaped links
    b. Mechanical coupling links
    c. Master link subassembly
    d. Coupling links

14. Which type of link serves best when attaching to crane hooks?
  a. Master link subassembly
  b. Oblong master links
  c. Pear-shaped links
  d. Coupling links

15. What type of rigging hardware is best for forming bridles?
  a. Oblong master link
  b. Pear master link
  c. Master rings
  d. Mechanical coupling link

16. Which ASME safety standard provides valuable information regarding links and rings used for hoisting and rigging operations?
  a. ASME B30.26
  b. ASME B30.5
  c. ASME B30.7
  d. ASME B30.22

17. What specialized overhead lifting device provides more headroom?
  a. Lifting beam
  b. Spreader bar
  c. Adjustable lifting bar
  d. Equalizer bar

18. What specialized overhead lifting device listed here is best suited to handle loads that need to be supported all the way throughout its length?
  a. Adjustable lifting bar
  b. Equalizer bar
  c. Spreader bar
  d. Lifting beam

19. Both spreader bars and lifting beams are designed to keep lifting slings below the units at or near what angle?
  a. 90 degree angle
  b. 85 degree angle
  c. 80 degree angle
  d. 75 degree angle

20. The ASME safety standard that covers spreader bars and lifting beams used by cranes for hoisting loads is
  a. ASME B30.5
  b. ASME B30.20
  c. ASME B30.22
  d. ASME B30.2

21. According to the ASME safety standard for metal mesh slings, a component will be removed from service when
  a. there is a lack of flexibility of the mesh fabric
  b. the sling does not have the correct size of handle eye
  c. there is a lack of chrome galvanized coating
  d. the sling is not constructed from Grade 80 metal overbearing surfaces

22. The maximum load weight of a sling that can be used during hoisting and rigging activities is
  a. its breaking strength
  b. its rated capacity
  c. its net load capabilities
  d. its maximum gross load

23. Identification tags attached to slings used in hoisting and rigging operations must include
  a. the number of threads in the fabric
  b. the resistance to ultraviolet rays rating
  c. the storage requirements when not in use
  d. the manufacturer's name

24. What is the maximum included angle for a shackle?
  a. 80 degrees
  b. 100 degrees
  c. 120 degrees
  d. 140 degrees

25. Synthetic slings must be inspected
  a. frequently
  b. periodically
  c. each day after use
  d. before each use

26. A chain sling should be inspected by a
  a. certified person
  b. certified rigger
  c. qualified person
  d. competent person

27. The ASME safety standard states that the minimum allowed chain grade that shall be used for overhead lifting is
  a. Grade 80
  b. Grade 70
  c. Grade 60
  d. Grade 50

28. When a new chain is received from the manufacturer, it should undergo a(n)

a. frequent inspection
b. periodic inspection
c. complete inspection
d. initial inspection

29. When slings used in hoisting and rigging operations can come into contact with loads with sharp edges, riggers must see to it that

a. the load is landed on adequate blocking
b. the lift is set up on ample tonnage so slings can be pulled out easily from beneath loads
c. the slings are protected with pads or softeners to safeguard and shield them
d. the loads are never stacked on top of each other to avoid cutting or crushing

30. According to the ASME B30.9 safety standards for slings, how many randomly distributed broken wires in one rope lay call for the removal of a wire rope sling from service?

a. 6 wires
b. 8 wires
c. 10 wires
d. 12 wires

Summary—and Then Some

In the crane hoisting and rigging industry, it can become very confusing trying to understand all the different regulations and standards and how they apply to different workers and parts of our operations. Both OSHA and ASME have specific, detailed definitions for all their terms that are frequently used around rigging hardware and how they apply to crane hoisting and rigging. All members of the hoisting and rigging team must be highly trained and qualified. Each should be familiar with OSHA and ASME regulations and standards and the descriptions of who does what, including these persons:

- **Competent person.** Someone who has the knowledge, skills, and authority to identify hazards and take corrective actions to ensure safety. Competent persons are not necessarily certified but do possess the necessary experience and training to recognize potential dangers and prevent accidents during lifting operations.

  While certification can be part of demonstrating competency, it's not a prerequisite to be a competent person. Experience and training can also establish competency. In comparison to the qualifications needed to be designated a *qualified person*, which generally requires a degree or official certifications, a competent person's qualifications are focused on practical experience and the ability to identify and address hazards.
- **Qualified person.** Someone who has demonstrated the ability to solve problems related to rigging, either through a recognized degree, certificate, or professional standing or through extensive knowledge, training, and experience. Qualified riggers are responsible for tasks like hooking, guiding loads, and making initial connections to components or structures. Qualifications are specific to the type of rigging being done. A person qualified for one type of rigging may not be qualified for another. OSHA regulations require qualified riggers for certain hoisting and rigging operations such as those involving fall zones and the assembly and disassembly of lattice booms and cranes.
- **Designated person.** Someone selected or assigned by an employer or an individual in charge to perform specific duties related to hoisting and rigging operations who is deemed qualified and competent to do so. This individual has the necessary training, knowledge, and experience for the specific tasks that the person is assigned. While the designated person's responsibilities do vary, the responsibilities normally involve ensuring the safe and correct execution of any hoisting and rigging operation. A group of people, rather than a single individual, may be used as a team of designated persons, each with specifically assigned duties within the overall hoisting and rigging operation.
- **Certified person.** A certified person in hoisting and rigging operations who has completed specific training and passed written and practical examinations, demonstrating the knowledge and skill needed in safely handling and moving loads using various rigging and lifting equipment, including cranes.

  Certification demonstrates a crane operator, lift director, rigger, or signal person's competence and knowledge in the person's profession. It offers assurance to other craft workers that the individual is qualified to handle equipment safely and can follow established protocols, understanding all relevant safety regulations and standards. In the hoisting and rigging industry, knowledge is a weapon. When the entire hoisting team is highly trained and knowledgeable, the hoisting operation performs like a symphony. If one individual is out of sync, the operation sputters, and production will suffer, all while compromising the highest level of safety. With the certification of the entire hoisting and rigging team, morale is at its highest, production increases, and the maximum level of safety is achieved. Be a professional in everything that you do, and you'll be a force to be reckoned with in your industry. Being highly trained and certified is a must, no matter what craft you work in.

Cranes on barges, as well as clamshell, dragline, and pile driving equipment, all complement mobile crane operations by extending their operational capabilities, improving their applicability in specific construction scenarios. These specialized attachments greatly extend a mobile crane's capabilities of tackling a broader range of construction applications, particularly in marine and foundation-related projects. The combined utilization of these tools is vital in maintaining high levels of safety, cost effectiveness, innovation, and production in complex, demanding construction environments.

Pullers and handheld lifting devices should never be used as a replacement for well-planned mobile crane hoisting and rigging operations. Using pullers safely and productively requires experience with good training. Handheld tools, such as pullers, come-alongs, and lever hoists are all valuable tools in the mobile crane hoisting and rigging industry. These types of equipment can enhance mobile cranes when performing hoisting that requires precise vertical or horizontal adjustments while setting a load into place. When it comes to efficiency, these types of tools can be the determining factor, and selecting the correct rigging equipment for a uniquely heavy lift requires specialized hoisting and pulling equipment. Every piece of rigging hardware and equipment plays an important role in the safe, productive movement of heavy loads. These types of specialized equipment play an important part where there are space constraints or load weights that make direct hoisting of the load impractical. These tools reduce physical effort, increase productivity, reduce fatigue, and contribute to smoother, more efficient hoisting and rigging operations.

21 Specialized Crane Operations, Tools, and Equipment

Mobile Cranes and Pile Driving

Crawler cranes play an important part in construction projects that involve pile driving. Lattice boom crawlers are the most popular cranes used for pile driving projects. Their long-reaching lattice booms allow them to pick and carry large, heavy loads while at the same time using heavy pile hammers. A hydraulic pile driver is a great tool in modern construction operations, providing a powerful method for driving and extracting piles. Regardless of whether you are building bridges or high-rise buildings, knowing how hydraulic pile drivers operate and their benefits can greatly improve your project's success. Hydraulic pile drivers, known as *vibratory pile drivers*, use vibrations to drive or extract piles into or from the ground. These machines consist of a vibratory hammer head and an attached pile, which can be made of wood, concrete, or steel. The vibrations reduce soil resistance, allowing the pile to be easily driven into the ground.

Mobile cranes can be set up with several types of pile driving hammers, including diesel, hydraulic, airstream, or vibrating hammers. Leads that guide structures for the hammer and pile can be operated fixed by a spotter, either with swinging leads secured to the ground or with swinging leads working offshore—a system used for marine construction.

Hydraulic pile drivers operate using a combination of high-frequency vibrations, including static weight to drive the piles into the ground.

Here is a breakdown of how mobile cranes are used in pile driving:

- The pile driver's large hammer is first fastened to the crane boom via a cable from the crane's winch or by leaders; a steel frame that connects to both the boom and its base, providing stability and control
- The crane is responsible for lifting the long piles into position and guiding them as the piles are driven into the ground.
- The crane's boom may include attachments for guiding a hammer, spotter, boiler, or moonbeam, depending on the specific pile driving method and project requirements.
- The hammer may be powered by hydraulics, steam, diesel, or air requiring a corresponding power pack or compressor to be mounted, potentially along with control valves, piping, and hoses.
- The crane operator in the superstructure controls the pile driving process using levers and foot pedals. This requires great skill and good judgment in order to manage multiple controls.

How do hydraulic pile drivers work? Here's how:

- **Vibration mechanism.** The core of the vibratory hammer is its vibration mechanism, which usually involves counter-rotating weights powered by electric motors. These weights generate high-frequency vibrations.
- **Reduction of soil resistance.** As the vibratory hammer is vibrating, it reduces the compaction of the pile by the surrounding soil. This allows the pile to penetrate the ground with minimal resistance.
- **Insertion or extraction.** Depending on the job, the pile is either driven into the ground or extracted; the vibrations loosen the soil, allowing the movement of the pile.
- **Control and precision.** Modern hydraulic pile drivers come with advanced control systems that allow the crane operators to adjust the frequency and amplitude of the vibrations, ensuring precise control over the pile driving process.

Factors to consider when selecting a crane for pile driving operations include:

- The charted capacity of the crane and the crane's boom length
- Construction site terrain and conditions
- Mobility and transportation needs
- Specific project requirements and features needed
- Safety features and regulatory compliance
- Cost and availability

Applications of hydraulic pile drivers can be used in a variety of construction projects including:

- **Foundation work.** Driving piles can create strong foundations for buildings and bridges.
- **Marine construction.** Installing piles for docks, piers, and other waterfront structures.

- **Earth retention systems.** Building retaining walls and other structures that require deep foundations.
- **Environmental applications.** Used in remediation projects where minimizing soil disturbance is important.

Benefits of using hydraulic pile drivers include:

- **Efficiency.** Pile driving speeds up projects.
- **Versatility.** Works well in many soil types, using different pile drivers.
- **Cost effective.** Efficiency can lead to lower project costs.
- **Precision.** Minimizes damage to piles and nearby structures.

In regard to safety, be sure to:

- Wear appropriate personnel protective equipment, and maintain a clear and organized work area. Housekeeping is always important.
- Store and handle materials safely.
- Inspect all equipment closely before use.

Types of pile driving hammers include:

- **The drop hammer.** This is the simplest type, which uses gravity to drive the pile.
- **The hydraulic hammer.** This hammer relies on hydraulic power to drive the pile, which offers precisely controlled driving.
- **The vibratory hammer.** It utilizes vibrations to install pile. It's very useful for sheet piles in challenging soil conditions.

Only qualified and certified experienced crane operators should operate cranes in pile driving operations. Prior to initiating a pile driving operation, a thorough risk assessment should be conducted, outlining all potential hazards and mitigation strategies, including the selection of appropriate equipment and all safety protocols.

Important safety devices include:

- Emergency stop buttons that provide immediate control to halt operations in case of a malfunction or hazard
- Pressure relief valves that prevent hose bursts and system failures by releasing excessive pressures in hydraulic systems
- Anti-two-block systems that automatically cut off crane functions to prevent the hook block from making full contact with the boom tip
- Load indicators and limit switches that display the load weight in real time, helping operators avoid overloading
- Limit switches that prevent crane movement beyond safe operational limits
- Anti-collision systems that prevent collisions between cranes on jobs with multiple cranes
- Safety harnesses and lanyards for all personnel working at heights that protect them from falling
- Lockout and tagout kits to ensure safety during maintenance and repairs by securing energy sources

By having good crane protection strategies in place and following safety protocols, risks during hydraulic pile driving can be minimized, ensuring safer and more productive operations.

In pile driving work involving cranes, inspections are highly important in ensuring safety and operational production. These inspections are usually categorized into frequent, daily to monthly, and periodic monthly to yearly inspections. Every four years, depending on the crane's usage and the severity of its service, pile driving inspections should be thorough and cover all aspects of the equipment:

- **Visual inspection.** Looking for signs of damage, wear, leaks, loose connections, and other potential issues on the exterior and accessible compartments.
- **Structural components.** Checking crane boom, mast, gantry, A-frame shoes, and revolving frame for cracks, bends, structural integrity, and excessive wear.
- **Wire rope and rigging.** Inspecting wire ropes for broken wires in the strands, corrosion, kinks, wear near wedges and socket assemblies, proper reeving, and sheave alignment. Special attention should be paid to potential issues like martensite development due to rope burning of tight sheaves.
- **Mechanical systems.** Evaluating the condition of the engine, hydraulic system, swing bearings and swing gear, propel machinery, and other critical mechanical equipment. This includes checking the levels of fluids such as coolant, oil, and fuel while inspecting hoses for leaks and ensuring that bearings and gears are properly lubricated.
- **Material verification.** Making sure piles are the correct size, type, and grade as specified for the project.
- **Pile driving check.** Inspecting piles for any damage, poor handling, and secure attachment of shoes (Figure 21.1) or helmets (Figure 21.2).

FIGURE 21.1 A pile shoe is located at the very bottom tip of the pile shaft designed to break through dense or obtrusive materials.

- **Hammer operation and alignment.** The inspectors that will be observing the hammer's operation must make sure the hammer is running properly, ensuring it's aligned correctly with the pile and not causing excessive stroke or blow count.
- **Pile alignment and plumbness.** Maintaining the pile's alignment, called *plumbness* or *batter*, during driving. This is very important. Inspectors should use templates and visually check for proper alignment.
- **Driving process monitoring.** Inspectors should watch the driving process for problems such as rebound, unexpected resistance, or damage to the pile. See Figure 21.3, which shows a crane performing pile driving operations.

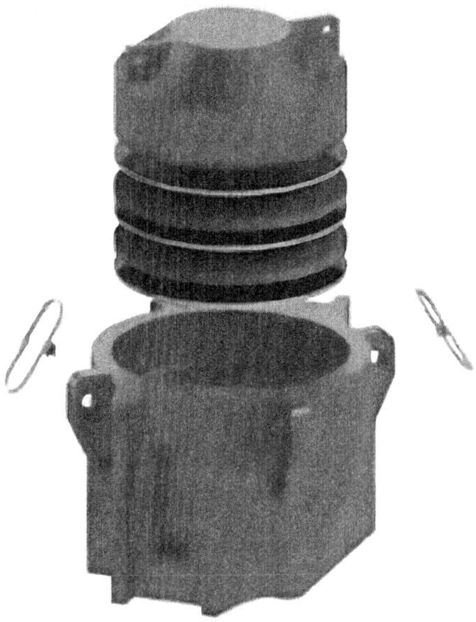

FIGURE 21.2 The helmet is a steel cap placed on top of the pile to protect it from the hammer's impact to distribute force.

FIGURE 21.3 The leads hanging vertically from the crane boom form a supporting framework that guides the pile hammer.

- **Splice inspection.** If splicing is necessary to achieve the planned pile length, making sure that the splices are constructed in accordance with the specified details—for example, making sure that welding is performed by a certified welder using approved procedures and materials.
- **Recording data.** Keeping good documentation of all parts of the driving process, including penetration rates, blow counts, or any problems. This is important for future reference and for calculating bearing capacity.

When troubleshooting and problem-solving, check for:

- **Excess blow count.** This may indicate problems with the driving system or soil conditions. Solutions include checking hammer efficiency, using a different hammer, or performing dynamic measurements to locate problems.
- **Driving out of alignment or location.** This can be addressed through alignment control measures like pile gates, templates, or fixed leads.
- **Pile head damage.** Make sure the helmet's size and cushioning meet specifications. Selecting the correct steel strength or adjusting the hammer stroke may prevent this problem.
- **Damage to piles.** This can be due to bending during driving, improper lifting, or manufacturer's defects. Inspections at various stages, including prior to and during driving, can help prevent or detect damage.

After pile driving, you should:

- Inspect the piles for any damage.
- Make sure that the piles are cut off to the specified elevation.
- Conduct dynamic pile testing in accordance with standards such as ASTM D4945 to be sure of the pile's capacity and integrity.
- Provide final certification confirming that the piles meet contract requirements.

Inspections are vital because they check for:

- **Safety.** Inspections that follow safety rules minimize the risks associated with heavy equipment work, ensuring a safe jobsite, which is paramount.
- **Quality control.** Inspections make sure the piles are installed properly, meeting design specifications and promoting the structural integrity of the foundation.
- **Cost efficiency.** Identifying and addressing issues early on can prevent costly delays, reworking, and possible structural failure.
- **Compliance.** Inspections make sure that work is completed within regulatory standards and building codes, avoiding potential penalties and legal issues.

A fantastic inspection program for pile driving operations provides for a more reliable, safer, and overall successful construction project.

Come-Alongs

Used in mobile crane hoisting and rigging operations, a come-along is a handheld operated ratchet lever winch used for pulling, lifting, or tensioning heavy loads (Figure 21.4). The original tool of this type was developed by Abraham Maasdam of Deep Creek, Colorado, about 1919, and it was later commercialized by his son, Felber Maasdam, about 1946. It has been copied by many manufacturers in recent decades.

A similar tool to the come-along is a cable puller, which does not have a drum and ratchet but directly grips the cable, allowing unlimited lengths of wire rope to be used. Some types of come-alongs use chains, but lever hoists are the preferred tool for lifting. Always make sure the come-along's capacity is sufficient for the load you are planning to lift. Come-alongs are generally used for small, more careful pulls or lifts compared with normal mobile crane heavy lifts. When gearing up for a load to be hoisted by a mobile crane, a come-along can be used to apply tension to slings and other rigging hardware and equipment to make sure they are properly seated and secure. They are great for making small adjustments and positioning a load, providing precise control. Come-alongs also work well to secure loads on trailers, and they work well on other platforms to stop shifting.

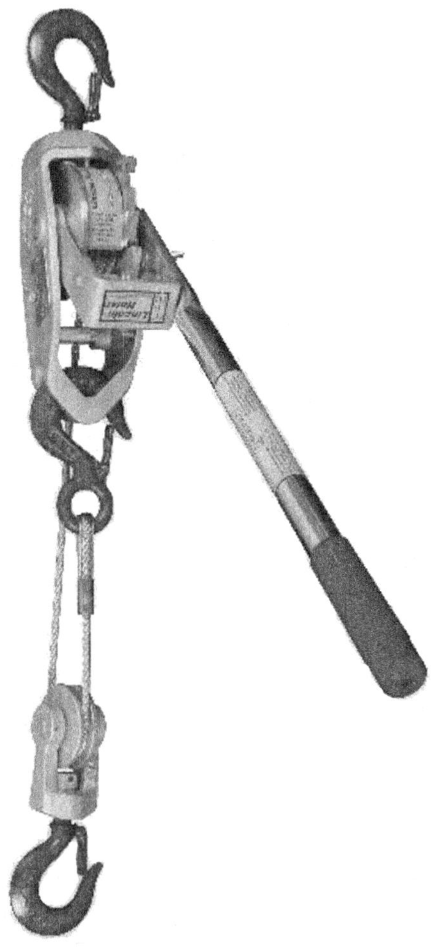

FIGURE 21.4 A come-along is a manually operated compact winch that uses a ratchet and cable to pull, lift, or stretch heavy loads.

How Does a Come-Along Work?

A wire rope puller operates very similarly to a lever block—the only difference being that wire rope is used instead of chain. To release the cable from the reel, make sure the mechanism is in a neutral position. These units are used for pulling only; they cannot be used for any lifting applications.

For Lifting Purposes

Every certified lever and chain block is manufactured to be capable of lifting a specific working load limit. The combined weight of the load being lifted, together with the rigging equipment the load is attached to, must never exceed this working load limit. The WLL of a lifting device will be clearly displayed on the side of the hoist, along with the name of the manufacturer and the serial number. Typically, the safety factor for come-alongs is 2:1 to 3:1 while pulling; when lifting, the safety factor is 5:1 to 7:1.

For Lashing Purposes

Cable pullers and lever hoists can be used for lashing. These are commonly used in the transport industry for tensioning or pulling a heavy load. The ASME safety standard B30.21 will provide more valuable information regarding come-along lever hoists.

Important notes regarding come-alongs: Come-alongs are intended for pulling a short distance or lifting a short distance, and they should never be used as a substitute for a sling. Exceeding a come-along's rated capacity can result in operational failure, potentially causing serious injuries, so always reference the manufacturer's specifications and the owner's manual for more information on operation safety if needed. Being properly trained and conducting safe work practices are also highly important while using come-alongs. Proper training and safe practice are highly important when using come-alongs.

Turnbuckles

Turnbuckles (Figure 21.5) are important tools used in mobile crane hoisting and rigging operations; they provide a safe and adjustable method for tensioning wire ropes and other rigging hardware and equipment. Turnbuckles feature two opposing threaded rods, one right-handed and the other left-handed. This allows tension to be adjusted by rotating its central body. Turnbuckles are common rigging devices used to add tension to, or reduce the slack in, wire a rope, cable, or similar tensioning assemblies.

Turnbuckles are a diverse type of rigging equipment and are used in a variety of applications across many different industries. They can be used in applications as simple as providing tension in a fence or perimeter cable, or in something as complicated as the construction of a suspension bridge. Like other rigging hardware products, turnbuckles are available in many different sizes, configurations, and types.

Using and selecting the right device is important in maintaining a safe and effective rigging program. Ensure you have the information necessary to help you determine the right type of turnbuckle for your application.

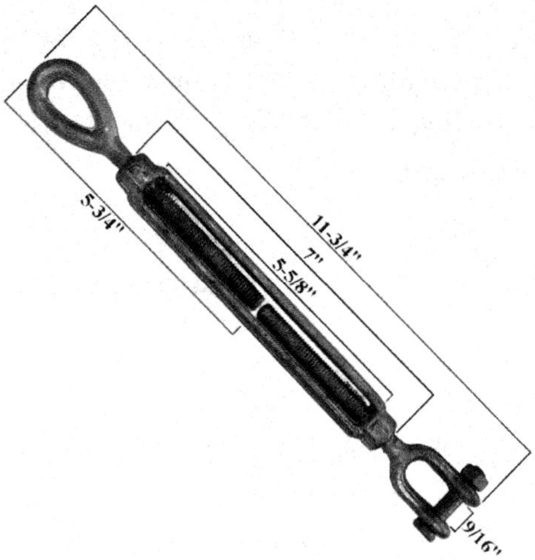

FIGURE 21.5 Turnbuckles are measured by their thread diameter and the body's maximum adjustment length.

The ASME B30.26 safety standard for rigging hardware describes a turnbuckle as an adjustable device consisting of a link with screw threads present on both ends; it can be turned to bring both ends closer together and is used for tightening a rod or stay. In crane hoisting and rigging operations, a turnbuckle is used to adjust the tension and length of different rigging systems, usually with the use of wire rope. It consists of a body with threaded ends and two end fittings. The fittings can be any combination of hooks, jaws, or eyes. The fittings screw into the body of the turnbuckle. This allows for pinpoint tension control when using different rigging equipment and hardware. It provides fine-tuning of the rigging equipment to ensure the loads rise balanced and are hoisted safely. Turnbuckles are also used to adjust slack out and provide equal load distribution when lifting loads. A turnbuckle has a central body with, as noted earlier, two threaded rods, left-handed at one end, and right-handed at the other. Rotating the central body of the turnbuckle moves these end fittings together, at the same time—in order to tighten or loosen the attached rigging equipment. The ASME safety standard B 30.26 provides more information about turnbuckles.

Types of End Fittings of Turnbuckles

There are three main types of end fittings that can be used together in any combination on either end of the turnbuckle:

- **Hook end.** A hook end fitting is used to create temporary connections because it is easy to quickly connect and disconnect. It should never be used in connections where tension could unexpectedly release.

- **Jaw end.** A jaw end fitting consists of a jaw and bolt secured together by a nut or a pin. This type of end fitting is used to connect two components that cannot be opened, such as eye bolts.
- **Eye end.** An eye end fitting is a closed loop that can be connected to a shackle or quick link.

The three main types of end fittings—hook end, jaw end, and eye end—are shown in Figure 21.6. OSHA 1926.251 mandates safety requirements for turnbuckles.

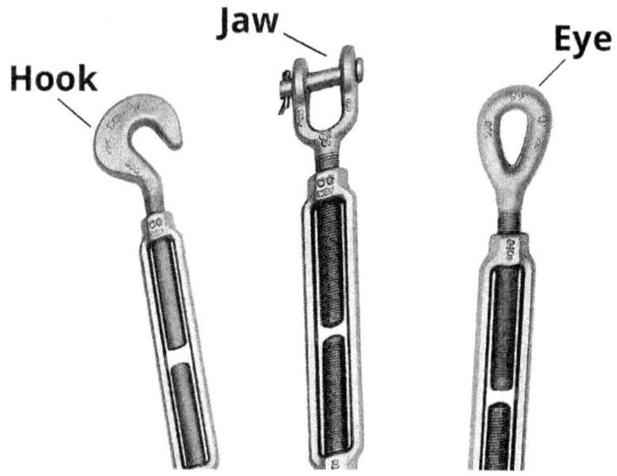

FIGURE 21.6 Select a turnbuckle with a working load limit that meets or exceeds your application's maximum load.

Turnbuckle Body

Open body turnbuckles allow for easy adjusting. Closed body turnbuckles protect the threaded rods inside the body from damage. The body itself turns to adjust tension by extending or retracting the end fittings. A jam nut can be used by placing it on the treads of the end fittings and tightening it against the turnbuckle body to stop the turnbuckle from becoming loose. See Figure 21.7, which shows a closed body turnbuckle and an open body turnbuckle.

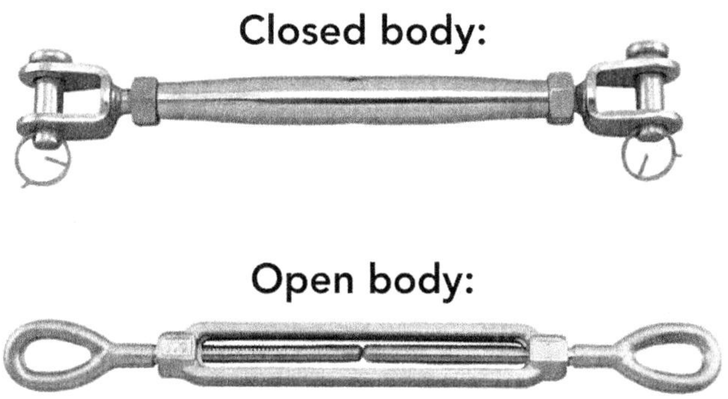

FIGURE 21.7 Closed body turnbuckles protect the threadline rods from becoming damaged.

Types of Turnbuckle Combinations

The most common types of end fitting combinations used with turnbuckles are:

- Hook and eye end fitting
- Jaw and jaw end fitting
- Eye and eye end fitting
- Eye and jaw end fitting

According to the ASME B30.26 safety standard for rigging hardware, turnbuckles must be clearly marked to show the manufacturer's name, the size of the turnbuckle, and the WLL. The three main types of end fittings can be used in any configuration to allow the unit to be attached to closed, open, temporary, or permanent connection points (Figure 21.8). Turnbuckles are available in stainless or galvanized steel. Always consider the WLL when making installations of turnbuckles.

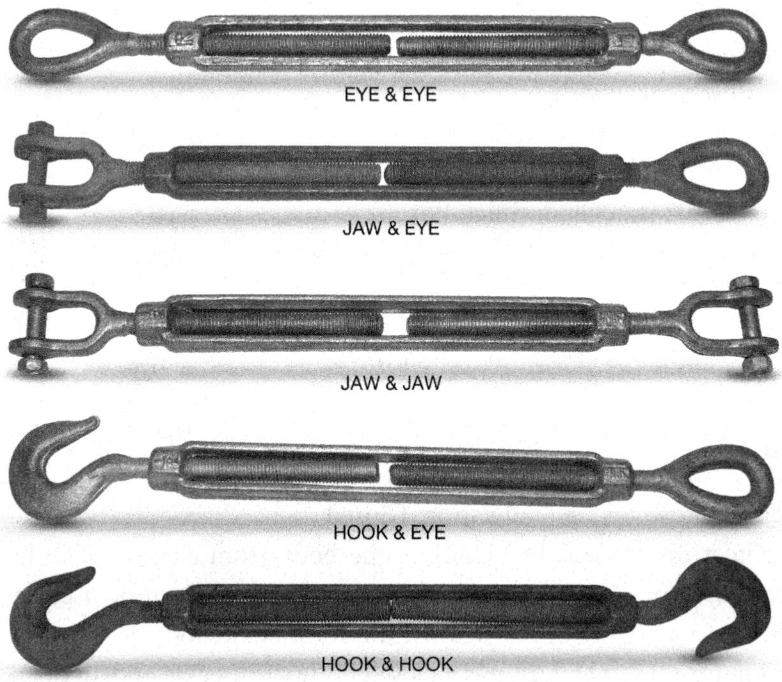

FIGURE 21.8 Ensure the body length is sufficient for your application for both the initial slack and final tensioning.

Using Turnbuckles Safely

To use turnbuckles safely:

- Secure turnbuckles to prevent accidental unscrewing.
- Always use turnbuckles in a straight in-line pull. Always avoid side loading.
- Avoid shock loading of the unit.
- Make sure the WLL of all turnbuckle parts in the rigging assembly is rated for the load being hoisted.
- Prevent obstructions from damaging the turnbuckle.

Inspections and Removing Turnbuckles from Service

ASME B30.26 clearly states that both daily visual inspections and periodic visual inspections of turnbuckles shall be made when used in crane hoisting and rigging operations. The inspections must be performed by a designated person. The inspection frequency depends on the type of hoisting activities and service conditions that the turnbuckle is being used for. Records should be kept on a periodic inspection basis at a minimum.

Turnbuckles must be removed from service if any of the following conditions is found:

- Excessive thread damage
- Missing or illegible identification tags or labels
- Unauthorized welding or brazing
- Heat damage from any source

- 10% reduction in dimension
- Excess pitting or corrosion
- Excessive nicks or gouges
- Bent, twisted, distorted, stretched, elongated, cracked, or broken load bearing components
- Any weld splatter or arc strikes

Always consult the most recent copy of ASME B30.26 for the most current information concerning turnbuckles. When using turnbuckles, it is very important to properly install and tighten them according to the manufacturer's guidelines. Overtightening or using them in improper applications can cause damage that affects their efficiency and safety.

Jack and Slide Systems

Many hoisting and rigging operations using mobile cranes start with a jack and slide system. The jack and slide system is a safe and productive way to move very large and often odd-shaped equipment. By supporting a heavy load on a track positioned low to the ground, this system allows its operators to keep advancing track sections ahead of the load as the load advances (Figure 21.9). Unlike a crane that hoists a load from above, the jack and slide system lifts from below the load and guides the load along the track.

FIGURE 21.9 Hydraulic cylinders push or pull the skid shoes that the load sits on, moving the entire load horizontally along the track.

These systems are versatile and productive, especially in tight, small places. Sections of track are laid out to form a road for the load to move on. The system uses hydraulic jacks to lift the load, and then by pushing or pulling the load, the load can be moved into place utilizing the system's skid shoes

(Figure 21.10). This type of operation is extremely useful when overhead clearance is low. It is perfect for moving very heavy loads that could be unsafe for cranes. Loads using this system are never as freely suspended as they would be if using a mobile crane, minimizing the risks of falling objects or loads. Unlike mobile cranes, which hoist their loads vertically, jack and slide systems move horizontally on a guided track that utilizes controlled friction, allowing very heavy loads to be moved smoothly with little effort.

FIGURE 21.10 The system's tracks and skid shoes are components that slide on the track, as shown here.

These systems are one of the safest and most efficient because they can be controlled remotely, and so no close-by supporting ground crew is needed when moving a load. For many operations, the jack and slide approach is great for some parts of the job; then the mobile crane is used for the final steps of hoisting the load and placing it. In other cases, the crane may handle the load first, and the jack and slide system finishes the job. Complementing cranes in many cases, jack and slide systems are used alongside cranes to form a very productive team.

Safety considerations while using jack and slide systems include the following:

- Never exceed the system's rated capacity or WWL.
- Good communication using standardized signals is important, especially when working closely with a crane.
- Ensure the jack and slide system is set firmly on stable, level ground that is capable of supporting the weight of the load.

Reviewing the industry guidelines on jack and slide systems, which can be referenced in the system operator's manual or the manufacturer's shop manual, is helpful in providing specifications and proper operating procedures. It's vital to carefully read and become familiar with these manuals before going to work.

While there isn't a single regulation specifically titled "Jack and Slide System Safety Standards," several OSHA regulations and best practices apply to the safe use of jacks and hydraulic systems for lifting and moving heavy loads. OSHA 1926.305's jacks, lever and ratchet, screw, and hydraulic outlines the general requirements for all types of jacks. A combination of OSHA regulations, ASME safety standards, and manufacturers' guidelines is very important to review to ensure the safe operation of jack and slide systems. Understanding load limits, maintaining equipment properly, and undergoing comprehensive training are vital in mitigating risks associated with heavy lifting and the movement of loads. When visually inspecting a jack, several areas should be closely looked over to make sure they are safe and ready to go. A jack and slide system in use is shown in Figure 21.11.

FIGURE 21.11 The load is never freely suspended, reducing the need for external restraints like winches.

Jack and Slide Inspections

OSHA's regulations and applicable ASME safety standards for the inspection of jack and slide systems are often used in tandem with one another when cranes are used for the hoisting and rigging of heavy loads. Understanding these OSHA regulations—especially those stressing the importance of regular, thorough inspections of equipment and the requirements of those inspections to ensure safe operation—is extremely important, and they should be reviewed as needed before any work is started.

Frequency of Inspections

A general inspection of jacks requires they be thoroughly inspected at intervals determined by service conditions, but no less than once every six months for constant or intermittent use at one location. Jacks sent out for special work or subjected to abnormal loads or shock must be inspected when sent out, when returned, and after each use.

When jacks are used with cranes, they also fall under the crane inspection requirements as set by OSHA regulations and ASME safety standard for that type of crane. These inspections must be conducted by a competent person.

Inspection details and procedures include the following:

- Make visual checks to examine signs of damage such as wear, defects, cracks, leaks, and corrosion.
- Conduct inspections that test functionality in order to evaluate and ensure the proper movement of the controls, safety devices, hydraulic systems, and lifting equipment.
- Check the jack body and lifting arm for any signs of structural damage.
- Assess the hydraulic hoses and fittings for leakage, wear, or deterioration.
- Check any control mechanisms for any necessary adjustments or corrections.
- Check the hydraulic fluid levels, the condition of the oil, and the hydraulic systems for leaks.
- Check all hooks for deformation or cracks.
- Check the hoist chain for wear, twist, or distortion.
- Check all the sheaves and clutch systems.
- Check any safety devices and operational aids.
- Check any structural members for any sign of cracks, deformation, or corrosion.

Maintenance, Repair, and Documenting Inspections

Any deficiencies found during inspections must be repaired or replaced before continued use. Repairs and modifications should be conducted in accordance with the manufacturer's specifications, using only manufacturer-approved parts and equipment, while keeping them properly lubricated at regular intervals. Jacks found to be defective must be tagged and removed from service until repairs are made. Always maintain records of inspection dates and defects found in any hoisting and rigging equipment.

Air Skates

Air skates are low-profile platforms equipped with air bearings. When compressed air is introduced into the bearing, a thin film of air is created between the skate and the ground's surface; this allows for nearly frictionless 360 degree travel and makes the final positioning of a heavy load in a congested area a much easier task. Also referred to as *air dollies, air bearings, air cushions*, or *machine skates*, these pieces of equipment enable ground personnel to move heavy loads affordably and safely, without tearing up floors or injuring the workers using them. Figure 21.12 shows how an air skate system works.

FIGURE 21.12 An air skate system uses a thin, pressurized film of air to move heavy equipment with minimal friction.

The Advantages of Using Air Skates

Advantages include:

- Air skates reduce the risk of injury to ground crew workers by reducing the need for manually lifting and moving heavy loads.
- Air skates reduce the risk of floor damage, making them a good choice for moving loads on sensitive flooring materials.
- Air skates may be more cost-effective than rollers or cranes.
- Air skates allow for quicker and more efficient movement of equipment, reducing downtime and increasing productivity.

Cranes fundamentally lift loads vertically. Air skates, on the other hand, are great at moving loads horizontally and allow for precise positioning once the load is on the ground while reducing or eliminating the need for other support equipment, such as forklifts. Air skates provide an air cushion that reduces vibrations, making them perfect for moving delicate, very expensive equipment, such as scientific instruments. Air skates provide a safe, efficient, and cost-effective means of transporting and positioning heavy loads in cooperation with cranes and hoists.

Required Inspections for Air Skates

Required inspections include visual and periodic inspections:

- **Visual inspection of air skates.** A competent person should conduct a visual inspection of the air skates; inspections must be conducted after each shift and as needed during use.
- **Periodic inspection of air skates.** A more thorough, deeper, and documented inspection is required every 12 months for equipment in regular use or subjected to harsh conditions. More frequent inspections are required on a quarterly or monthly basis, or as conditions may require.

And then there's the initial inspection:

- **Initial inspection of air skates.** Before using new air skates, you should perform an initial inspection to ensure that they match the original purchase order and are correctly rated for the work you intend to perform. Identifying labels must include a fireproof plate or sticker permanently adhered to the frame of the air skate. These identification labels must include the name of the manufacturer, the type or model of the air skate, the serial number, and the skate's rated load capacity. Familiarize yourself with the advisory warning labels.

For any inspection, the manufacturer's specifications must be consulted in the operator's manual and must be strictly followed. These documents contain critical operating parameters and inspection criteria and specify how often inspections should take place. You should be cross-referencing these documents when conducting any visual inspections of air skates and their components.

Every component of the air skate requires its own combination of close, visual inspection. Check the hoses and fittings by examining the lines for signs of abrasion, punctures, bulges, or kinking. Air lines can be inspected by applying pressure to the system while listening for sounds that might indicate air leaks or vibrations that possibly indicate a faulty piston. Applying a soap-and-water solution, while looking for bubbles present near the hoses or fittings, will also indicate the location of a leak.

Functional testing of the pneumatic system must be conducted after first making sure that it's been depressurized. Check around valve bodies where they connect to the airlines for any seals that may be compromised or displaced. Inspect the regulator for cracks, rust, corrosion, or any other damage, and ensure the regulator is properly installed with the air flow direction arrow pointing in the appropriate direction. Check the regulator filter for debris, clogging, or contamination while making sure the gasket seals and O-rings are also in good condition. Inspect the face of the pressure gauge to ensure it is free of cracks or other structural damage, as well as the break chamber for loose mountings or bent push rods, making sure that both the filtration and regulation systems are working properly. Ensure that a clean, dry air supply is being provided as specified, and frequently drain the air compressor's tank to remove accumulated oil and moisture.

The surface that will be bearing the weight of the load and air skate must also be inspected. Be sure that the air skate is operating on a smooth, level surface. Avoid warped or uneven surfaces that might potentially affect performance or safety. Be sure that the surface is strong enough to handle the bearing weight of the load.

Inspection of the smaller components is equally important. Be sure that all springs, pins, and other parts are in place and free of debris.

The air casters, or torus bags, are what will contain the pressurized air used by the skate. It's imperative that they are inspected for damage that could cause the air to leak while moving a load. Visually check the torus bag for cracks, abrasions, punctures, tears, bulging, or any other signs of possible separation. Keeping the bag clean—free of oil, grease, or dirt—can prevent compromising the bag's integrity over time.

When conducting wheel and bearing inspections, you must disconnect the air supply before performing any maintenance. Visually inspect each wheel for signs of wear and tear, such as flat spots, cracks, or chunks missing from the wheel. Look the wheels over for uneven wear, which could indicate misalignment or issues with load distribution. Remove any dirt or debris from the wheels to help enhance a good inspection. Check for loose bearings by gently wiggling the individual wheels side to side, looking for excessive movement. Spin the wheels by hand. A good bearing will rotate smoothly and freely. Listen for noises as the wheel spins for any unusual sounds. Grinding or crunching sounds indicate debris inside the bearing. Squeaking or hissing could mean the bearing is dry and needs lubrication. Check for rust on the bearing surface, which is a major sign of moisture damage.

Fasteners and connections should include a careful, systematic visual check to identify obvious problems. Inspect all fasteners and welds for any signs of rust or corrosion, which may weaken the fasteners. Look over the frame and connection points for any cracks, especially around welds. Hairline cracks can indicate metal fatigue that could cause structural failure. Inspect for any bent, warped, or distorted framework that could cause early failure or premature wear. Scan for any missing nuts, bolts, lock washers, or other parts. Inspect for fretting and signs of movement between parts that should be held stationary. Check the tightness of fasteners, and always use a calibrated torque wrench. Never overtighten. Lightly tap welds or apply pressure to check for any signs of cracking or separation. Any cracks, regardless of how small, are cause for immediate repair.

Additional general inspection areas include:

- **Identification labels.** Ensure that all required identification labels are legible and are adhered to the equipment.
- **Manufacturer's specifications.** Verify that the equipment is being used in accordance with the manufacturer's specifications and operating requirements.

Powered Saddle Rollers

Powered saddle rollers are specialized, precision rigging equipment used for moving and positioning heavy loads. They work very well in areas with little headroom.

Advantages of using powered saddle rollers to move heavy loads include the following:

- They are perfect for jobs with overhead obstructions and low headroom.
- They require less equipment and ground support personnel.
- They are fast and easy to assemble and disassemble.
- They have remote-controlled steering with jacking and lifting capabilities.
- Specialized tracks aren't necessary with saddle rollers, and this saves time, labor, and expense.

Unlike some other systems that require tracks or rails for movement, powered saddle rollers operate without them, making setup easier and reducing work crew effort and labor. They often include remote steering and jacking controls, allowing the system operator to adjust and guide the rollers with precision. Many systems have built in fail safety for both traveling and lifting operations. Powered saddle rollers provide robust solutions for moving very large and very heavy equipment where other means are difficult or impossible.

Operational safeguards and inspection concerns to be aware of include:

- **Structural integrity.** Look for signs of deformation, cracks, and excessive wear to any structural members.
- **Components and fastener.** Check for loose or missing guards, fasteners covers, or stops.
- **Labels.** Verify that all operating control markings are legible and in good condition.
- **Control and drive mechanisms.** Look for proper adjustment, excessive wear, and contamination.
- **Alignment.** Ensure the rollers are properly aligned to prevent wear.
- **Lubrication.** Inspect for proper lubrication of the rollers, shafts, and other moving parts.
- **Fasteners.** Inspect all bolts, rivets, and other items for looseness, stripping, or corrosion.
- **Sheaves and drums.** Look for excessive wear and cracks.
- **Bearings.** Inspect for wear and damage.
- **Air casters.** Inspect the air system for sufficient clean, dry air and the hoses for proper length and diameter. See Figure 21.13.

FIGURE 21.13 For the inspection of air casters, visually assess the air diaphragms for any cuts or other damage while verifying that the air supply is at the correct pressure.

Frequency of Inspection

Inspection criteria for powered saddle rollers are governed by the ASME B30.20 safety standards as well as the manufacturer's specifications and consist of three main categorizations:

- **Regular inspection.** Regular inspection occurs before, during, and after each shift. The equipment operator must visually inspect the device before using it and during the lift for any signs of damage or failure.

- **Frequent inspection.** Frequent inspections are conducted visually by the system operator or other designated persons at intervals based on the severity of use. For normal usage, conduct inspections every month; for heavy usage, conduct inspections every week; and for severe usage, conduct daily inspections.
- **Periodic inspection.** *Periodic inspection* refers to a complete inspection performed by a qualified inspector with records documented at intervals in accordance with OSHA regulations and ASME safety standards.

Temporary Trunnions

Temporary trunnions increase safety and improve productivity during complex lifts involving the removal of old vessels with temporary or nonfunctioning trunnions. Temporary trunnions allow for lifting at multiple points at specific locations during the length of the vessel. A temporary trunnion is used as a hitch point in cases where the loads do not have hitch points or where the hitch points that are currently in place no longer function. Installation of a trunnion depends on the needs of the operation and can be a less expensive, more effective option, compared with fabricating welded lifting lugs. Temporary trunnions are often used in operations involving the upending or rolling of vessels and in operations in which the existing hitch points are not suitable for the operation. This scope of work provides a secure point for slings and other rigging hardware and equipment used for a lift.

FIGURE 21.14 A lifting trunnion is either cast or fabricated and is bolted to an object being lifted, such as a tank or vessel.

Temporary trunnions are also very helpful during hoisting operations like tailing. This is when the vessel is lifted from a horizontal load to a vertical position, much like flipping a load from one orientation to another. Temporary trunnions can be used on flat-sided surfaces or shells, often at spots for both vertical lifting and tailing operations. They are a great tool in heavy hoisting and rigging operations, creating secure attachment points for cranes and other hoisting equipment when traditional options are inadequate. Figure 21.14 shows a trunnion installation and lift in action.

When installing temporary trunnions in hoisting and rigging operations, carefully follow these important requirements:

- Check for compliance with ASME BTH-1. Points of attachment must be designed and tested in accordance with all applicable safety standards, as well as OSHA regulations.
- Check for inadequate design. Poor welding, missing components, inadequate fastening, improper welding, or poor materials create unsafe lifting points and should not be used at hitch points.

- Look for damage and defects such as cracks, scratches, corrosion, or deformation in the trunnions themselves, including bolts, pins, spacers, and attachments.

If trunnions are in heavy or extreme service, consider inspecting before each use or more often as conditions require. Some operations may justify multiple inspections each shift. During the hoisting and rigging operation, review your inspection of previous lifts, and discuss them with a present qualified person; then schedule your future inspections accordingly, based on your findings. A trunnion installation is shown in Figure 21.15.

FIGURE 21.15 Trunnions are especially important for lifts that involve rotating or a "tailing load."

A review of welding procedure specifications should occur as needed, while also following current welding safety standards and regulations. Initial checks start with a thorough visual inspection before and after each weld and include checking:

- **Weld appearance.** Examine the weld bead for uniformity, smoothness, and consistent width.
- **Surface defects.** Look for common visual defects like cracks, porosity, undercut, slag inclusions, spatter, and overlap:
  - **Cracks.** These are serious defects that require immediate attention and repair.
  - **Porosity.** *Porosity* refers to tiny holes caused by trapped gas or lack of filler.
- **Weld profile.** Use weld gauges to check the weld's flatness, convexity, and concavity.
- **Documentation.** Document any defects found, including the location, size, and type.

A few more things to note about initial inspections:

- Relevant safety standards, such as ASME safety standards B31.3, Table 341.3.2, specify the acceptable limits of weld defects.
- A combination of visual inspection and nondestructive testing methods is often the most effective approach for meeting the surveyor's and client's requirements.
- Maintaining structural integrity, especially for lifting trunnions involving temporary installations, is highly important.

Codes and Standards for Temporary Welded Trunnions

Always follow relevant welding codes and ASME safety standard B31.3 to define the acceptable limits for weld defects. Remember, a combination of a visual inspection and nondestructive testing methods is often the most effective approach for ensuring weld quality and integrity, especially applications involving temporary trunnions.

Cranes and Clamshell Operations

A clamshell is a specialized bucket attachment designed for cranes. Clamshell buckets are normally used for digging, dredging, loading, and unloading with general use in the handling or transferring of bulk materials such as sand, gravel, soil, debris, and organic and nonorganic waste. It gets its named from the hinged, shell-like bucket with an open and closing motions that resemble those of a clam. Figure 21.16 shows a hydraulic clamshell bucket at work.

FIGURE 21.16 A lattice boom crawler crane mounted on a barge conducts dredging operations with a clamshell.

A clamshell bucket's mechanisms mainly fall into two categories: mechanical and hydraulic. Clamshell attachments are normally used by conventional-type cranes with lattice booms, which are considered the standard for two-line cable operations needed for clamshell work and require a minimum of two hoist drums. Clamshells are also often used for dredging and other general material-handling purposes. Depending on the type of operation, a variety of six different clamshell buckets are available. With buckets made from high-strength steel, they offer the highest levels of wear and impact resistance.

The most common types of clamshell buckets include:

- Hydraulic clamshells
- Single-rope clamshells
- Electrohydraulic clamshells

- Rehandling clamshells
- Digging clamshells
- Ore clamshells

Clamshell Design

As mentioned earlier, a clamshell bucket consists of two "shells" that resemble a clam's shells. The bucket's hinged halves operate with an open-and-closing action to collect or release material. This action may be controlled either by hydraulic cylinders or by wire rope if used by a conventional crane. Figure 21.17 shows a clamshell bucket picking up another clamshell bucket.

FIGURE 21.17 A clamshell dredge coming up from the bottom of a harbor.

Applications of Clamshell Buckets

Clamshell bucket attachments for cranes are valuable tools in many industries for the handling of a wide variety of materials:

- **Construction.** Excavating, trenching, and managing construction debris
- **Mining.** Handling extracted minerals and ore
- **Dredging.** Removing sedimentary debris from waterways
- **Maritime shipping.** Loading and unloading bulk materials and cargo when shipping vessels are in ports and harbors
- **Waste management.** Handling and disposing of various waste materials
- **Power plants.** Managing materials such as coal and biomass

Safe Inspection of Clamshell Buckets

Whenever performing maintenance or inspecting a clamshell bucket, it is important to have the bucket in the closed position and sitting securely on the ground to prevent any accidental movement or reopening of the jaws during inspection. Figure 21.18 shows a crawler crane–clamshell operation.

FIGURE 21.18 A duty cycle crawler crane is performing clamshell work by the ocean.

If the clamshell bucket is hooked up to a hydraulic system, make sure that it is de-energized and that all the hydraulic pressure is relieved before inspecting the bucket. If there is any chance of the bucket moving while the inspection process is taking place, you should block and chock it to secure its position. A clamshell bucket is very heavy and can cause serious injury if it were to move unexpectedly. A stable clamshell bucket will provide for a more complete inspection of all its parts, including any hydraulic cylinders, bushings, pins, teeth, edges, or hinges.

Proper positioning makes full access to all parts of the clamshell bucket easy for inspection or any adjustments or repairs. Make sure that the bucket is nonoperational and placed on the ground and that it is 25 feet from any edges or declines, with the crane shut down. If inspections require working at heights, always wear the proper fall protection equipment. Areas to inspect include:

- Pins and bushings for wear and proper lubrication.
- Bucket teeth for wear. Rotating the position of these teeth from time to time will lengthen their service life.
- Mounting hinges for cracks or any other signs of damage.
- All leading edges or corners for thinning or wear.
- Bucket edges for signs of structural damage.

It is very important to consult the correct OSHA standards, such as CFR 1926.1400, for construction work. No single OSHA standard is dedicated solely to clamshell work in the general industry.

Several standards can apply, depending on the specific equipment used and the nature of the work being performed.

It's important to know that the entire project will operate around specific OSHA standards and regulations that depend on the type of equipment being used, the materials handled, and the surrounding work being performed. Always consult the specific OSHA standards and regulations relevant to your workplace and operations. Also refer to any applicable manufacturer's guidelines for the equipment in use. The ASME B30 safety standard series will also be a fantastic resource for additional information.

Always read the clamshell manufacturer's operational and shop manuals for more specific information regarding inspection schedules. By thoroughly inspecting the areas noted above, you will help locate problems before they turn into failures. That will help protect all those working around or near the clamshell operation, preventing jobsite injuries and saving the lives of your coworkers.

Dragline Operations and Cranes

A dragline is a duty cycle crane with an attached open dragline bucket designed to pick up, hold, and transfer large amounts of materials above and below water. The dragline bucket is sometimes heavily perforated. The bucket is suspended from the front by two chain lengths, as shown in Figure 21.19. Both chains are attached to the hoisting and lowering wire ropes of the crane. The dragline hoisting rope is routed over the boom head and operated by the first wire rope hoist drum. The dragline wire rope runs head-on through the base of the boom and is connected to the second wire rope hoist drum. Both hoist drums have free-fall capabilities.

FIGURE 21.19 Always know and follow the capacity of both the bucket and the crane.

The cranes used for duty cycle operations, such as draglines and clamshells, are built with superstructures and undercarriages strong enough to withstand the immense stresses of continuous, hard work. *Duty cycle* refers to the repetitive lifting, swinging, and digging cycles of work done by the crane. Dragline operations can be found in mining, quarrying, and material-handling operations for excavating materials like gravel, sand, coal, and clay. A dragline–duty cycle crane operates by lowering the bucket, which is suspended from the crane boom, via a hoist rope toward the ground in the area where the digging will soon start. The operator will begin to drag the bucket; the dragrope connected to the bucket is pulled, causing the bucket to scrape along the mass of material until it is filled. The hoist rope then lifts the bucket as the operator swings the boom until the load is positioned over the dumping area. The dragrope is then released as the operator tilts the bucket, emptying the load of material.

Conventional lattice boom crawler cranes are widely chosen for this type of work due to their robust durability. These cranes, especially crawlers, are built to withstand the highly dynamic range of stresses present while conducting duty cycle dragline operations. Most manufacturers require the charted load capacities of a crane to be derated by 20% while conducting dragline and clamshell operations, known as *cycle work*. The tracked undercarriages of crawler cranes also provide a stable platform when working on soft or uneven ground. Duty cycle cranes are equipped with free-fall winches that are fantastic for casting and retrieving the bucket being used in a controlled, productive fashion. Conventional lattice boom cranes can hoist heavy material capacities with a longer boom reach, making them perfect workhorses for material handling and similar types of work.

Hydraulic cranes are also incredibly powerful machines in their own right. However, they have limitations when performing these types of operations. The needs and demands of dragline or clamshell work, such as the need for continuous duty cycle operations, heavy shock loading, and heavy material movement, often exceed the limitations ofl hydraulic operating systems. See Figure 21.20, which shows a lattice boom crawler crane performing dragline operations.

FIGURE 21.20 A duty cycle crawler crane casts spoils from a dragline into a stockpile.

Conducting dragline work using a conventional crawler crane requires that intensive precautionary inspections and strict procedural guidelines be in place to help protect the crane operator and other ground support personnel.

Procedure for the Safe Inspections of Dragline Buckets

Steps in the pre-inspection process include:

1. Lower the bucket to the ground. The bucket should be placed on firm, level ground.
2. Set the bucket at least 25 feet from any edges or declines.
3. Make sure there is slack in the drag chains. This safety measure allows the bucket to sit freely.
4. Suspend the spreader bar above the bucket using its hoist chains.
5. Clear the dump rope so it is free of contact with the arch for safe accessibility during inspection.
6. Power off all controls to prevent any accidental movement of the bucket during the inspection process.
7. Ensure that lockout and tagout procedures are in place as specified in your company safety plan.

For the main inspection areas:

- Check the overall condition of the entire bucket for signs of damage, deformation, cracks, bends, and excessive wear.
- Check the bottom of the bucket for wear and integrity.
- Inspect the side panels for wear and tear, and verify the side panels' thickness.
- Check all welds for integrity, looking for cracks, rewelds, or any other defects that could compromise the bucket's structural integrity.
- Inspect the teeth of the bucket:
  - Check for any missing, loose, or worn teeth.
  - Look for any missing bolts that secure the teeth.
  - Conduct a visual scan to look for uneven wear patterns that may indicate improper alignment or usage.
  - Inspect for cracks, chips, or significant wear.
  - Ensure that the teeth are properly attached and tightened.
- Inspect the adapters to ensure they are securely fastened to the bucket, and check for signs of wear or damage around the locking mechanism.
- Check the bucket's cutting edge for signs of scalloping, wear, or damage.
- Inspect the drag chains for wear, corrosion, cracks, or damage that may weaken them.
- Check the tension of the drag chains.
- Inspect the wire rope strands for any broken wires, especially near the wedge and socket assembly.
- Check to make sure the dump ropes are free from contact with the arch.
- Inspect dump ropes for excessive roughness, looking for cuts, bulges, flat areas, bumps, discoloration, or lumps that may indicate internal damage.
- Inspect the spreader bar; carefully look for chains that might be bent, twisted, gouged, or pitted or are showing signs of any weld spatter or heat damage.
- Ensure that the WLL is clearly marked on all hardware associated with the bucket.

- Look for excessive wear of the hoist block or overhaul ball and any support shackles.
- Check for structural damage, cracks, or excessive wear of the spreader bar.
- Look for any loose or missing guards, fasteners, covers, stops, or manufacturer's nameplates.
- Check rigging hardware for wear, corrosion, or any other damage; make sure load capacity markings are clear and easy to read.
- Make sure the correct size and rated pins are installed on the shackles.
- Ensure that the boom and dipper arm are free of cracks, bends, dents, or other damage.
- Evaluate the hydraulic system's hoses for breaks or bulges while making sure the couplings are in good condition.
- Inspect the cylinders for leaks, dents, or scratches.
- Evaluate the fairlead, checking for cracked members or welds, damage in the swivel frame, or wear on the guide rollers; and ensure that the sheaves have clearance, allowing free rotation.
- Check that the automatic lube system is functioning properly. Lubrication must be present.
- Check all the lube lines to ensure they are in good condition

Always read the dragline's operation manual for specific inspection schedules. By properly inspecting these areas, you will help locate potential problems early, preventing failures. That will protect the safety of the supporting ground crew personnel working around or near the dragline operation.

Those Qualified to Conduct Inspections

Those qualified to conduct inspections include qualified persons and competent persons:

- **Qualified persons.** Qualified persons must perform all annual inspections. They must be familiar with crane design as well as any assembly or dismantling procedures. In some jurisdictions, this person must be certified by a nationally accredited organization.
- **Competent persons.** Competent persons make any daily to monthly inspections. They must be knowledgeable regarding all crane components and be able to identify any potential hazards and be authorized to take corrective actions.

Documentation

Full documentation, including the information required for monthly inspections listed in the manufacturer's operations manual, must be kept on file in the crane's log book. Follow OSHA when working in construction or MSHA (Mine Safety and Health Administration) when working in mining, as well as the state requirements for all safety standards or regulations that apply locally. All are used to make sure crane operators, riggers, and ground support personnel are safe when performing dragline work within their different jurisdictions. Always refer to, and comply with, applicable OSHA standards 29 CFR 1926.180 and the crane manufacturer's instructions for inspections and maintenance.

Mobile Cranes Working on Barges

There are several different cranes that can be used in marine construction, including floating cranes, crawler cranes, and hydraulic cranes. Barge-mounted cranes are much like floating cranes but are used for close-by tasks such as lifting piles, constructing piers, and installing coastal defenses. This type of crane is mounted on a barge that can move to specific work areas, sometimes requiring the assistance of a tugboat. Barge-mounted cranes require special attention to maintain balance and stability when lifting heavy loads over open water. Make sure to use the correct weight distribution to prevent accidents when hoisting loads over water. Anchoring systems or spuds that are hydraulically operated legs are used to stabilize the barge during work, reducing the chance of tipping. Crawler cranes work best when used for both onshore and nearshore marine construction for lifting piles, assembling components, and building bridges.

The tracks of crawler cranes help the cranes move over the soft, muddy ground often found at construction projects near water. Hydraulic cranes can move heavy loads with accuracy, making them great for work in marine environments where careful movements are required. These cranes can be useful for work on shore or on barges. Telescopic cranes work best in confined or hard-to-reach areas in marine construction, such as between bridge pylons or around structures where space is limited. Figure 21.21 shows a lattice boom crane working on a barge.

FIGURE 21.21 A barge crane is used to replace a railroad bridge.

Mobile cranes working on barges are important tools in many marine construction operations where land-based access is not available or is limited. Mobile cranes are a good solution for lifting and placing heavy loads on or near water. Spud barges are built to support mobile cranes, often featuring a boxed stern for maximum flotation. They use spud poles driven into the seabed to anchor the barge, providing good stability for pile driving and other associated crane work. Spud barges are specially constructed to support a working crane. A reinforced keel is generally provided under the crane working area by use of extra bulkheads, internal trusses, or deck frames. Wooden crane mats under the crane help distribute the concentrated load over a larger area and also provide traction for the crane when moving is necessary. The operation of the spuds is provided by the crane on the barge, or a spud winch can be installed as well.

Mooring is completed by a four-point mooring system that provides the stable working environment needed for crane work. Here are some considerations and challenges of cranes on barges to think about:

- **Stability.** Maintaining a barge's stability when lifting heavy or off-center loads is highly important to prevent capsizing or tip-overs. Factors like wind, waves, and tides can significantly affect stability. Ensuring correct load distribution and securing the crane on the barge are very important.
- **Reduced charted lifting capacity.** The lifting capacity of a land crane is usually reduced when used on a barge to account for the barge's stability and movement. Check the manufacturer's specifications for the crane being used. Typically, a 20% reduction of the crane's charted capacity is required.
- **Environmental factors.** Adverse weather conditions can hinder crane operations. Always be aware, and pay special attention to weather reports in the area.

Safety Standards, Regulations, and Inspections

Operating cranes on water provides big risks and requires specialized safety requirements. Good, strong safety protocols, visual inspections, and the use of personnel protective equipment is very important to prevent accidents.

- **Communication.** Clear communication between the crane operator, barge crew, and all other supporting personnel is highly important. Refer to the ASME B30.5 safety standards for more information.
- **Regulations.** The ASME B30.8 Floating Cranes and Floating Derricks standard applies to cranes and derricks that are mounted on barges or pontoons and are used for vertical lifting and lowering of suspended loads.

OSHA 29 CFR 1926.1437 covers the use floating cranes, derricks, land cranes, and derricks on barges in construction work.. Operating mobile cranes on barges introduces additional complexities and requires specific protocols beyond land-based crane inspections. Here are some key points to consider.

- Cranes on barges are subjected to frequent, periodic, and annual inspections, the same as land cranes are. However, cranes used for shipyard employment require thorough examinations by OSHA-accredited persons both initially and then every 12 months. Retesting should occur every four years.

- Barge vessel inspections are subjected to shift inspections, monthly inspections, and annual inspections by a qualified person with expertise in vessels. Flotation devices should be used to inspect the external part of the barge or vessel for these problems:
  - Significant wear or corrosion or deformation of cleats, bits, chokes, fenders, capstans, ladders, and sanctions
  - Any external evidence of leaks or structural damage
- Barge vessel inspections should include four-corner draft readings, measuring the depth of the barge's hull at each corner while in the water. This process is typically done by a qualified surveyor.
- Firefighting equipment should be inspected to ensure it's functioning and in good condition.
- Every four years, a marine engineer, marine architect, licensed surveyor, or other qualified person must survey the internal portion of the barge, pontoons, vessels, or means of flotation.

Stability and Load Requirements

Key points to consider:

- The barge flotation device must be able to sustain changing loading conditions without excessive heeling or turning over, maintaining stability.
- When lifting a load onto, or from, a water vessel, weight must be distributed evenly across its deck, both laterally and vertically. Uneven loading creates instability, so the center of gravity must be as low as possible.
- For operations involving cranes with heavy loads, a stability assessment from an accredited marine surveyor may be necessary in order to ensure safety.
- The rated capacity of land cranes on flotation devices must be reduced to account for instability factors such as list, trim, wave action, and wind.

Documentation

Key points to consider:

- Documentation of the required inspections, including monthly, annually, and quadrennially, must be kept on file for at least four years. Keep documentation up-to-date, and include any manifests or safety certifications.
- Documentation should include the items inspected, the results, the name of the person who's conducted the inspection, and the date on which the inspection occurred. Figure 21.22 shows an inspector carefully looking over a crane.
- Have a comprehensive emergency response plan in place that includes evaluations and procedures for spills or accidents. All personnel should be educated and trained on these plans.

Make sure that all personnel involved in the operation are qualified, trained, and certified. Mobile cranes on barges offer a versatile and valuable solution for many marine and construction projects. However, good planning, following strict safety requirements and a well-rounded understanding of the special challenges posed by marine environments, is essential. Safety is paramount when working on barges; the dynamic dangers present are extremely hazardous.

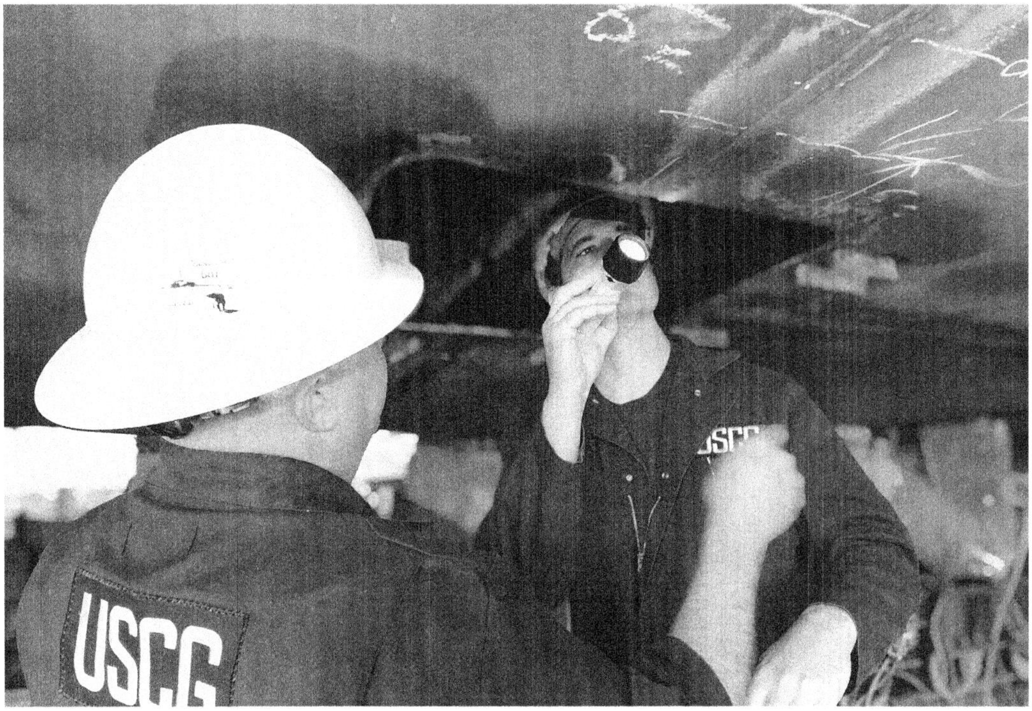

FIGURE 21.22 It is important for barge inspectors to look over the structural integrity of the hull and check for signs of corrosion.

Chapter 21 Review Quiz

1. What ASME safety standard pertains to the use of come-alongs for pulling or lifting loads?
   a. ASME B30.21
   b. ASME B30.22
   c. ASME B30.25
   d. ASME B30.26

2. What ASME safety standard pertains to the use of turnbuckles for tightening down loads?
   a. ASME B30.21
   b. ASME B30.22
   c. ASME B30.25
   d. ASME B30.26

3. What is the typical safety factor for come-alongs, cable pullers, and hand winches when lifting loads?
   a. 1:1 to 2:1
   b. 2:1 to 3:1
   c. 3:1 to 4:1
   d. 4:1 to 5:1

4. What is the typical safety factor for come-alongs, cable pullers, and hand winches when pulling loads?
   a. 1:1 to 3:1
   b. 2:1 to 5:1
   c. 3:1 to 4:1
   d. 5:1 to 6:1

5. The OSHA safety standard that pertains to turnbuckles is
   a. 1926.550
   b. 1926.1400
   c. 1926.251
   d. 1926.1411

6. Why is it important to inspect the come-along's cable, hooks, and chains before every use?
   a. To ensure the unit is intact and free of rust
   b. To make sure all labels are properly placed and display the correct WWL
   c. To identify wear, deformation, kinks, or cracks that could lead to a sudden failure while under load
   d. To properly lubricate the internal ratchet mechanism with a quality lubricant

7. What ASME safety standard pertains to the safe use of come-alongs, lever hoists, and chain lever hoists?
   a. ASME B30.26
   b. ASME B30.21
   c. ASME B30.10
   d. ASME B30.12

8. Manufacturers should design turnbuckles with a minimum safety factor of
   a. two times the working load limit
   b. three times the working load limit
   c. four times the working load limit
   d. five times the working load limit

9. What ASME safety standard pertains to the use of jacks?
   a. ASME B30.1
   b. OSHA 1926.1400
   c. OSHA 1926.1305
   d. There isn't a single regulation or standard titled "Jack and Slide System Safety."

10. Many mobile crane manufacturers require the crane operator to derate the load chart when conducting duty cycle operations. The amount needed to derate the chart is
   a. 20%
   b. 25%
   c. 30%
   d. 35%

11. The primary OSHA regulation that is titled Floating Cranes, Derricks, and Cranes on Barges is
   a. 29 CFR 1926.1400
   b. 29 CFR 1926.1404
   c. 29 CFR 1926.1406
   d. 29 CFR 1926.1437

12. The primary ASME safety standard covering mobile land cranes on barges is
   a. ASME B30.8
   b. ASME B30.3
   c. ASME B30.5
   d. ASME B30.6

13. The ASME standard that applies to clamshell work with a mobile crane is
   a. ASME B30.5
   b. ASME B30.8
   c. ASME B30.13
   d. ASME B30.20

14. The ASME standard that applies to the clamshell bucket itself that is used for clamshell work is

a. ASME B30.5
b. ASME B30.8
c. ASME B30.13
d. ASME B30.20

15. What is the hoisting task called when a vessel is lifted from a horizontal load to a vertical position, much like flipping a load from one orientation to another?

a. Tailing
b. Shadowing
c. Topping
d. Following

16. Temporary trunnions are specialized rigging attachments that are used to provide

a. streamlined rigging processes when lifting vertically
b. secure hoisting or lifting points on loads that have none
c. engineered locations along the length of an object
d. pivot points to ensure a stable lift when hoisting horizontally

17. Why are conventional lattice boom crawler cranes widely chosen for dragline work?

a. They have long, exceptionally strong booms.
b. They have torque converters that work well for quick starts and slow stops.
c. They have crawlers that are better suited for getting the crane level.
d. They have heavy-duty construction to withstand high dynamic stresses that come with duty cycle dragline operations.

18. Hydraulic pile drivers used in pile driving operations use a combination of static weight and high frequency. They also use which of the following to move piles into the ground?

a. Thump
b. Oscillation
c. Pulsation
d. Vibration

19. A top safety concern when operating land cranes on barges is

a. the center of gravity
b. proper mooring
c. maintaining stability
d. hardwood timber mats to sit on

20. A mobile land crane when working on water cannot exceed a minimum wind speed of

a. 10 miles per hour
b. 20 miles per hour
c. 30 miles per hour
d. 40 miles per hour

21. When a mobile land crane is working on a barge, what type of indicator must be clearly visible to the crane operator based on the crane's rated capacity to ensure the barge deck remains out of the water?

a. Boom angle indicator
b. Boom length indicator
c. Anti-two-block indicator
d. List or trim indicator

22. When inspecting a dragline or clamshell bucket, be sure the equipment is on level ground and kept away from any edges by at least

a. 15 feet
b. 20 feet
c. 25 feet
d. 30 feet

Summary—and Then Some

Using specialized tools and equipment in mobile crane operations provides some major advantages in safety, efficiency, and overall project quality. This specialized equipment is also imperative for the protection not only of workers, but of the load and surrounding jobsite as well. It allows crane operators to hoist complex, heavy loads with a level of precision and control that would not be possible with the standard equipment available to the industry alone. Reducing setup time while streamlining complicated hoisting and rigging tasks, along with utilizing specialized equipment, results in significant reductions in completion time and cost savings. Emerging technology in specialized hoisting and rigging equipment focuses on advancements in automation, sustainability, and data-driven intelligence to increase safety, efficiency, and quality of product.

With trending ideas, such as artificial intelligence, advanced robotics, eco-friendly power sources, and immersive training, finishing large construction projects takes all the guesswork out of making heavy lifts. But there is still no room for taking chances. The best way to support the crane, and its operator, is to use qualified workers who know how to correctly use specialized equipment, which can take a lot of the anxiety out of executing heavy lifts. Using specialized equipment with qualified ground support and a skilled crane operator makes a shift run safely, smoothly, and productively, while at the same time inspiring confidence among the entire rigging team and ground personnel.

It all goes well until the lifting team becomes complacent about the crane inspections required by OSHA. This includes inspection of all specialized equipment, support machinery, rigging hardware, and slings. Failure to make these inspections can lead to faulty components, structural weakness, or malfunctioning safety devices that could cause catastrophic failure of the crane's components or of specialized equipment. These types of failures can cause disastrous accidents like boom failures, dropped loads, turnovers, and much more. Unfortunately, these types of accidents can lead to serious injury and fatalities of ground support personnel and the innocent public. The traumatic stress of these events causes immense physical, emotional, and financial suffering for the victims and their families. In some cases, serious accidents can be followed by serious criminal penalties for those responsible for the safety violations that caused the accident. Serious or repeated violations can result in the suspension or revocation of a company's certification to operate cranes.

By prioritizing maintenance and inspections, a crane company shows its commitment to the safety of workers, improving morale and building a strong culture of safety among all personnel. Preventive maintenance programs anticipate and address potential problems before they lead to unexpected breakdowns. Making frequent inspections helps spot the minor issues—such as worn wire ropes, hydraulic leaks, or faulty brakes—before they lead to major breakdowns or catastrophic accidents. Thoroughly inspect the crane by examining its structural components, such as booms and jibs, for cracks, corrosion, and other damage that may compromise safety. A company with a verifiable track record of strong safety procedures and equipment maintenance can often qualify for lower insurance premiums. It is your job to never become complacent in your duty to complete all required inspections diligently, so that everyone involved goes home happy and safe.

A Crane Terminology and Definitions

This chapter presents terms and definitions that you should know and serves as a reference if you come across a term you don't know and want to look up.

TERM	DEFINITION
Abrasion	Functional surface wear on wire rope.
Accessory	A secondary part or assembly of parts that contributes to the overall function and usefulness of the equipment.
Angle indicator	An accessory that measures the angle between the boom base and the horizontal.
Anti-two-block device	A device that, when activated, disengages all equipment functions whose movement can cause two-blocking.
Auxiliary hoist	A secondary mechanism used either in conjunction with, or independently of, the main hoist.
Axis of rotation	The vertical axis around which the crane's superstructure rotates; also called the *center of rotation* or *swing axis*.
Axle	The shaft or spindle with which or about a wheel rotates. It refers to a type of axle assembly including housings, gearing, differential, bearings, and mounting appurtenances.
Back-hitch gantry	A structure, fixed or adjustable in height, forming part of the upper structure of a crane, to which the lower spreader (carrying the live boom suspension ropes) is anchored; also called the *A-frame* or *gantry*.
Backstay cable	The cable running from the jib mast to the main boom. Used to maintain the jib's offset to the boom.
Backward stability	The ability of a crane to resist overturning in the direction opposite the boom while in a loaded or unloaded condition.
Ballast	Weight used to supplement the weight of the machine in providing stability for lifting loads (the term *ballast* is normally associated with locomotive cranes).
Barrel	The body part of a rope drum hoist.
Base	The traveling base, or carrier, on which the rotating superstructure is mounted. It transmits loads to the ground or other supporting surface.
Base (butt) section	The first section of boom closest to the crane draw works, beginning at the boom base pins. The lowest section of a telescopic boom; it does not telescope but contains the pin mountings of the boom foot and the upper end mountings of the boom hoist cylinder.
Basic boom	The minimum length of a sectional lattice boom that can be mounted and operated, usually consisting of a boom base and tip section only.
Becket	An end attachment permitting wire rope installation; sometimes referred to as a *wedge and socket*.
Birdcage	The look of wire rope linearly compressed, with the outer strands forming a cage and at times displacing the core.
Block	Term applied to wire rope sheaves (pulleys) enclosed in aside plates and fitted with some attachments for lifting a load such as a hook or shackle.
Boom (crane)	A member hinged to the rotation superstructure and used for supporting the hoisting tackle; also called the *jib* (European).
Boom angle	The angle above or below the horizontal of the longitudinal axis base boom section.
Boom angle indicator	An accessory that measures the angle of the boom to the horizontal.
Boom base	The lowest section of a sectional lattice boom having the attachment or boom foot pins mounted at its lower end; also called the *boom butt* or *butt section*.
Boom deflection	The bend in the boom due mainly to boom-tip loading, causing compression and tension.

TERM	DEFINITION
Boom guy line	A fixed-length rope forming part of the boom suspension system; also called the *hog line*, *boom stay*, *standing line*, *stay rope*, or *pendant*.
Boom head (tip) section	The last section of a boom containing the head sheaves.
Boom hoist	The rope drum(s), drives(s), and reeving controlling the derricking motion of the boom; also called the *topping lift*.
Boom hoist cylinder	The hydraulic ram used instead of a rope boom suspension; the most common means of derricking (raising or lowering) telescopic booms.
Boom hoist mechanism	Means for supporting the boom and controlling the boom angle.
Boom inserts	Center sections of a sectional lattice boom.
Boom length indicator	A device providing the crane and/or operator with boom length information. Typically for telescopic booms.
Boom point	The outer extremity of the crane boom, containing the hoist sheave assembly.
Boom point sheave assembly	An assembly of sheaves and pins built as an integral part of the boom point.
Boom stop	A device (mechanical or electrical) used to limit the angle of the boom at the highest recommended position.
Boom suspension	A system of ropes and fittings, either fixed or variable in length, that supports the boom and controls the boom angle.
Brake	A device (other than a motor) used for retarding or stopping motion by friction or power means.
Breaking strength	The load required to cause tensile failure in tested samples, as with wire rope.
Bridle	A frame, forming part of the boom suspension, supporting sheaves for the live suspension ropes and attached to the fixed suspension ropes (pendants); also called a *spreader*, *upper spreader*, *live spreader*, *spreader bar*, or *floating harness*.
Bridle sling	A multi-leg wire rope, fabric, or chain sling arrangement.
Cab	A housing that covers the rotating superstructure machinery or the operator's cab or driver's station.
Cabling	Hook block twisting beyond ½ revolution, 180 degrees, relative to the head sheaves.
Cheek weights	Overhauling weights attached to the side plates of a lower load block.
Choker	A sling that may be formed as a slip noose around an object to be lifted or moved.
Clearance	The minimum distance from any part of the crane to the point of the nearest obstruction.
Clevis (shackle)	A U-shaped rigging device with a pin across the open end to connect a load to a pulling device.
Clutch	A means for engagement or disengagement of power.
Commercial truck vehicle	A commercial motor vehicle designed primarily for the transportation of property in connection with business and industry.
Competent person	A person who is capable of identifying existing and predictable hazards in surroundings or working conditions that are unsanitary, hazardous, or dangerous to employees, and who has authority to take prompt, corrective measures to eliminate them.

TERM	DEFINITION
Controlling entity	A "prime" contractor, general contractor, construction manager, or other legal entity that has the overall responsibility for the construction of the project.
Counterweight	The weight used to supplement the weight of the machine in providing stability for lifting working loads.
Crane	A power-operated machine capable of lifting or lowering loads and moving them horizontally, which uses wire rope. The hoisting mechanism is an integral part of the machine.
Crane carrier	The undercarriage of a wheel-mounted crane specifically designed for transporting the rotating crane superstructure. It may or may not provide its own travel mechanism. It is distinguished from a commercial truck vehicle in that it is not designed to transport personnel, materials, or equipment other than the crane rotating superstructure.
Crane configuration	The post-assembly arrangement of the crane including the boom, supporting base, counterweight(s), jib(s), extension(s), and attachments.
Critical lift	A hoisting or lifting operation that has been determined to present an increased level of risk beyond normal lifting activities. For example, increased risk may relate to personnel injury, damage to property, interruption of plant production, delays in schedule, release of hazards to the environment, or other jobsite factors.
Crossover points	In multiple-layer spooling of rope on a drum, those points or rope contact where the rope crosses the preceding rope layer.
Dead end	The point of fastening of one rope end in a running rope system, the other (live) end being fastened to the hoist drum.
Dedicated spotter	A qualified signal person who is in continuous contact with the operator and whose sole responsibility is to verify that the specified clearance is maintained.
Dog	A mechanical device used with a ratchet wheel built into one flange (side) of a hoist drum to lock the drum, preventing it from rotating backward.
Drifting	Pulling a suspended load laterally to change its horizontal position.
Drum	The cylindrical member around which a rope is wound for lifting and lowering loads.
Dynamic test	An operational test with the machine under full load and operating all motions in turn.
Ensure	A term used when the meaning "take steps to see that" or "make sure" is intended.
Equalizer sheave	The sheave at the center of a rope system over which no rope movement, other than equalizing, occurs.
Extension cylinder	A hydraulic ram used to extend, or retract, a section(s) of a telescopic boom.
Fair lead	A guide mounted in front of a drum, through which runs a drag cable, for cranes configured as draglines.
Fall	A number of running ropes supporting a load; also called *parts of line*.
Fatigue (wire rope)	A progressive fracture process due to the bending of individual wires, hurried by lubrication shortage.
Flange point	The point of contact between the rope and the drum flange where the rope changes layers on a hoist drum.
Fleet angle	The angle the rope, leading onto a hoist drum, makes with the line perpendicular to the drum's rotating axis when the rope is making a wrap against a flange.

TERM	DEFINITION
Fleeting sheave	A sheave mounted on a shaft parallel to the hoist drum shaft so arranged that it can slide laterally as the rope spools to help promote proper spooling.
Fly section	On a telescopic boom, the outermost powered telescoping section.
Forestay cable	The wire rope connecting the jib tip to the jib mast.
Forward stability	The ability of the crane to resist overturning in the direction of the boom point while in a loaded or unloaded condition.
Free fall	Lowering the hook (load) or boom by gravity where the lowering speed is controlled only by a retarding device such as a brake.
Gantry (A-frame)	A structural frame, extending above the superstructure, to which the boom support ropes are reeved.
Gate block	A single- or double-sheave block so arranged that one or more of the cheek plates can be opened, permitting the block to be reeved; also called a *snatch block*.
Grade	Wire rope or strand classification by tensile strength and/or steel type.
Gradeability	The slope that a machine can climb, expressed as a percentage.
Guy rope (line)	A fixed-length supporting rope intended to maintain a nominally fixed distance between the two points of attachment; also called a *stay rope* or *pendant*.
Hand line	A line (usually fiber rope) tied to a suspended load to control or limit its rotation tendencies.
Headache ball	A weighted hook for single-line service.
Hog line	A fixed-length rope forming part of the boom suspension system; also called a *standing line* or *pendant line*.
Hoist	To lift an object; the mechanism used for lifting.
Hook radius	The maximum reach of the crane hook from the crane's center of rotation.
Horse	A short strut or frame mounted on the boom head to provide the means for attaching the jib support rope; also called a *jib strut*, *rooster*, *jib mast*, or *A-frame*.
Independent wire rope core (IWRC)	A wire rope used as the core around which the main strands are laid.
Jib	An extension attached to the boom point to provide added boom length for lifting specified loads. The jib may be in line with the boom or offset to various angles in the vertical plane of the boom.
Jib angle	The angle of the jib relative to the horizontal.
Jib backstop	A device that will restrain the jib from turning over backward.
Jib offset	An included angle (in degrees) between a line drawn as an extension of the main boom's centerline and the jib's centerline.
Jib point	The outer extremity of the crane jib, containing the hoist sheave assembly.
Jobsite	Area within the boundaries of the project where the equipment is to be operated.
Lagging	Removable shells (optional) for use on a hoist rope drum to produce a change in line pull or line speed.
Lang lay rope	Wire rope in which the lay of the wires in the strand is in the same direction as the lay of the strand in the rope.

TERM	DEFINITION
Lattice boom	A boom constructed of four longitudinal corner members, called *chords*, assembled with transverse and/or diagonal members, called *lacings*, to form a trusswork in two directions. The chords carry the axial boom forces and bending moments, while the lacings resist the shears.
Layer	A series of wraps of wire rope around the hoist drum barrel, extending full from flange to flange.
Lead line	The primary rope line from the power source or hoist drum to the first sheave in a block and tackle arrangement or crane boom head.
Line pull (available)	The maximum pulling force attainable in a rope leading off a hoist drum or lagging at a particular pitch diameter drum, for first layer.
Line pull (permissible)	A line pull restricted by rope strength and other mechanical limitations along the hoist train. The force also being a function of the number of parts of line.
Line speed	The maximum speed attainable in a rope leading off a hoist drum or lagging of a particular pitch diameter, or as a function of the number of parts of line.
Live boom	A boom that is operated by a hoist mechanism that will unwind without restrictions.
Live mast	A gantry mounted on the front of a crane, which changes angle relative to the boom.
Load block	The assembly of hook, swivel, bearings, sheaves, pins, and frame suspended from the hoisting rope(s).
Load hoist	A hoist drum and rope reeving system used for lifting and lowering loads.
Load radius	The horizontal distance from the center (axis) of rotation of the crane to the center of gravity of a suspended load.
Load ratings	Crane ratings in pounds (kilograms) established by the manufacturer in accordance with ASME B30.5.
Loading	An external action that causes forces in members of a structure. It may be in the form of a direct physical action (i.e., weight put on the structure or pressure applied by the wind or motion).
Lower load block	The assembly of hook or shackle, swivel sheaves, pins, and frame suspended by hoisting rope.
Lower spreader	A frame, forming part of the boom suspension supporting sheaves for the suspension ropes and attached to the gantry or upper structure.
Luffing	Changing the boom angle by varying the length of the boom suspension ropes or hydraulic boom hoist cylinder; also called *derricking, booming up or down,* or *topping*.
Luffing attachment	A front-end attachment for a mobile crane that uses an upper working boom or jib, which is capable of changing angle during operation and is mounted on top of the lower boom.
Machine resisting moment	The moment of the deadweight of the crane, less the boom weight, about the tipping axis (fulcrum); hence the moment that resists overturning.
Manual insert	An optional nonpowered section in a telescopic boom that forms the outermost boom section when provided and is usable either fully extended or fully retracted.
Manufacturer's recommendations	The recommendations supplied by the original equipment manufacturer for the application of the given parts or components.

TERM	DEFINITION
Mast (boom)	A frame hinged at or near the boom hinge for use in connection with supporting a boom. The head of the mast is usually supported and raised or lowered by the boom hoist ropes.
Mast (jib)	A frame hinged at or near the boom point for use in connection with supporting the jib.
Maximum horizontal reach	The maximum horizontal distance the crane configuration can reach during load lifting or boom raising/lowering operations.
Midpoint suspension	An additional set of boom suspension lines attached to the main boom at some point between the main suspension attachment and the boom foot.
Mid section(s)	On a telescopic boom, the intermediate powered telescoping section(s), inner and outer, mounted between the base and head, or fly, sections.
Minimum breaking force	The minimum load at which a new and unused wire rope will break when loaded to destruction in direct tension.
Moment	A term used to measure the tendency to produce motion about a point of axis (fulcrum). Measurements are usually given in foot-pounds or inch-pounds (newton-meters).
Mounting base	The traveling base on which the rotating superstructure is mounted.
Multi-crane lifting	A team of cranes lifting a single object.
Multiple load line operation	The simultaneous use of two or more lines reeved over sheaves on a single shaft or multiple shafts of a crane with multiple load drums to lift, rotate, or hold a single load.
Normal operational conditions	Conditions during which equipment is performing functions within the manufacturer's operating recommendations. Under these conditions, the operator is at an operator control station described in the instructions for the equipment; no other persons, except those designated, are to be on the equipment.
Operational aid	An accessory that provides information to facilitate the operation of a crane or that takes control of particular functions without action of the operator when a limiting condition is sensed. Examples of such devices include, but are not limited to, the following: anti-two-block device, rated capacity indicator, rated capacity (load) limiter, boom angle or radius indicator, lattice boom, hoist disconnect device, boom length indicator, crane level indicator, drum rotation indicator, load indicator, and wind speed indicator.
Original language(s)	Language(s) used by the manufacturer to develop product instructions and manual(s).
Outriggers	Extendable or fixed members attached to the mounting base, which rest on supports at the outer ends used to support the crane.
Overhauling weight	Weight added to a hoist rope to overcome resistance and permit unspooling at the hoist drum when no live load is being applied; also called the *headache ball.*
Overturning moment	The tendency for a machine to overturn.
Pads	The plates that transmit the crane weight to the ground.
Parts of line (falls)	The number of times a continuous rope runs between the boom or jib head and the hook or hook block.
Pawl (dog)	A device that engages a ratchet to prevent rotation.
Paying out	Giving slack to a line or relieving load on a line by letting (spooling) out rope.
Payload	The load or loads being transported by the commercial truck chassis from place to place.

TERM	DEFINITION
Pendant	A rope or strand of specified length with fixed end connections.
Pin-connected boom	Crane boom sections that are connected with pins rather than bolts.
Pitch diameter	The diameter of a sheave or hoist drum measured at the centerline of the rope.
Power-controlled lowering	A system or device in the power train, other than the load hoist brake, that can control the lowering rate of speed of the load hoist mechanism.
Pre-forming	A process where each wire and strand is permanently formed to the helical shape it will assume in the finished rope.
Preventer	In rigging practice, a means, usually consisting of, but not limited to, a wire rope, for preventing an unwanted movement or occurrence by acting as a saving device when an anchorage fails.
Quadrants of operation	A crane's prescribed working arc relevant to its center of rotation.
Qualified person	A person who, by possession of a recognized degree in an applicable field or certificate of professional standing, or who, by extensive knowledge, training, and experience, has successfully demonstrated the ability to solve or resolve problems relating to the subject matter and work.
Radius	The distance from the axis of rotation of the superstructure to the center of gravity of the load.
Rail clamp	A tong-like metal device mounted on a locomotive crane car, which can be connected to the track.
Range diagram	A diagram showing an elevation view of a crane with circular arcs marked off to show the luffing path of the tip(s) for all boom and jib lengths and radial lines marking boom angles. A vertical scale indicates height above the ground, while a horizontal scale is marked with the operating radii. The diagram can be used to determine lift heights, clearance of the load from the boom, and clearance for lifts over obstructions.
Rated capacity	The maximum allowable load the equipment can lift at any given radius.
Rated capacity indicator	A device that automatically monitors the radius, load weight, and load rating and warns the crane operator of an overload condition.
Rated capacity (load) limiter	A device that automatically monitors radius, load weight, and load rating and prevents movements of the crane, which would result in an overload condition.
Reeving	A rope system in which the rope travels around drums and sheaves to provide a mechanical advantage for a hoist.
Reeving diagram	A diagram showing the path of the rope through a system of sheaves or blocks.
Repetitive pick-up point	When operating on a short cycle operation, the rope being used on a single layer and being spooled repetitively over a short portion of the drum.
Rooster	A common, everyday term used for one or more struts at the top of a boom and mast, such as a jib strut or a tower crane top tower.
Root diameter	The diameter of a sheave or grooved hoist drum measured at the base of the groove, or the diameter of a smooth barrel on a hoist drum; also called the *tread diameter*.
Rope	Refers to wire rope unless otherwise specified.
Rope drum	Part of a hoist that consists of a rotating cylinder with side flanges on which hoisting rope is spooled or wrapped; also called a *hoist drum*.

TERM	DEFINITION
Rotation-resistant wire rope	Stranded wire rope consisting of at least two layers of strands where the outer layer of strands is laid opposite to the underlying layer. The design results in a reduction in load-induced torque.
Running rope	A rope that travels around sheaves or drums.
Safe working load (SWL)	The safe working load = catalog strength over the factor or safety, i.e., rigging ropes.
Safety devices	Any electrical or mechanical device fitted to a crane whose prime function is to automatically prevent malfunctions and/or mishaps due to potential overloading.
Shall	A word indicating a requirement.
Sheave	A wheel or pulley with a circumferential groove designed for a particular size of rope; used to change direction of a running line.
Should	A word indicating a recommendation.
Side loading	A loading applied at any angle to the vertical plane of the boom, caused by wind, sudden starting or stopping of the swing motion, or dragging of an unsuspended load.
Signals	Audible or visual hand communication to direct the crane.
Slewing	A crane function where the boom or load-supporting member rotates about a vertical axis (axis of rotation); also called *swing*.
Spread bar	In rigging, a device used in the rigging of a load with multi-leg slings to prevent the slings from squeezing or crushing the load.
Stabilizers	Extendable or fixed members that are attached to the mounting base to increase the stability of the crane, but that may not have the capability of relieving all the weight from the wheels or tracks.
Standby crane	A crane not in regular service that is used occasionally or intermittently as required.
Standing rope	A supporting rope that maintains a constant distance between the points of attachment to the two components connected by the rope.
Strand	A group of individual wires helically wound together to form a part of a wire rope's construction.
Structural competence	The ability of the equipment and its components to withstand the stresses imposed by applied and dynamic loads.
Superstructure	The rotating upper frame structure of the machine and the operating machinery mounted thereon.
Swage fitting	A 100% efficient end fitting into which wire rope is inserted, then permanently cold-pressed (swaged).
Swing	Rotation of the superstructure for movement of loads in a horizontal direction about the axis of rotation.
Swing (house) lock	A mechanical device (latch/pin) to lock the superstructure to the carrier, preventing swing motion.
Swing mechanism	The machinery involved in providing horizontal rotation of the superstructure.
Swivel	A load-carrying member with thrust bearing to permit rotation under load in a plane perpendicular to the direction of the load.
Tackle	An assembly or ropes and sheaves arranged for lifting, lowering, or pulling.
Tag line (winder)	Spring-loaded, boom-fitted reel, from which a cable is attached to a clamshell bucket minimizing spinning.

TERM	DEFINITION
Tail swing	Clearance distance from the center of rotation to maximize the rear extension of the revolving superstructure.
Tailing crane	In a multiple-crane operation in which a long object is being erected from a horizontal starting position to a vertical position, the crane controlling the base end of the object.
Taking up	The process of removing slack from a line or drawing in.
Telescoping boom	A boom from which one or more boom extensions are telescoped to varying lengths.
Tipping	A point in the overturning process when there is no safe return.
Transit	The moving or transporting of equipment from one jobsite to another.
Travel	The function of the equipment moving under its own power from one location to another on a jobsite.
Two-block damage prevention feature	A system that will stall when two-blocking occurs without causing damage to the hoist rope or crane machinery components.
Two-block warning feature	A warning device to alert the operator of an impending two-blocking condition.
Two-blocking	The condition in which the lower load block or hook assembly comes in contact with the upper load block or boom point sheave assembly.
Upper load block	The assembly of shackle, swivel, sheaves, pins, and frame suspended from the boom point.
Wedge socket	A wire rope end attachment where a wedge within a rope loop is seated into a socket (becket) or winch drum.
Whip line (runner or auxiliary)	A secondary rope system usually of lighter load capacity than that provided by the main rope system.
Winch	A hoisting mechanism made up of one or more hoist drums. Can be operated by air, hydraulics, or mechanical friction controls; also called a *hoist*, *drum hoist*, or *hoisting mechanism*.
Winch head	A power-driven spool for handling loads by means of friction between fiber or wire rope and the spool.
Wind moment	The tendency for a machine to overturn as a result of excessive wind force.
Working load	The external weight applied to the equipment, including the weight of the attaching equipment such as the lower load block, shackle, and slings.
Wrap	One full revolution of a wire rope around a hoist drum.
Working load limit (WLL)	The maximum allowable force or weight that lifting equipment such as slings, shackles, and hooks can safely handle under normal service conditions, as defined by the manufacturer. It incorporates a safety factor, often 4:1 to 7:1, to account for variables such as like shock localing or abrasion.

B Mobile Crane Nomenclature and Identification

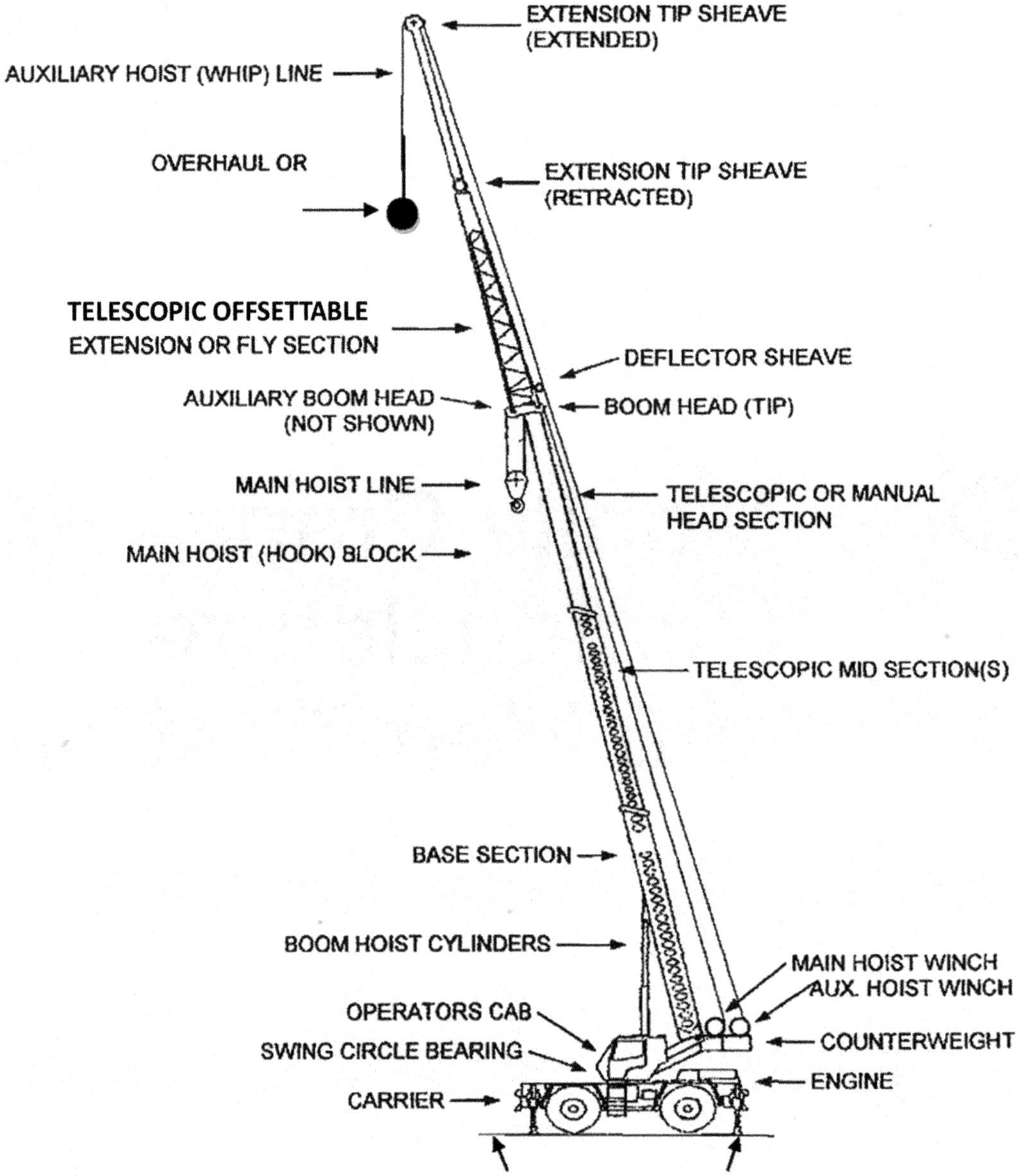

FIGURE B.1 Offsettable typical wheel-mounted rough-terrain crane.

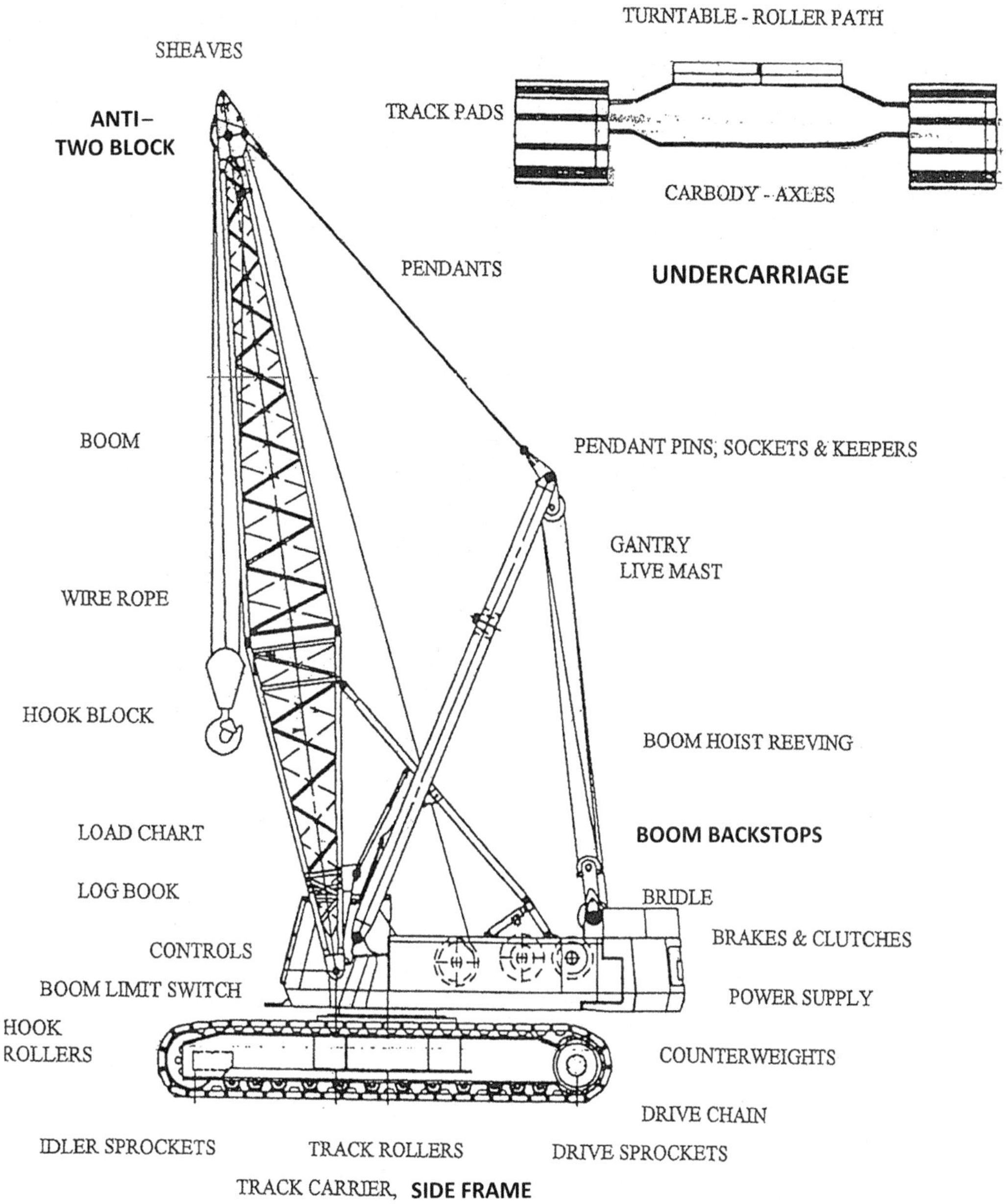

FIGURE B.2 Typical crawler-mounted lattice boom crane.

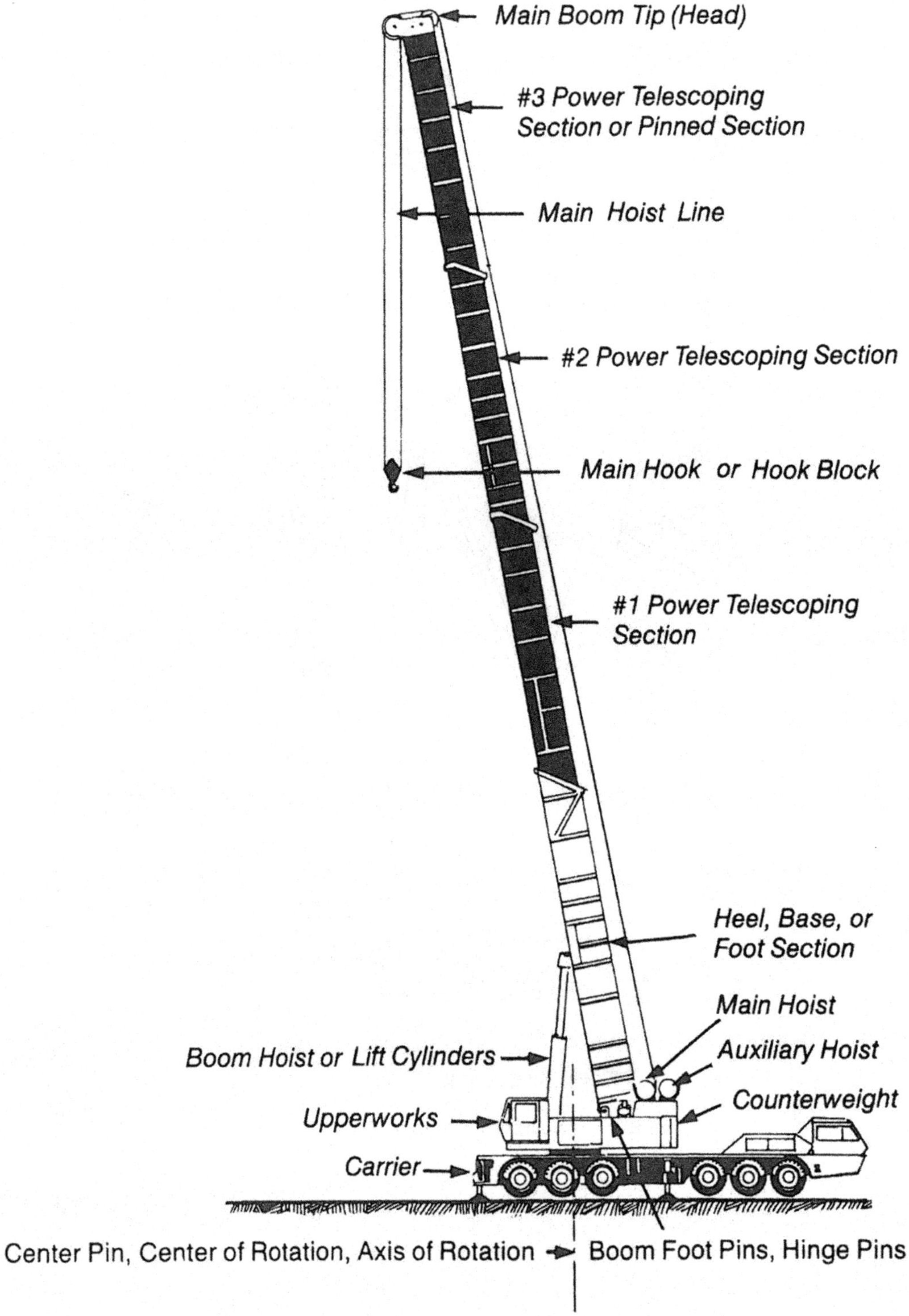

FIGURE B.3 Carrier-mounted telescopic boom cranes.

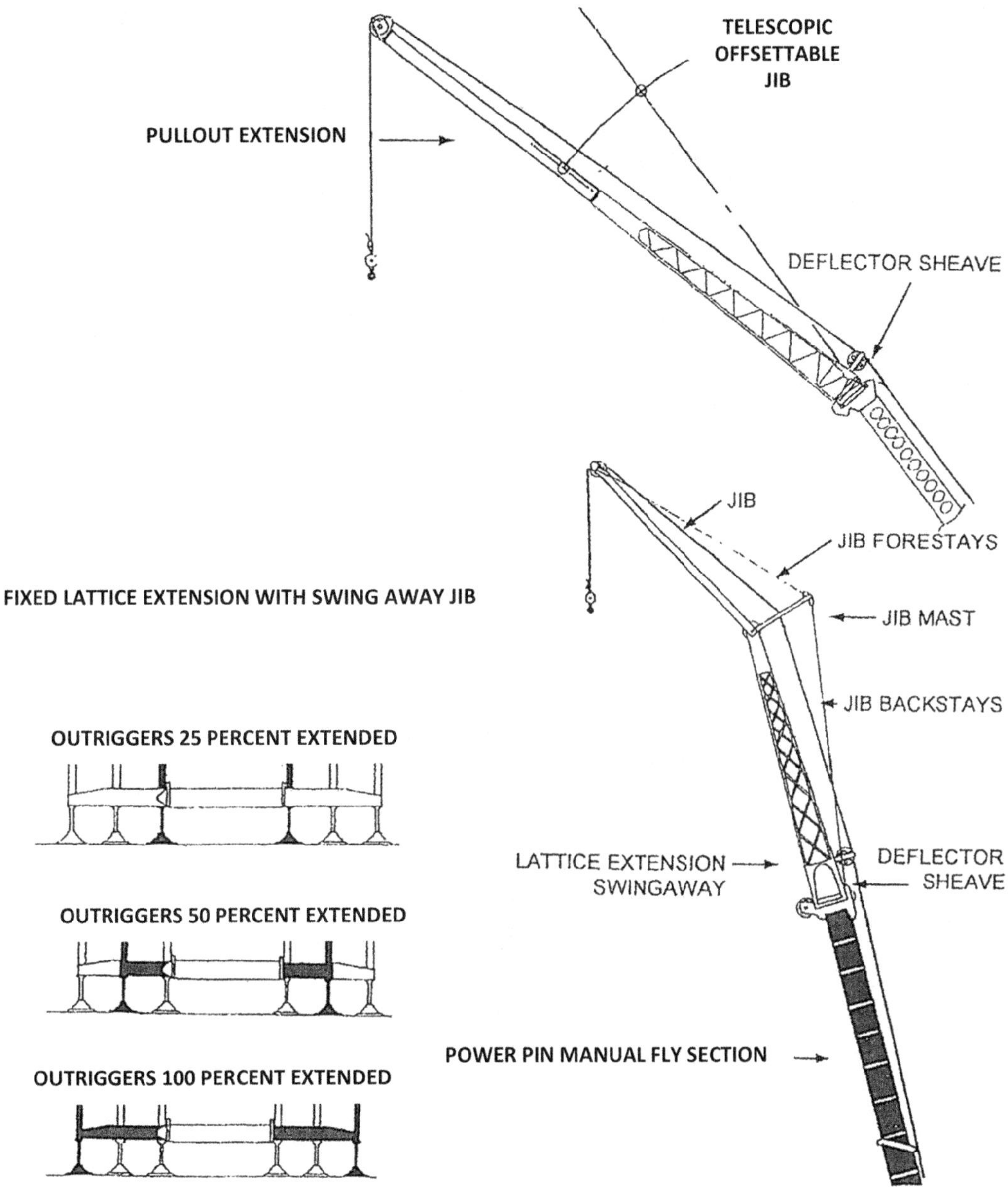

FIGURE B.4 A partial and intermediate position view. On top is a telescopic offsettable jib and below is a look at a power pin manual fly section with a lattice extension swingaway.

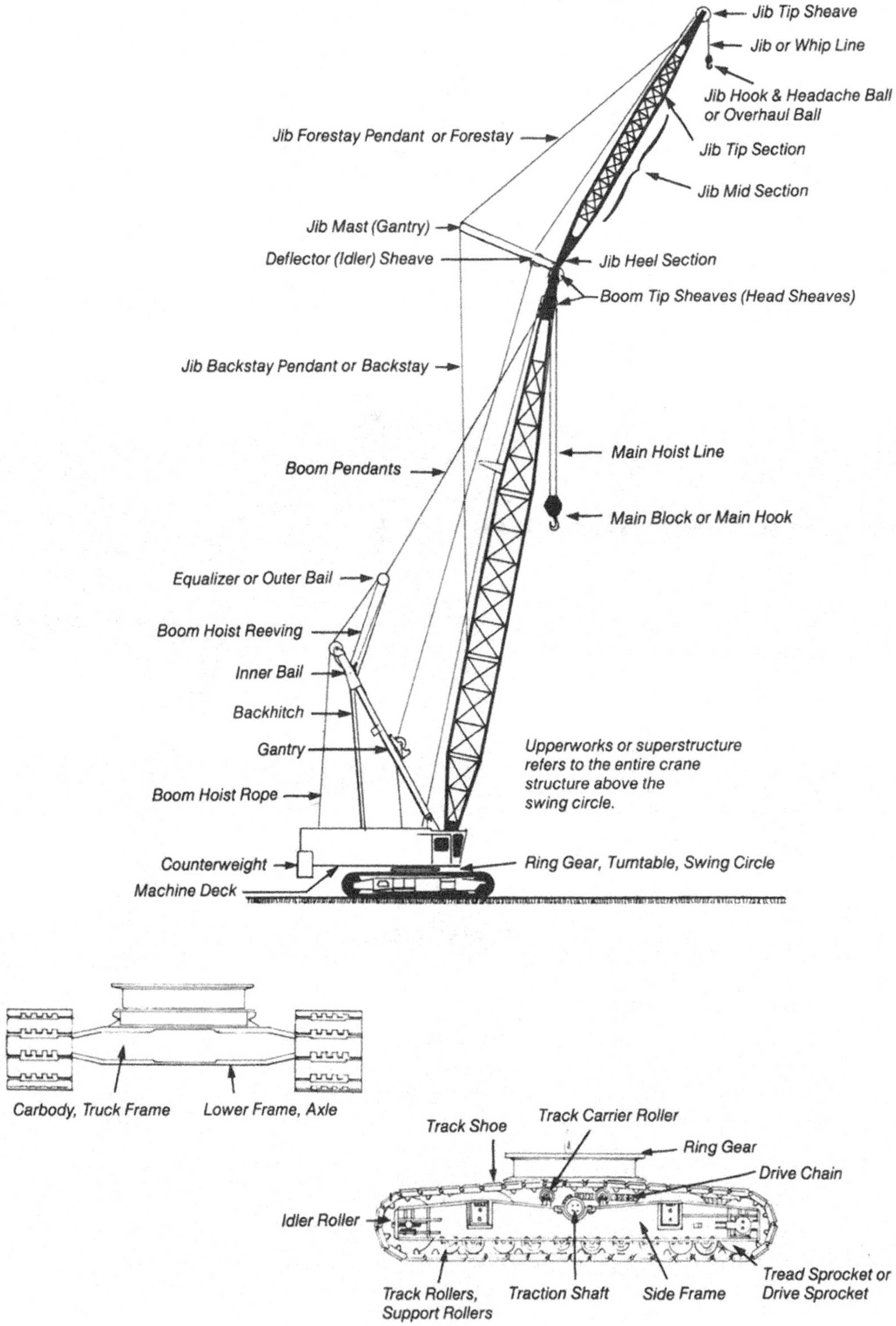

FIGURE B.5 Crawler-mounted lattice boom cranes

Review Quiz Answers

Chapter 1

1. The first crane-type tool was used for irrigation and was called the
   a. shadoof crane
2. Who gets credit for inventing the very first crane used for construction purposes?
   a. The Greeks
3. The tongs used by medieval cranes to lift heavy blocks are called
   d. Lewis tongs
4. Who gets credit for making great improvements to the treadwheel crane, thereby increasing its functionality and mechanical advantage?
   a. The Romans
5. In the Middle Ages, operating treadwheel cranes was so dangerous, owners resorted to hiring
   a. blind people because they could not see the dangers

Chapter 2

1. OSHA analysis of crane-related accidents between 2011 and 2017 identified the major causes of accidents. Nearly 45% of all cases reported were the result of
   d. crane boom contact with energized power lines

2. From 2011 to 2017, in a span of 7 years, the census of Fatal Occupational Injuries reported 297 crane-related deaths. On average, how many people died each month during this time period?
   a. 42 people died.

3. Can a face shield be used instead of safety glasses?
   c. Yes, they can be worn if used in conjunction with safety glasses or goggles

4. Who is responsible for seeing that each employee uses appropriate eye or face protection?
   d. The employer

5. OSHA sets legal limits on noise exposure in the workplace based on a worker's exposure over what time-weighted average?
   d. 8 hours—TWA noise level of 90 dba

6. Can you wear headphones while operating heavy equipment?
   c. No, OSHA has issued citations for such practices.

7. What class of hard hat is required to be worn on construction projects?
   b. Class B helmets

8. How often should hard hats be replaced?
   b. No more than 2 years of regular use or 5 years from the date of manufacture

9. Are gloves required to be worn on construction sites?
   c. Yes, employers must provide the correct hand protection and gloves should be worn when exposed to hazards.

10. Are steel-toed boots required to be worn on construction sites?
    a. There is nothing in the rules that requires steel-toed shoes.

11. Is the employer required to pay for non-specialty toe protection footwear?
    d. No, the employer is not required to pay for nonspecialty safety toe protection, including steel-toed boots

12. What is the main purpose of wearing foot protection?
    a. To maintain balance, preventing trips or slips

13. The responsibility for the safe operation of the crane belongs to the
    c. crane operator

14. According to ASME safety standards, crane operators are required to have a physical examination at least every
    c. 3 years

15. Who has custodial control of the crane by virtue of a lease or ownership?
d. The crane owner

16. Who arranges the crane's presence on a worksite and controls its use there?
c. The crane user

17. The person that directly oversees the work being performed by a crane and the associated rigging crew is the
d. lift director

18. The person who selects, configures, and assembles the rigging equipment for attachment, support, control, and detachment of the load during lifting activities is the
c. rigger

19. What person would exercise supervisory control over the worksite on which a crane is being used and over the work that is being performed on that site?
d. The site supervisor

20. What individual directly controls the crane's functions?
d. The crane operator

Chapter 3

1. The OSHA regulation pertaining to the use of mobile cranes used in construction is
d. 1926.1400

2. What OSHA standard pertains to crawler, locomotive, and truck cranes?
d. 1926.180

3. What OSHA standard pertains to crane lattice boom assembly and disassembly?
a. 1926.1404

4. What OSHA standard pertains to qualified riggers?
a. 1926.1401

5. What OSHA standard pertains to signal person qualifications?
d. 1926.1428

6. What ASME safety standard pertains to mobile and locomotive cranes?
a. ASME B30.5

7. What ASME safety standards pertain to crane hooks?
d. ASME B30.10

8. What ASME safety standard refers to all below-the-hook lifting devices?
b. ASME B30.20

9. What ASME safety standard pertains to personal lifting systems?
c. ASME B30.23

Chapter 4

1. Rigging personnel are allowed to ride on a load to be lifted
   d. never; no personnel are ever allowed.

2. Crane operators are allowed to hoist and carry loads over the heads of personnel
   d. never; loads should never be hoisted and moved over personnel.

3. Mobile crane operators should carefully plan
   d. each lift before hoisting the load

4. The rigging shall be secured and balanced before the load is lifted more than
   a. a few inches

5. Before making a lift, the crane operator must first determine the
   b. weight of the load

6. The center of gravity of a load depends on its
   b. mass

7. When planning to lift a heavy load, the crane's hook must be located
   a. directly over the center of gravity

8. When lifting a heavy load and the crane hook is positioned directly over the center of gravity, what must the crane operator be aware of after lifting the object?
   a. Swinging out of the load, increasing the load radius

9. The person who directly oversees the work being performed by a crane and the associated rigging crew is the
   d. lift director

10. What person would exercise supervisory control over the worksite on which a crane is being used and over the work that is being performed on that site?
    d. The site supervisor

11. Who has the custodial control of the crane by virtue of a lease or ownership
    d. the crane owner

12. The responsibility for the safe operation of the crane belongs to the
    c. certified crane operator

13. According to OSHA, crane operators working in construction must be certified by a nationally accredited organization and be re-certified every
    d. five years

Chapter 5

1. The radius is defined as the horizontal distance from the
   b. center of rotation of the crane to the center of gravity of the suspended load

2. The net load is described as the
   a. actual weight of the load itself

3. What best describes the center of rotation of the crane?
   b. The vertical axis around which the superstructure rotates

4. Which of the following best describes the term *tipping axis*?
   d. The point around which the crane tips

5. Which best describes the term *load moment*?
   b. The force that causes the crane to rotate around its tipping axis

Chapter 6

1. Before setting a periodic inspection schedule for any crane, you should
   b. refer to the manufacturer's manual

2. During routine inspection of the crane, running rope should be
   a. visually inspected

3. Crane inspections that take place daily to monthly are classified as
   d. frequent

4. On a monthly basis, you should inspect several items on your crane, one of which would include
   c. sheaves and drums

5. An inspection of all running rope in service on your crane should occur
   a. daily

6. All inspections, tests, repairs, maintenance, and modifications to the crane must be recorded in the
   c. equipment log book

7. If there is a warning sign on the switch or engine starting controls, the operator shall not shut off or start the engine until the sign has been removed by the
   a. authorized person

8. Load tests that are conducted on cranes must not exceed
   b. 100%

9. It is acceptable to load a crane beyond the rated load
   b. never

10. Before adjustments and repairs are started on a crane's hydraulic components,
    d. relieve the hydraulic oil pressure from all circuits and components

Chapter 7

1. According to OSHA regulations, what is the minimum accessible area of operation a crane must be barricaded?
    c. swing radius of the rear of the superstructure
2. You are setting up an all-terrain crane with all three outriggers extended. In this configuration, when are you allowed to place blocking under the extended outrigger beam?
    a. You are never allowed.
3. You are setting up an all-terrain crane on outriggers. How should the wheels on your crane be positioned in this configuration?
    c. Tires high enough off the supporting surface enough to see daylight
4. Blocking under outrigger floats should cover an area approximately
    d. three times the area of the outrigger floats
5. Before placing the crane in a new work area, the operators should confirm the ground support capabilities by
    b. checking with the site supervisor
6. The reason for using blocking under crane outrigger floats is to
    b. decrease the applied ground pressure
7. The stability safety margin for crawler-mounted cranes is greater than that of truck-mounted cranes on outriggers because
    c. they are more difficult to get level
8. You are operating a 30-ton all-terrain crane with the outriggers fully extended and set. What quadrant of operation would produce the *maximum* ground bearing pressure?
    b. Directly over the corner
9. You are operating an all-terrain crane in a newly constructed housing development. What would cause you the most concern?
    c. The location of the recent backfill and the amount of compaction
10. When operating a crane on outriggers, for safety purposes, it is important to check
    b. that the outrigger floats are properly secured to the jacks

Chapter 8

1. Load tests that are conducted on mobile cranes must not exceed
    b. 100%

2. The crane's load rating chart must be
   c. accessible to the crane operator

3. A crane boom should never be used to
   d. drag a load sideways

4. The capacities shown on the crane's load chart are
   a. the gross capacities

5. Capacities listed above the solid line on the load chart are based on
   d. the crane's structural limitations

6. Specific details regarding the effective weights of attached crane components such as jibs are found in
   a. the load chart data

7. When the actual boom length being used falls between those listed on the load chart, use the capacity listed for
   d. the next longer or shorter boom length, whichever is less

8. Increasing the number of parts of line to the crane's hoist block
   c. increases the mechanical advantage

9. Deflection of the crane boom when making a lift will increase
   a. the load radius

10. When a jib is erected but not used for the lift, the operator must make a deduction from
    c. the effective weight of the jib

11. When the jib is erected from the stowed position and not being used, its effective weight
    d. increases

12. The side loading of the crane boom is permitted
    d. only with freely suspended loads

13. The stability safety margin for crawler mounted cranes is greater than truck-mounted cranes on outriggers because
    c. they are harder to get level

14. When making a lift with a crawler crane, ground pressures would be highest when
    c. lifting over the corner of the tracks

15. Mobile crane stability is based on the
    c. principles of leverage

16. As you boom down, the leverage of the load on your mobile crane will
    d. increase

17. When operating your crane, rapid acceleration or sudden stopping of the load would
    c. produce load forces greater than the load's actual weight

18. When is the rated capacity of the rigging or attachment a limiting factor for a lift?

c. when the rated work load is less than the crane's capacity

19. When the crane's gross capacity is between two stated values on the load chart, always

c. use the next lower value

20. When the crane is equipped with a manual boom extension that isn't being used, the operator must

c. deduct weight of the extension from the crane's gross capacity

21. When an all-terrain crane makes a lift off the ground on rubber, what effect does it have on the crane?

b. The boom angle will decrease

22. You have just changed the jib angle to the ground on your crane from 65 degrees to 35 degrees. You have now

b. decreased the capacity of the jib

23. You are operating a lattice boom crawler crane. The boom angle on your load chart is measured between the

c. loaded boom and the horizontal

24. The large telescopic boom crane you are operating is set up on firm, level ground with outriggers fully extended and set. You are operating with 70 feet of boom at a 53-foot radius. Your load chart indicates a capacity of 11,950 pounds at a 50-foot radius, followed by a capacity of 10,500 pounds at a 55-foot radius. What is your gross capacity at the 53-foot radius?

a. 10,500 pounds

25. What is the net capacity of a truck crane with 40 feet of boom extended, unless the manufacturer indicates otherwise?

a. The capacity indicated on the load chart and subtract the weight of any rigging, jibs, balls and attachments

Telescopic Boom Crane - Fixed Cab (TSS)

Outriggers	Fully Extended & Set	**Extension /Jib**	23 – 40 ft Jib Stowed
Counterweight	NA	**Extension /Jib Offset**	NA
Tires	NA	**Block**	NA
Crawlers	NA	**Overhaul Ball**	5 TON
Main Boom	Fully Extended	**Rigging**	10 lbs
Main Boom Angle	NA	**Wire Rope (main hoist)**	9/16" Rotation Resistant
Radius	MAXIMUM	**Wire Rope (auxiliary)**	NA
Other	NA	**Other**	NA

Note: For line in use, when wire rope deductions are required, the rope length shall be measured from the boom tip to the ground.

1. Your net capacity at maximum operating radius over the "side" using the necessary hook block could be.

A. 1400

B. 790

C. 1080

D. 2880

Gross Capacity	=	1400 lbs
Deductions		
Stowed Jib	-	190 lbs
Overhaul Ball	-	120 lbs
Rigging	-	10 lbs
Net Capacity	=	1080 lbs

Use for Workspace

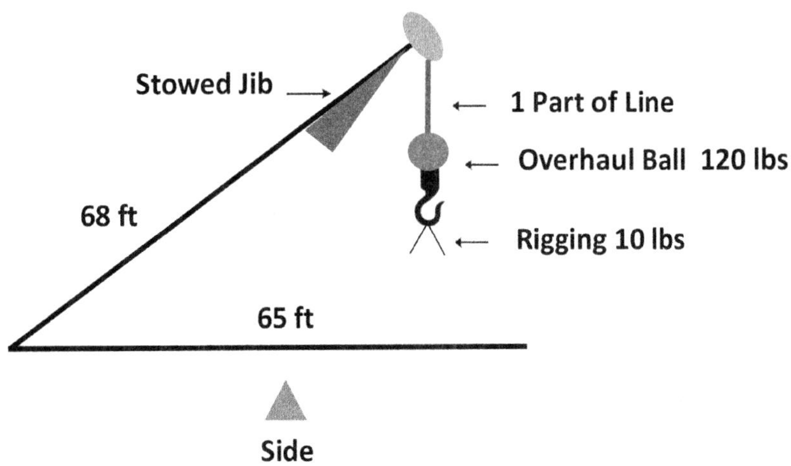

OPERATING RADIUS FT	LOADED BOOM ANGLE ∠	BOOM LENGTH 26 FT	∠	38 FT	∠	48 FT	∠	58 FT	∠	68 FT	OPERATING RADIUS FT	LOADED BOOM ANG ∠	23 FT. JIB FOR ALL BOOM LENGTH SEE WARNING NOTE* RATED LOAD IN POUNDS	LOADED BOOM ANG ∠	40 FT. JIB FOR ALL BOOM LENGTH SEE WARNING NOTE* RATED LOAD IN POUNDS
5	77	34000													
8	70	24000	77	21500											
10	66	20110	74	18460	78	14600					10				
12	61	17360	71	16010	76	12680	79	11810			12				
15	52	14310	66	13320	72	10500	76	9830	78	9200	15				
20	36	10490	57	10350	65	8160	70	7690	74	7280	20	78	3500		
25			47	8310	58	6570	65	6240	70	5910	25	75	3060	78	1940
30			36	6610	51	5420	59	5200	65	4940	30	72	2700	75	1690
35			17	4560	42	4470	53	4390	60	4200	35	69	2400	72	1490
40					32	3580	47	3720	55	3610	40	65	2150	70	1320
45					16	2380	39	3120	50	3110	45	62	1950	67	1180
50							29	2500	43	2670	50	58	1770	64	1060
55							15	1610	36	2240	55	55	1550	61	960
60									28	1780	60	50	1350	58	870
65									14	1400	65	46	1160	55	790
70											70	41	1010	51	730
75											75	36	830	48	670
80											80			44	610
85											85			40	570
90											90			35	520
95											95				
100											100				
	480 LBS.		330 LBS.		260 LBS.		220 LBS.		190 LBS.		DEDUCTIONS FOR STOWED JIB				

Table heading: LOAD RATINGS IN LBS WITH OUTRIGGERS AND STABILIZERS EXTENDED — JIB LOAD RATINGS WITH OUTRIGGERS AND STABILIZERS EXTENDED (boom sections: BASE, 1ST, 2ND)

DEDUCTIONS

Auxiliary Block	50 lb	*22.68 kg*
Overhaul Ball	120 lb	*54.43 kg*
Single-Sheave Load Block	260 lb	*117.93 kg*
Double-Sheave Load Block	350 lb	*158.76 kg*
Hose Reel	190 lb	*86.18 kg*

Swing-Around Jib (Stowed) .... See Load Rating Chart

WARNING

Lifting off the main boom point while the swing-around jib is erected is not intended or approved.

AREA OF OPERATION

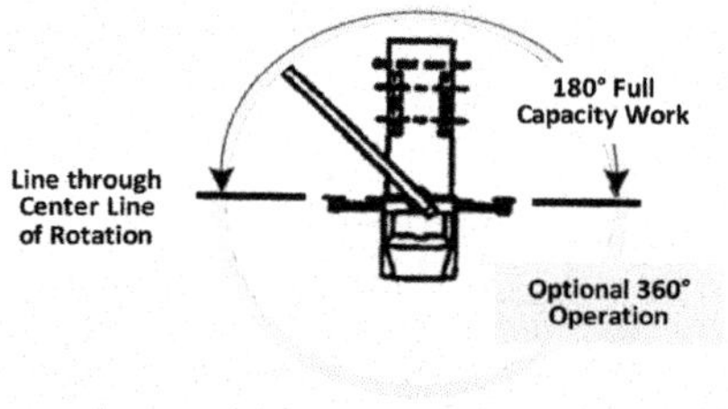

Telescopic Boom Crane - Fixed Cab (TSS)

Outriggers	Fully Extended & Set	Extension /Jib	40 ft Erected
Counterweight	NA	Extension /Jib Offset	NA
Tires	NA	Block	NA
Crawlers	NA	Overhaul Ball	5 - Ton
Main Boom	48 ft	Rigging	50 lbs
Main Boom Angle	75	Wire Rope (main hoist)	9/16" Rotation Resistant
Radius	NA	Wire Rope (auxiliary)	NA
Other	Hose Reel Installed	Other	NA

Note: For line in use, when wire rope deductions are required, the rope length shall be measured from the boom tip to the ground.

2. What is the net capacity when a lift is made on the extension over the "side" of the crane.

A. 1380

B. 1520

C. 1330

D. 1690

Gross Capacity	=	1690 lbs
Deductions		
Hose Reel	-	190 lbs
Overhaul Ball	-	120 lbs
Rigging	-	50 lbs
Net Capacity	=	1330 lbs

Use for Workspace

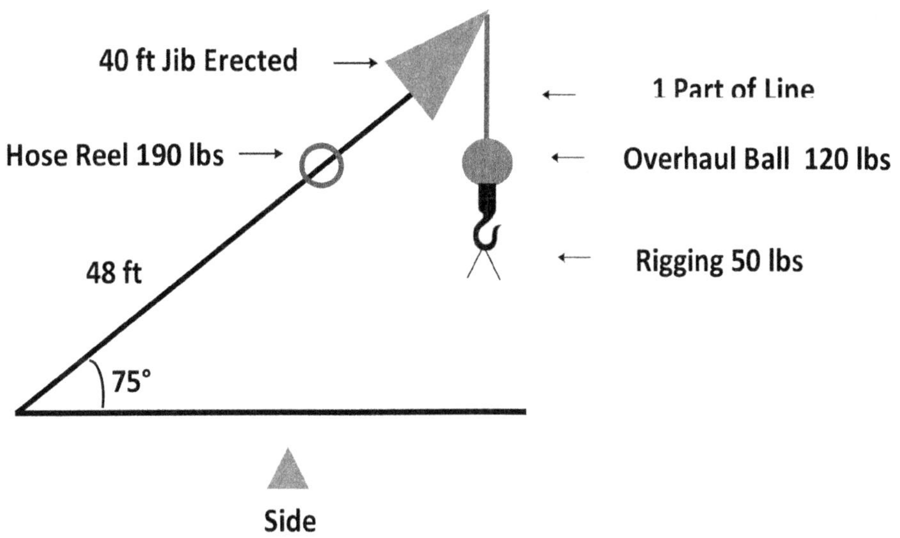

LOAD RATINGS IN LBS WITH OUTRIGGERS AND STABILIZERS EXTENDED										
OPERATING RADIUS FT	LOADED BOOM ANGLE ∠	BOOM LENGTH 26 FT	∠	38 FT	∠	48 FT	∠	58 FT	∠	68 FT
5	77	34000								
8	70	24000	77	21500						
10	66	20110	74	18460	78	14600				
12	61	17360	71	16010	76	12680	79	11810		
15	52	14310	66	13320	72	10500	76	9830	78	9200
20	36	10490	57	10350	65	8160	70	7690	74	7280
25			47	8310	58	6570	65	6240	70	5910
30			36	6610	51	5420	59	5200	65	4940
35			17	4560	42	4470	53	4390	60	4200
40					32	3580	47	3720	55	3610
45					16	2380	39	3120	50	3110
50							29	2500	43	2670
55							15	1610	36	2240
60									28	1780
65									14	1400
70										
75										
80										
85										
90										
95										
100										
	480 LBS.		330 LBS.		260 LBS.		220 LBS.		190 LBS.	

(Boom diagram labels: BASE, 1ST, 2ND, ∠)

JIB LOAD RATINGS WITH OUTRIGGERS AND STABILIZERS EXTENDED				
OPERATING RADIUS FT	LOADED BOOM ANG ∠	23 FT. JIB FOR ALL BOOM LENGTH SEE WARNING NOTE* RATED LOAD IN POUNDS	LOADED BOOM ANG ∠	40 FT. JIB FOR ALL BOOM LENGTH SEE WARNING NOTE* RATED LOAD IN POUNDS
10				
12				
15				
20	78	3500		
25	75	3060	78	1940
30	72	2700	75	1690
35	69	2400	72	1490
40	65	2150	70	1320
45	62	1950	67	1180
50	58	1770	64	1060
55	55	1550	61	960
60	50	1350	58	870
65	46	1160	55	790
70	41	1010	51	730
75	36	830	48	670
80			44	610
85			40	570
90			35	520
95				
100				
DEDUCTIONS FOR STOWED JIB				

DEDUCTIONS

Auxiliary Block	50 lb	22.68 kg	
Overhaul Ball	120 lb	54.43 kg	
Single-Sheave Load Block	260 lb	117.93 kg	
Double-Sheave Load Block	350 lb	158.76 kg	
Hose Reel	190 lb	86.18 kg	

Swing-Around Jib (Stowed) .... See Load Rating Chart

WARNING

Lifting off the main boom point while the swing-around jib is erected is not intended or approved.

AREA OF OPERATION

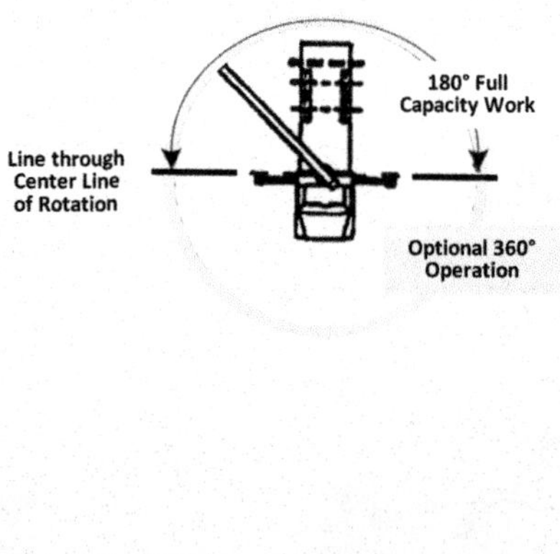

Telescopic Boom Crane - Fixed Cab (TSS)

Outriggers	Fully Extended & Set	**Extension /Jib**	23 – 40 ft Jib Stowed
Counterweight	NA	**Extension /Jib Offset**	NA
Tires	NA	**Block**	NA
Crawlers	NA	**Overhaul Ball**	5 - Ton
Main Boom	Maximum	**Rigging**	75 lbs
Main Boom Angle	NA	**Wire Rope (main hoist)**	9/16" Rotation Resistant
Radius	Maximum	**Wire Rope (auxiliary)**	NA
Other	Hose Reel Installed	**Other**	Spreader Bar 110 lbs

Note: For line in use, when wire rope deductions are required, the rope length shall be measured from the boom tip to the ground.

3. What is the net capacity of this crane for a lift over the rear the with the main boom.

A. 715

B. 1400

C. 905

D. 825

Gross Capacity	=	1400 lbs
Deductions		
Hose Reel	-	190 lbs
Stowed Jib	-	190 lbs
Overhaul Ball	-	120 lbs
Rigging	-	75 lbs
Spreader Bar	-	110 lbs
Net Capacity	=	715 lbs

Use for Workspace

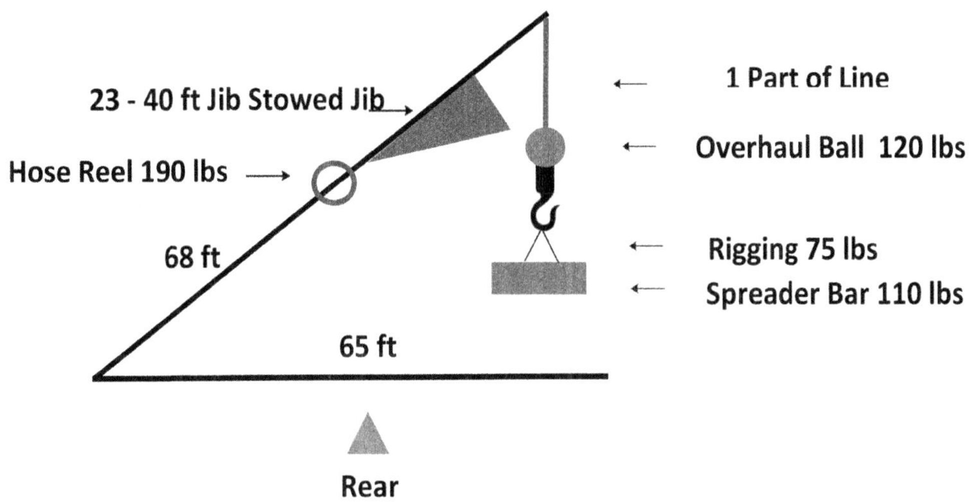

LOAD RATINGS IN LBS WITH OUTRIGGERS AND STABILIZERS EXTENDED										
OPERATING RADIUS FT	LOADED BOOM ANGLE ∠	BOOM LENGTH 26 FT	∠	38 FT	∠	48 FT	∠	58 FT	∠	68 FT
5	77	34000								
8	70	24000	77	21500						
10	66	20110	74	18460	78	14600				
12	61	17360	71	16010	76	12680	79	11810		
15	52	14310	66	13320	72	10500	76	9830	78	9200
20	36	10490	57	10350	65	8160	70	7690	74	7280
25			47	8310	58	6570	65	6240	70	5910
30			36	6610	51	5420	59	5200	65	4940
35			17	4560	42	4470	53	4390	60	4200
40					32	3580	47	3720	55	3610
45					16	2380	39	3120	50	3110
50							29	2500	43	2670
55							15	1610	36	2240
60									28	1780
65									14	1400
70										
75										
80										
85										
90										
95										
100										
	480 LBS.		330 LBS.		260 LBS.		220 LBS.		190 LBS.	

BASE, 1ST, 2ND, ∠

JIB LOAD RATINGS WITH OUTRIGGERS AND STABILIZERS EXTENDED				
OPERATING RADIUS FT	LOADED BOOM ANG ∠	23 FT. JIB FOR ALL BOOM LENGTH SEE WARNING NOTE* RATED LOAD IN POUNDS	LOADED BOOM ANG ∠	40 FT. JIB FOR ALL BOOM LENGTH SEE WARNING NOTE* RATED LOAD IN POUNDS
10				
12				
15				
20	78	3500		
25	75	3060	78	1940
30	72	2700	75	1690
35	69	2400	72	1490
40	65	2150	70	1320
45	62	1950	67	1180
50	58	1770	64	1060
55	55	1550	61	960
60	50	1350	58	870
65	46	1160	55	790
70	41	1010	51	730
75	36	830	48	670
80			44	610
85			40	570
90			35	520
95				
100				
DEDUCTIONS FOR STOWED JIB				

DEDUCTIONS

Auxiliary Block	50 lb	*22.68 kg*
Overhaul Ball	120 lb	*54.43 kg*
Single-Sheave Load Block	260 lb	*117.93 kg*
Double-Sheave Load Block	350 lb	*158.76 kg*
Hose Reel	190 lb	*86.18 kg*

Swing-Around Jib (Stowed) .... See Load Rating Chart

WARNING

Lifting off the main boom point while the swing-around jib is erected is not intended or approved.

AREA OF OPERATION

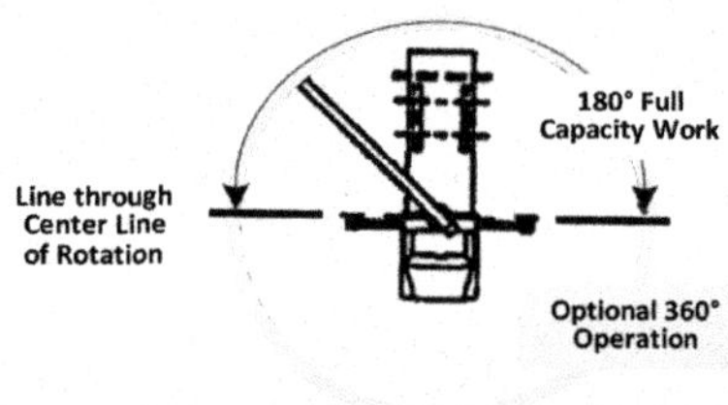

Telescopic Boom Crane - Fixed Cab (TSS)

Outriggers	Fully Extended & Set	**Extension /Jib**	23 – 40 ft Jib Stowed
Counterweight	NA	**Extension /Jib Offset**	NA
Tires	NA	**Block**	Operators Choice
Crawlers	NA	**Overhaul Ball**	Operators Choice
Main Boom	26 ft	**Rigging**	150 lbs
Main Boom Angle	NA	**Wire Rope (main hoist)**	9/16" Rotation Resistant
Radius	10 ft	**Wire Rope (auxiliary)**	NA
Other	NA	**Other**	NA

Note: For line in use, when wire rope deductions are required, the rope length shall be measured from the boom tip to the ground.

4. Lifting from the main boom over the "side" of the crane using the necessary hook block or ball, your **maximum** net load could be.

A. 19,000

B. 19,170

C. 19,630

D. 21,000

Gross Capacity	=	20,110 lbs
Deductions		
Auxiliary Block	-	50 lbs
Stowed Jib	-	480 lbs
Single Sheave Load Block	-	260 lbs
Rigging	-	150 lbs
Net Capacity	=	19,170 lbs

Use for Workspace

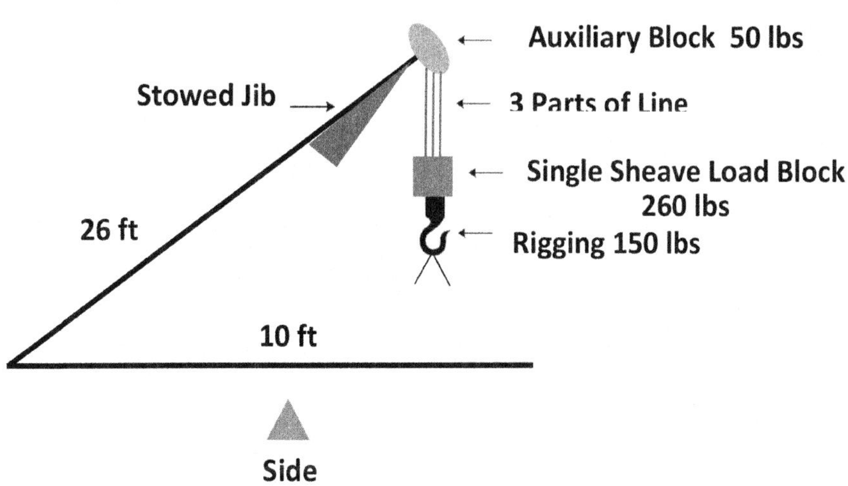

LOAD RATINGS IN LBS WITH OUTRIGGERS AND STABILIZERS EXTENDED											JIB LOAD RATINGS WITH OUTRIGGERS AND STABILIZERS EXTENDED				
OPERATING RADIUS FT	LOADED BOOM ANGLE	BOOM LENGTH 26 FT	∠	38 FT	∠	48 FT	∠	58 FT	∠	68 FT	OPERATING RADIUS FT	LOADED BOOM ANG ∠	23 FT. JIB FOR ALL BOOM LENGTH SEE WARNING NOTE* RATED LOAD IN POUNDS	LOADED BOOM ANG ∠	40 FT. JIB FOR ALL BOOM LENGTH SEE WARNING NOTE* RATED LOAD IN POUNDS
5	77	34000													
8	70	24000	77	21500											
10	66	20110	74	18460	78	14600					10				
12	61	17360	71	16010	76	12680	79	11810			12				
15	52	14310	66	13320	72	10500	76	9830	78	9200	15				
20	36	10490	57	10350	65	8160	70	7690	74	7280	20	78	3500		
25			47	8310	58	6570	65	6240	70	5910	25	75	3060	78	1940
30			36	6610	51	5420	59	5200	65	4940	30	72	2700	75	1690
35			17	4560	42	4470	53	4390	60	4200	35	69	2400	72	1490
40					32	3580	47	3720	55	3610	40	65	2150	70	1320
45					16	2380	39	3120	50	3110	45	62	1950	67	1180
50							29	2500	43	2670	50	58	1770	64	1060
55							15	1610	36	2240	55	55	1550	61	960
60									28	1780	60	50	1350	58	870
65									14	1400	65	46	1160	55	790
70											70	41	1010	51	730
75											75	36	830	48	670
80											80			44	610
85											85			40	570
90											90			35	520
95											95				
100											100				
	480 LBS.		330 LBS.		260 LBS.		220 LBS.		190 LBS.		DEDUCTIONS FOR STOWED JIB				

BASE, 1ST, 2ND

DEDUCTIONS

Auxiliary Block	50 lb	*22.68 kg*
Overhaul Ball	120 lb	*54.43 kg*
Single-Sheave Load Block	260 lb	*117.93 kg*
Double-Sheave Load Block	350 lb	*158.76 kg*
Hose Reel	190 lb	*86.18 kg*

Swing-Around Jib (Stowed) .... See Load Rating Chart

WARNING
Lifting off the main boom point while the swing-around jib is erected is not intended or approved.

ALLOWABLE LINE PULL

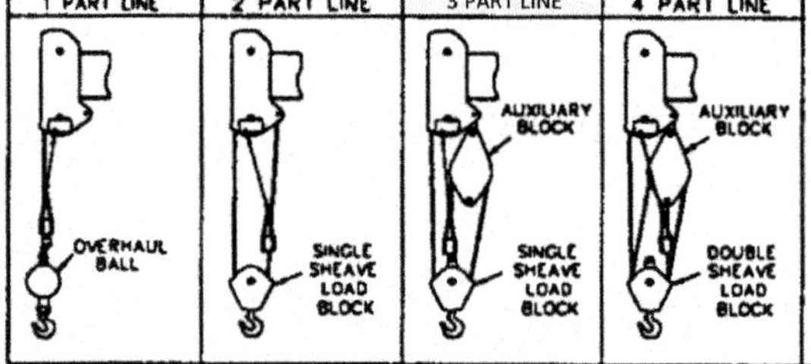

1 PART LINE	2 PART LINE	3 PART LINE	4 PART LINE	
8500 lb 3856 kg	17000 lb 7711 kg	25500 lb 11567 kg	34000 lb 15422 kg	9/16" (14.29 mm) 6x25 IWRC (3.5:1 SF). 29750 lb (13494 kg) Minimum breaking strength.
7400 lb 3357 kg	14800 lb 6313 kg	22200 lb 10070 kg	29600 lb 13426 kg	9/16" (14.29 mm) Rot resistant (5.0:1 SF). 37000 lb (16783 kg) Minimum breaking strength.

WARNING

Anti-Two-Block system must be in good operating condition before operating crane. Refer to Owner's Manual.

Keep at least three wraps on load line on drum at all times.

AREA OF OPERATION

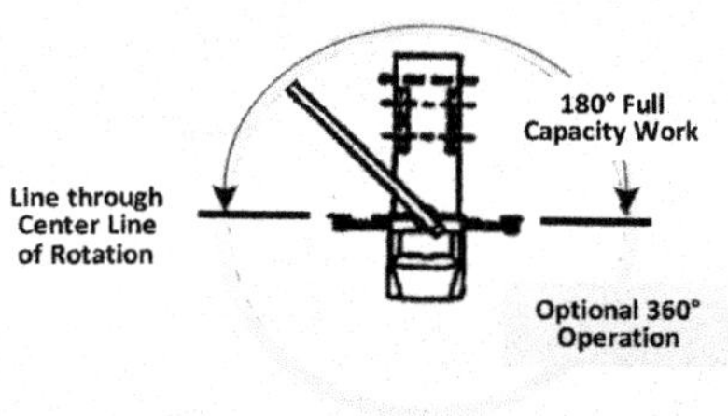

Telescopic Boom Crane - Fixed Cab (TSS)

Outriggers	Fully Extended & Set	Extension /Jib	23 – 40 ft Jib Stowed
Counterweight	NA	Extension /Jib Offset	NA
Tires	NA	Block	Single Sheave
Crawlers	NA	Overhaul Ball	NA
Main Boom	36 ft	Rigging	45 lbs
Main Boom Angle	71	Wire Rope (main hoist)	9/16" Rotation Resistant
Radius	NA	Wire Rope (auxiliary)	NA
Other	NA	Other	Lifting Beam 110 lbs

Note: For line in use, when wire rope deductions are required, the rope length shall be measured from the boom tip to the ground.

5. What is the net capacity of this crane when a lift is made off the main boom over the "side".

A. 15,215

B. 15,165

C. 19,630

D. 15,275

Gross Capacity	=	16,010 lbs
Deductions		
Auxiliary Block	-	50 lbs
Stowed Jib	-	330 lbs
Single Sheave Load Block	-	260 lbs
Rigging	-	45 lbs
Lift Beam	-	110 lbs
Net Capacity	=	15,215 lbs

Use for Workspace

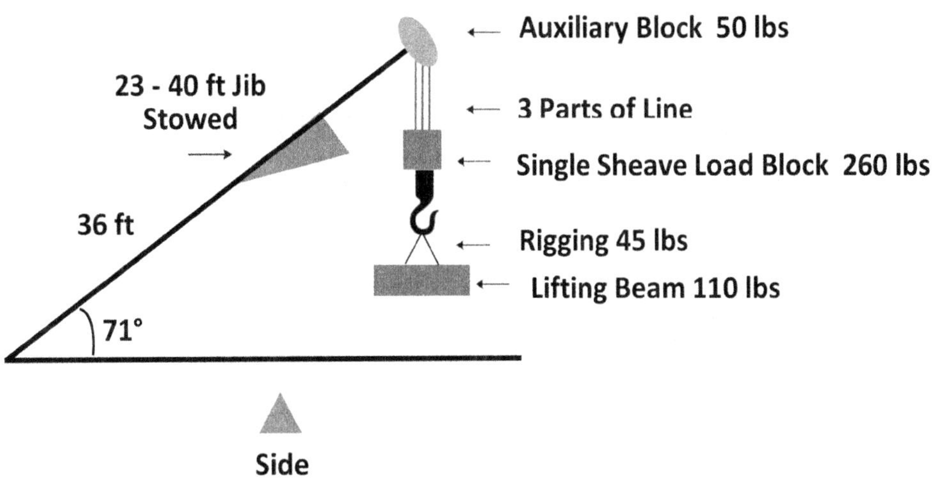

LOAD RATINGS IN LBS WITH OUTRIGGERS AND STABILIZERS EXTENDED											JIB LOAD RATINGS WITH OUTRIGGERS AND STABILIZERS EXTENDED				
OPERATING RADIUS FT	LOADED BOOM ANGLE ∠	BOOM LENGTH 26 FT	∠	38 FT	∠	48 FT	∠	58 FT	∠	68 FT	OPERATING RADIUS FT	LOADED BOOM ANG ∠	23 FT. JIB FOR ALL BOOM LENGTH SEE WARNING NOTE* RATED LOAD IN POUNDS	LOADED BOOM ANG ∠	40 FT. JIB FOR ALL BOOM LENGTH SEE WARNING NOTE* RATED LOAD IN POUNDS
5	77	34000													
8	70	24000	77	21500											
10	66	20110	74	18460	78	14600					10				
12	61	17360	71	16010	76	12680	79	11810			12				
15	52	14310	66	13320	72	10500	76	9830	78	9200	15				
20	36	10490	57	10350	65	8160	70	7690	74	7280	20	78	3500		
25			47	8310	58	6570	65	6240	70	5910	25	75	3060	78	1940
30			36	6610	51	5420	59	5200	65	4940	30	72	2700	75	1690
35			17	4560	42	4470	53	4390	60	4200	35	69	2400	72	1490
40					32	3580	47	3720	55	3610	40	65	2150	70	1320
45					16	2380	39	3120	50	3110	45	62	1950	67	1180
50							29	2500	43	2670	50	58	1770	64	1060
55							15	1610	36	2240	55	55	1550	61	960
60									28	1780	60	50	1350	58	870
65									14	1400	65	46	1160	55	790
70											70	41	1010	51	730
75											75	36	830	48	670
80											80			44	610
85											85			40	570
90											90			35	520
95											95				
100											100				
	480 LBS.		330 LBS.		260 LBS.		220 LBS.		190 LBS.		DEDUCTIONS FOR STOWED JIB				

DEDUCTIONS

Auxiliary Block	50 lb	22.68 kg
Overhaul Ball	120 lb	54.43 kg
Single-Sheave Load Block	260 lb	117.93 kg
Double-Sheave Load Block	350 lb	158.76 kg
Hose Reel	190 lb	86.18 kg

Swing-Around Jib (Stowed) .... See Load Rating Chart

WARNING

Lifting off the main boom point while the swing-around jib is erected is not intended or approved.

ALLOWABLE LINE PULL

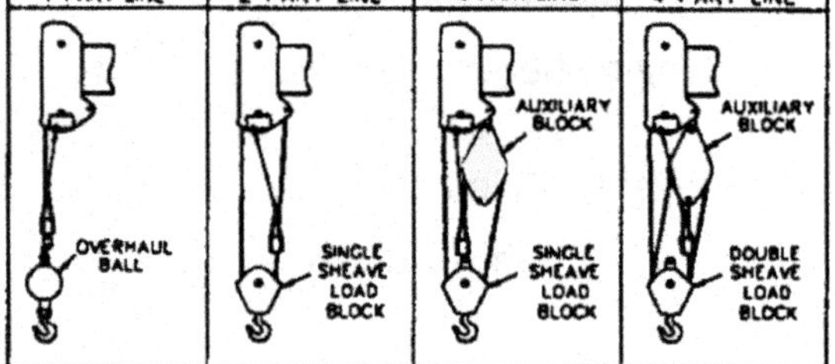

1 PART LINE	2 PART LINE	3 PART LINE	4 PART LINE	
8500 lb 3856 kg	17000 lb 7711 kg	25500 lb 11567 kg	34000 lb 15422 kg	9/16" (14.29 mm) 6x25 IWRC (3.5:1 SF). 29750 lb (13494 kg) Minimum breaking strength.
7400 lb 3357 kg	14800 lb 6313 kg	22200 lb 10070 kg	29600 lb 13426 kg	9/16" (14.29 mm) Rot resistant (5.0:1 SF). 37000 lb (16783 kg) Minimum breaking strength.

WARNING

Anti-Two-Block system must be in good operating condition before operating crane. Refer to Owner's Manual.

Keep at least three wraps on load line on drum at all times.

AREA OF OPERATION

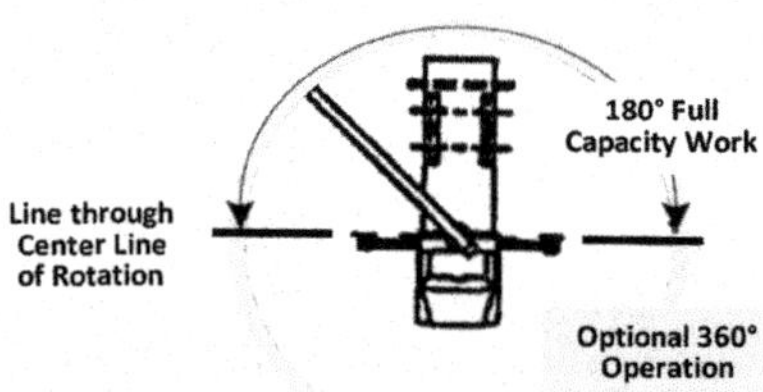

Telescopic Boom Crane - Fixed Cab (TSS)

Outriggers	Fully Extended & Set	**Extension /Jib**	40 ft Extended
Counterweight	NA	**Extension /Jib Offset**	NA
Tires	NA	**Block**	NA
Crawlers	NA	**Overhaul Ball**	5 - Ton
Main Boom	68 ft	**Rigging**	75 lbs
Main Boom Angle	69	**Wire Rope (main hoist)**	9/16" Rotation Resistant
Radius	NA	**Wire Rope (auxiliary)**	NA
Other	Spreader Bar 110 lbs	**Other**	Hose Reel Installed

Note: For line in use, when wire rope deductions are required, the rope length shall be measured from the boom tip to the ground.

6. What is the gross capacity when the lift is made on the extension over the "side" of the crane.

A. 1320

B. 875

C. 685

D. 1180

Gross Capacity = 1180 lbs

Use for Workspace

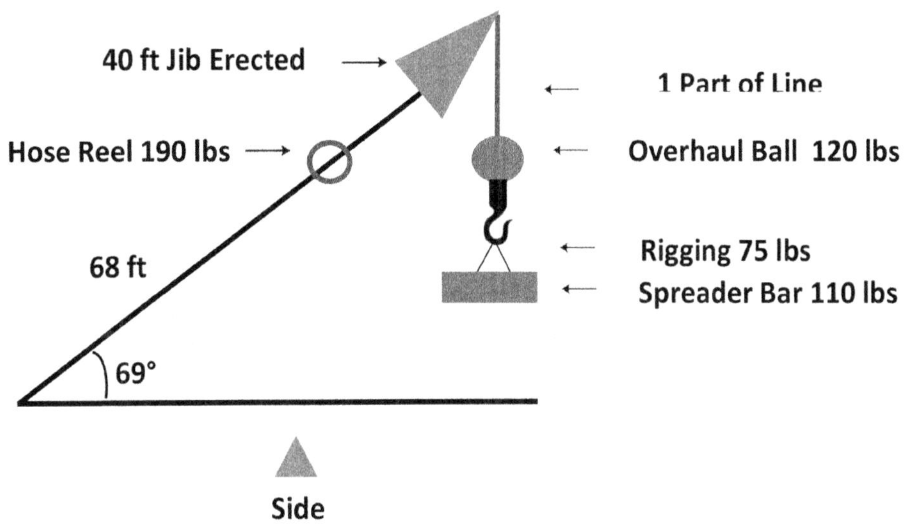

LOAD RATINGS IN LBS WITH OUTRIGGERS AND STABILIZERS EXTENDED											JIB LOAD RATINGS WITH OUTRIGGERS AND STABILIZERS EXTENDED				
OPERATING RADIUS FT	LOADED BOOM ANGLE	BOOM LENGTH 26 FT	∠	38 FT	∠	48 FT	∠	58 FT	∠	68 FT	OPERATING RADIUS FT	LOADED BOOM ANG ∠	23 FT. JIB FOR ALL BOOM LENGTH SEE WARNING NOTE* RATED LOAD IN POUNDS	LOADED BOOM ANG ∠	40 FT. JIB FOR ALL BOOM LENGTH SEE WARNING NOTE* RATED LOAD IN POUNDS
5	77	34000													
8	70	24000	77	21500											
10	66	20110	74	18460	78	14600					10				
12	61	17360	71	16010	76	12680	79	11810			12				
15	52	14310	66	13320	72	10500	76	9830	78	9200	15				
20	36	10490	57	10350	65	8160	70	7690	74	7280	20	78	3500		
25			47	8310	58	6570	65	6240	70	5910	25	75	3060	78	1940
30			36	6610	51	5420	59	5200	65	4940	30	72	2700	75	1690
35			17	4560	42	4470	53	4390	60	4200	35	69	2400	72	1490
40					32	3580	47	3720	55	3610	40	65	2150	70	1320
45					16	2380	39	3120	50	3110	45	62	1950	67	1180
50							29	2500	43	2670	50	58	1770	64	1060
55							15	1610	36	2240	55	55	1550	61	960
60									28	1780	60	50	1350	58	870
65									14	1400	65	46	1160	55	790
70											70	41	1010	51	730
75											75	36	830	48	670
80											80			44	610
85											85			40	570
90											90			35	520
95											95				
100											100				
	480 LBS.		330 LBS.		260 LBS.		220 LBS.		190 LBS.		DEDUCTIONS FOR STOWED JIB				

DEDUCTIONS

Auxiliary Block	50 lb	22.68 kg
Overhaul Ball	120 lb	54.43 kg
Single-Sheave Load Block	260 lb	117.93 kg
Double-Sheave Load Block	350 lb	158.76 kg
Hose Reel	190 lb	86.18 kg

Swing-Around Jib (Stowed) .... See Load Rating Chart

WARNING
Lifting off the main boom point while the swing-around jib is erected is not intended or approved.

AREA OF OPERATION

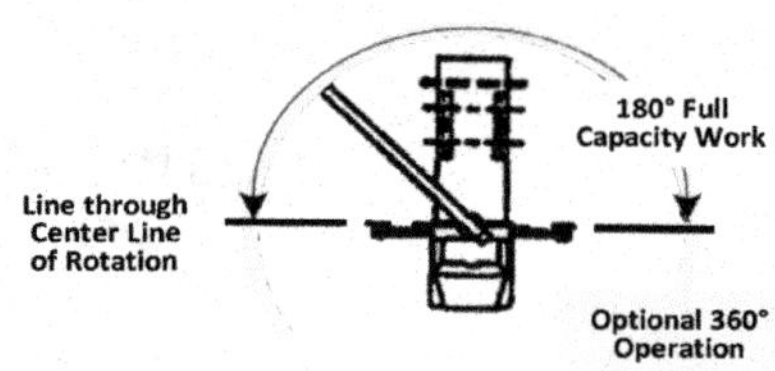

LOAD RATINGS IN LBS WITH OUTRIGGERS AND STABILIZERS EXTENDED											JIB LOAD RATINGS WITH OUTRIGGERS AND STABILIZERS EXTENDED				
OPERATING RADIUS FT	LOADED BOOM ANGLE ∠	BOOM LENGTH 26 FT	∠	38 FT	∠	48 FT	∠	58 FT	∠	68 FT	OPERATING RADIUS FT	LOADED BOOM ANG ∠	23 FT. JIB FOR ALL BOOM LENGTH SEE WARNING NOTE* RATED LOAD IN POUNDS	LOADED BOOM ANG ∠	40 FT. JIB FOR ALL BOOM LENGTH SEE WARNING NOTE* RATED LOAD IN POUNDS
5	77	34000													
8	70	24000	77	21500											
10	66	20110	74	18460	78	14600					10				
12	61	17360	71	16010	76	12680	79	11810			12				
15	52	14310	66	13320	72	10500	76	9830	78	9200	15				
20	36	10490	57	10350	65	8160	70	7690	74	7280	20	78	3500		
25			47	8310	58	6570	65	6240	70	5910	25	75	3060	78	1940
30			36	6610	51	5420	59	5200	65	4940	30	72	2700	75	1690
35			17	4560	42	4470	53	4390	60	4200	35	69	2400	72	1490
40					32	3580	47	3720	55	3610	40	65	2150	70	1320
45					16	2380	39	3120	50	3110	45	62	1950	67	1180
50							29	2500	43	2670	50	58	1770	64	1060
55							15	1610	36	2240	55	55	1550	61	960
60									28	1780	60	50	1350	58	870
65									14	1400	65	46	1160	55	790
70											70	41	1010	51	730
75											75	36	830	48	670
80											80			44	610
85											85			40	570
90											90			35	520
95											95				
100											100				
	480 LBS.		330 LBS.		260 LBS.		220 LBS.		190 LBS.		DEDUCTIONS FOR STOWED JIB				

DEDUCTIONS

Auxiliary Block	50 lb	*22.68 kg*
Overhaul Ball	120 lb	*54.43 kg*
Single-Sheave Load Block	260 lb	*117.93 kg*
Double-Sheave Load Block	350 lb	*158.76 kg*
Hose Reel	190 lb	*86.18 kg*

Swing-Around Jib (Stowed) .... See Load Rating Chart

WARNING

Lifting off the main boom point while the swing-around jib is erected is not intended or approved.

AREA OF OPERATION

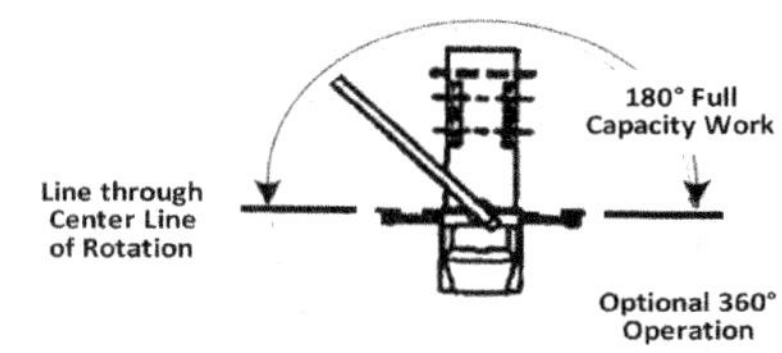

Telescopic Boom Crane—Swing Cab (TLL)			
Outriggers	Fully Extended & Set	Extension/Jib	34 ft Fly Erected
Counterweight	Full	Extension/ Jib Offset	15° Offset
Tires	N/A	Block	N/A
Crawlers	N/A	Overhaul Ball	8.5 Ton
Main Boom	110 ft (Mode B)	Rigging	150 lbs
Main Boom Angle	67.5°	Wire Rope (Main Hoist)	18 X 19 Rotation Resistant 1 lb per ft
Radius	65 ft	Wire Rope (Auxiliary)	18 X 19 Rotation Resistant 1 lb per ft
Other	N/A	Other	22 ft. Fly Stowed

1. What ***maximum*** net load can be lifted with the offset fly over the "side" of the crane

A. 6490 lbs

B. 7000 lbs

C. 7510 lbs

D. 6190 lbs

Gross Capacity	=	7000 lbs
22 ft Fly Tip Stowed	-	300 lbs
85 ton Hook Ball	-	360 lbs
Rigging	-	150 lbs
Net Capacity	=	6190 lbs

Note: For line in use, when wire rope deductions are required, the rope length shall be measured from the boom tip to the ground.

BOOM MODE "B"

Maximum Allowable Lifting Capacities
Rated Lifting Capacities In Pounds
On Fully Extended Outriggers
See Set Up Note 2.

110 ft Main Boom +34 Ft. Offset Fly

Load Radius In Feet	1° Offset		15° Offset		30° Offset		Load Radius In Feet
	Loaded Boom Angle (Deg.)	360°	Loaded Boom Angle (Deg.)	360°	Loaded Boom Angle (Deg.)	360°	
35	76.5	10,500					35
40	74.5	10,500					40
45	72.5	10,500	76.0	9,800			45
50	70.5	9,800	74.0	9,000	77.0	8,300	50
55	68.5	8,900	71.5	8,200	75.0	7,700	55
60	66.5	8,200	69.5	7,600	72.5	7,100	60
65	64.0	7,500	67.5	7.00	70.5	6,600	65
70	62.0	6,900	65.0	6,500	68.0	6,200	70
75	59.5	6,400	63.0	6,100	65.5	5,800	75
80	57.5	6,000	60.5	5,700	63.0	5,500	80
85	55.0	5,600	58.0	5,300	60.5	5,100	85
90	52.5	5,100	55.5	5,000	58.0	4,800	90
95	49.5	4,700	53.0	4,700	55.5	4,600	95
100	47.0	4,200	50.0	4,300	52.5	4,300	100
105	43.5	3,600	47.0	3,900	49.5	4,000	105
110	40.5	3,100	43.5	3,400	46.0	3,600	110
115	37.0	2,600	40.5	2,900	42.5	3,100	115
120	33.5	2,200	36.5	2,400	38.5	2,600	120
125	29.5	1,800	32.5	2,000	34.0	2,100	125
130			27.5	1,600	28.5	1,700	130

⚠ WARNING

Do Not Lower 34 Ft. Offset Fly In Working Position Below 26 Degrees Unless Main Boom Length Is 98 Ft. Or Less, Since Loss Of Stability Will Occur Causing A Tipping Condition.

Load Handling Equipment	Weight (lb)
Auxiliary Head Attached	150
60 Ton Hook Block (See Hook Block For Actual Weight)	1,100
40 Ton Hook Block (See Hook Block For Actual Weight)	720
8.5 Ton Hook Ball (See Hook Ball For Actual Weight)	360

22 Ft. Fly Tip Stowed On Boom Base	300
22 Ft. Tip Erected But Not Used	**PROHIBITED**
22 Ft. Tip Stowed On 34 Ft. Offset Fly	**PROHIBITED**

Note: Capacity deductions are for manufacturer-supplied equipment only.

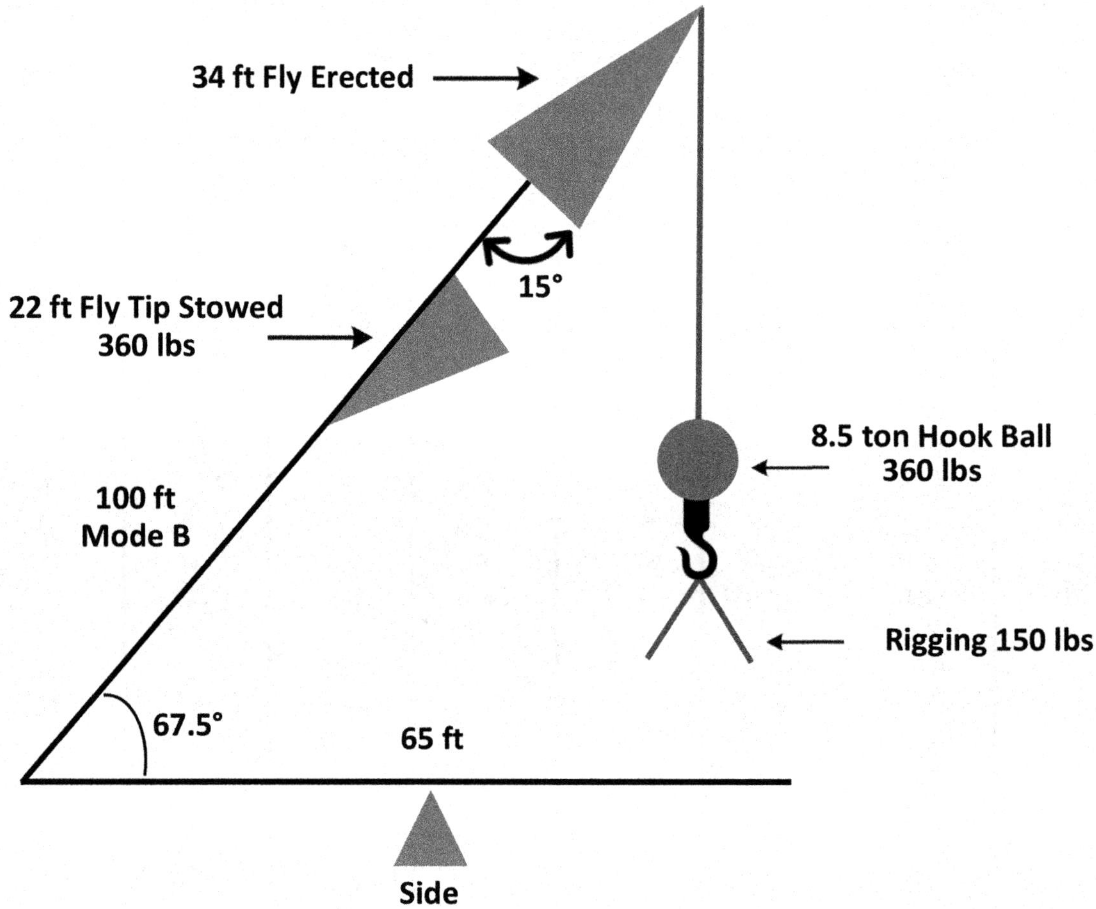
34 ft Fly Erected
15°
22 ft Fly Tip Stowed
360 lbs
8.5 ton Hook Ball
360 lbs
100 ft
Mode B
Rigging 150 lbs
67.5°
65 ft
Side

Telescopic Boom Crane—Swing Cab (TLL)			
Outriggers	Fully Extended & Set	Extension/Jib	34 ft Fly Erected
Counterweight	Full	Extension/ Jib Offset	15°
Tires	N/A	Block	N/A
Crawlers	N/A	Overhaul Ball	8.5 Ton
Main Boom	Length 110 ft (Mode B)	Rigging	150 lbs
Main Boom Angle	72°	Wire Rope (Main Hoist)	18 X 19 Rotation Resistant 1 lb per ft
Radius	N/A	Wire Rope (Auxiliary)	18 X 19 Rotation Resistant 1 lb per ft
Other	Auxiliary Boom Head Installed	Other	22 ft. Fly Stowed

2. What is the net capacity when a lift is made off the fly over the "side" of the crane

A. 8860 lbs

B. 8200 lbs

C. 6640 lbs

D. 7240 lbs

Gross Capacity	=	8200 lbs
22 ft Fly Tip Stowed	-	300 lbs
85 ton Hook Ball	-	360 lbs
Rigging	-	150 lbs
Auxiliary Head	-	150 lbs
Net Capacity	=	7240 lbs

Note: For line in use, when wire rope deductions are required, the rope length shall be measured from the boom tip to the ground.

BOOM MODE "B"

Maximum Allowable Lifting Capacities
Rated Lifting Capacities In Pounds
On Fully Extended Outriggers
See Set Up Note 2.

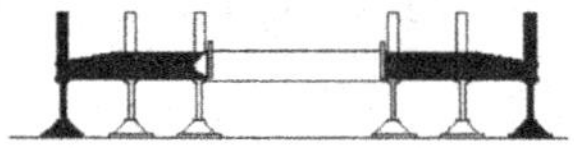

110 ft Main Boom +34 Ft. Offset Fly

Load Radius In Feet	1° Offset		15° Offset		30° Offset		Load Radius In Feet
	Loaded Boom Angle (Deg.)	360°	Loaded Boom Angle (Deg.)	360°	Loaded Boom Angle (Deg.)	360°	
35	76.5	10,500					35
40	74.5	10,500					40
45	72.5	10,500	76.0	9,800			45
50	70.5	9,800	74.0	9,000	77.0	8,300	50
55	68.5	8,900	71.	8,200	75.0	7,700	55
60	66.5	8,200	69.5	7,600	72.5	7,100	60
65	64.0	7,500	67.5	7,000	70.5	6,600	65
70	62.0	6,900	65.0	6,500	68.0	6,200	70
75	59.5	6,400	63.0	6,100	65.5	5,800	75
80	57.5	6,000	60.5	5,700	63.0	5,500	80
85	55.0	5,600	58.0	5,300	60.5	5,100	85
90	52.5	5,100	55.5	5,000	58.0	4,800	90
95	49.5	4,700	53.0	4,700	55.5	4,600	95
100	47.0	4,200	50.0	4,300	52.5	4,300	100
105	43.5	3,600	47.0	3,900	49.5	4,000	105
110	40.5	3,100	43.5	3,400	46.0	3,600	110
115	37.0	2,600	40.5	2,900	42.5	3,100	115
120	33.5	2,200	36.5	2,400	38.5	2,600	120
125	29.5	1,800	32.5	2,000	34.0	2,100	125
130			27.5	1,600	28.5	1,700	130

⚠ WARNING

Do Not Lower 34 Ft. Offset Fly In Working Position Below 26 Degrees Unless Main Boom Length Is 98 Ft. Or Less, Since Loss Of Stability Will Occur Causing A Tipping Condition.

Load Handling Equipment	**Weight (lb)**
Auxiliary Head Attached	150
60 Ton Hook Block (See Hook Block For Actual Weight)	1,100
40 Ton Hook Block (See Hook Block For Actual Weight)	720
8.5 Ton Hook Ball (See Hook Ball For Actual Weight)	360

22 Ft. Fly Tip Stowed On Boom Base	300
22 Ft. Tip Erected But Not Used	**PROHIBITED**
22 Ft. Tip Stowed On 34 Ft. Offset Fly	**PROHIBITED**

Note: Capacity deductions are for manufacturer-supplied equipment only.

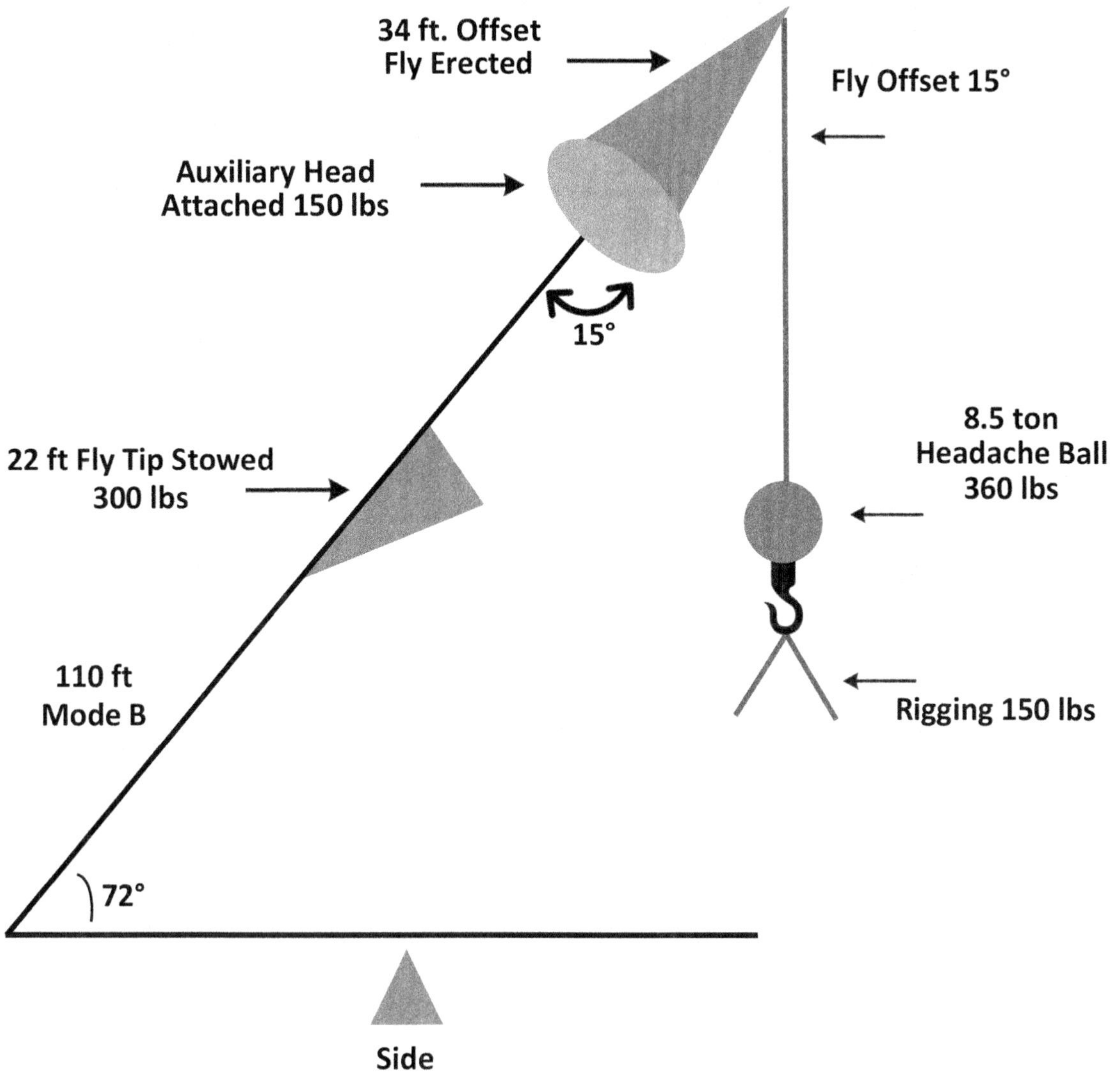
34 ft. Offset
Fly Erected
Fly Offset 15°
Auxiliary Head
Attached 150 lbs
15°
22 ft Fly Tip Stowed
300 lbs
8.5 ton
Headache Ball
360 lbs
110 ft
Mode B
Rigging 150 lbs
72°
Side

Telescopic Boom Crane—Swing Cab (TLL)			
Outriggers	Fully Extended & Set	Extension/Jib	34 ft Offset Fly Stowed
Counterweight	Full	Extension/ Jib Offset	N/A
Tires	N/A	Block	N/A
Crawlers	N/A	Overhaul Ball	8.5 Ton
Main Boom	Length 110 ft (Mode B)	Rigging	200 lbs
Main Boom Angle	59.5°	Wire Rope (Main Hoist)	18 X 19 Rotation Resistant 1 lb per ft
Radius	N/A	Wire Rope (Auxiliary)	18 X 19 Rotation Resistant 1 lb per ft
Other	N/A	Other	N/A

3. What is the net capacity of this crane for a lift over the "rear" with the main boom

A. 10,440 lbs

B. 11,440 lbs

C. 11,000 lbs

D. 9900 lbs

Gross Capacity at 57°	=	11,900 lbs
34 ft Offset Fly Stowed	-	900 lbs
8.5 ton Hook Ball	-	360 lbs
Rigging	-	200 lbs
Net Capacity	=	10,440 lbs

Note: For line in use, when wire rope deductions are required, the rope length shall be measured from the boom tip to the ground.

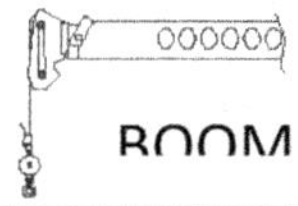

BOOM MODE "B"

Maximum Allowable Lifting Capacities
Rated Lifting Capacities In Pounds
On Fully Extended Outriggers
See Set Up Note 2.

95 Ft. To 110 Ft. Main Boom

Load Radius In Feet	95 Ft.			105 Ft.			110 Ft.			Load Radius In Feet
	Loaded Boom Angle (Deg.)	360°	Over Front	Loaded Boom Angle (Deg)	360°	Over Front	Loaded Boom Angle (Deg.)	**360°**	Over Front	
20	77.5	31,800	31,800							20
25	74.5	28,300	28,300	76.0	25,700	25,700	77.0	22,500	22,500	25
30	71.0	25,300	25,300	73.5	23,100	23,100	74.5	22,200	22,200	30
35	68.0	22,900	22,900	70.5	20,900	20,900	72.0	20,100	20,100	35
40	64.5	20,800	20,800	67.5	19,000	19,000	69.0	18,300	18,300	40
45	61.5	19,000	19,000	65.0	17,400	17,400	66.0	16,700	16,700	45
50	58.0	16,500	17,500	61.5	15,900	15,900	63.5	15,200	15,200	50
55	54.0	13,800	15,900	58.5	13,900	14,700	60.5	13,900	13,900	55
60	50.0	11,800	13,500	55.0	11,900	13,600	57.	11,900	12,500	60
65	45.5	10,000	11,700	51.5	10,100	11,800	54.0	10,200	11,200	65
70	41.0	8,600	10,000	48.0	8,700	10,100	50.5	8,700	10,100	70
75	36.0	7,300	8,700	43.5	7,400	8,800	47.0	7,500	8,800	75
80	30.0	6,300	7,500	39.5	6,400	7,600	43.0	6,400	7,700	80
85	23.0	5,400	6,500	34.5	5,500	6,600	38.5	5,500	6,700	85
90	12.0	4,500	5,600	29.0	4,700	5,700	34.0	4,700	5,800	90
95				22.0	4,000	4,900	28.5	4,000	5,000	95
100				11.5	3,300	4,200	22.0	3,400	4,300	100
105							11.0	2,800	3,700	105
Min. Boom Angle/ Cap.	0°	3,100	3,100	0°	2,100	2,100	0°	1,700	1,700	Min. Boom Angle/ Cap.

Load Handling Equipment	**Weight (lb)**
Auxiliary Head Attached	150
60 Ton Hook Block (See Hook Block For Actual Weight)	1,100
40 Ton Hook Block (See Hook Block For Actual Weight)	720
8.5 Ton Hook Ball (See Hook Ball For Actual Weight)	360

Lifting From Main Boom With:	
22 Ft. Fly Tip Stowed On Boom Base	300
34 Ft. Offset Fly Stowed On Boom Base	900
34 Ft. Offset Fly Erected But Not Used	4,400
56 Ft. Offset Fly Stowed On Boom Base	1,200
56 Ft. Offset Fly Erected But Not Used	7,800

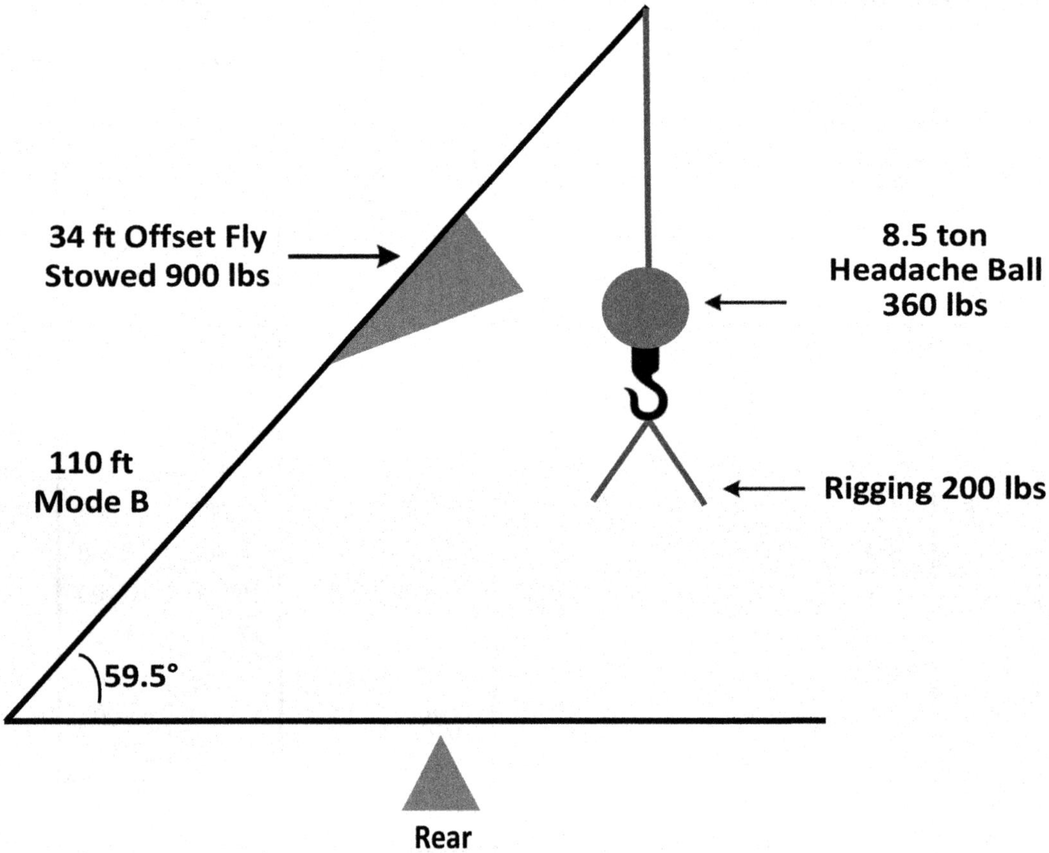
34 ft Offset Fly
Stowed 900 lbs
8.5 ton
Headache Ball
360 lbs
110 ft
Mode B
Rigging 200 lbs
59.5°
Rear

Telescopic Boom Crane—Swing Cab (TLL)			
Outriggers	Fully Extended & Set	Extension/Jib	56 ft Fly Stowed
Counterweight	Full	Extension/ Jib Offset	N/A
Tires	N/A	Block	40 Ton Reeved to 6 Parts
Crawlers	N/A	Overhaul Ball	8.5 Ton Hanging 10 ft below ABH
Main Boom	Length 75 ft (Mode B)	Rigging	450 lbs
Main Boom Angle	N/A	Wire Rope (Main Hoist)	18 X 19 Rotation Resistant 1 lb per ft
Radius	25 ft	Wire Rope (Auxiliary)	18 X 19 Rotation Resistant 1 lb per ft
Other	Auxiliary Boom Head Installed	Other	N/A

4. What is the net capacity of this crane when a lift is made off the main boom over the "side"

A. 37,100 lbs

B. 39,630 lbs

C. 38,650 lbs

D. 39,120 lbs

Gross Capacity	=	41,700 lbs
56 ft Offset Fly Stowed	-	1200 lbs
40 ton Hook Block	-	720 lbs
8.5 ton Hook Ball	-	360 lbs
Rigging	-	450 lbs
Excess Wire Rope	-	160 lbs
Excess Wire Rope	-	10 lbs
Auxiliary Head	-	150 lbs
Net Capacity	=	38,650 lbs

Note: For line in use, when wire rope deductions are required, the rope length shall be measured from the boom tip to the ground.

65 Ft. To 85 Ft. Main Boom										
Load Radius In Feet	65 Ft.			75 Ft.			85 Ft.			Load Radius In Feet
	Loaded Boom Angle (Deg.)	360°	Over Front	Loaded Boom Angle (Deg.)	360°	Over Front	Loaded Boom Angle (Deg.)	360°	Over Front	
12	77.0	42,000	42,000							12
15	74.5	42,000	42,000	77.0	42,000	42,000				15
20	70.0	42,000	42,000	73.0	42,000	42,000	75.5	36,000	36,000	20
25	65.5	42,000	42,000	69.0	41,700	41,700	72.0	31,500	31,500	25
30	60.5	40,700	40,500	65.0	37,100	37,100	68.5	28,200	28,200	30
35	55.0	30,700	35,100	60.5	30,900	32,500	64.5	25,400	25,400	35
40	49.0	24,200	27,600	56.0	24,400	27,800	61.0	23,000	23,000	40
45	43.0	19,500	22,300	51.0	19,700	22,600	57.0	19,900	21,100	45
50	35.5	16,000	18,400	46.0	16,300	18,700	52.5	16,400	18,800	50
55	27.0	13,300	15,400	40.0	13,600	15,600	48.0	13,700	15,800	55
60	13.5	11,100	12,900	33.5	11,500	13,200	43.0	11,700	13,400	60
65				25.0	9,700	11,300	38.0	9,900	11,500	65
70				12.5	8,200	9,700	31.5	8,400	9,900	70
75							24.0	7,200	8,500	75
80							12.0	6,100	7,300	80
Min. Boom Angle/ Cap.	0°	8,000	8,000	0°	5,900	5,900	0°	4,300	4,300	Min. Boom Angle/ Cap.

Load Handling Equipment	Weight (lb)
Auxiliary Head Attached	150
60 Ton Hook Block (See Hook Block For Actual Weight)	1,100
40 Ton Hook Block (See Hook Block For Actual Weight	720
8.5 Ton Hook Ball (See Hook Ball For Actual Weight)	360
Lifting From Main Boom With:	
22 Ft. Fly Tip Stowed On Boom Base	300
34 Ft. Offset Fly Stowed On Boom Base	900
34 Ft. Offset Fly Erected But Not Used	4,400
56 Ft. Offset Fly Stowed On Boom	1,200
56 Ft. Offset Fly Erected But Not Used	7,800

WORKING RANGE DIAGRAM

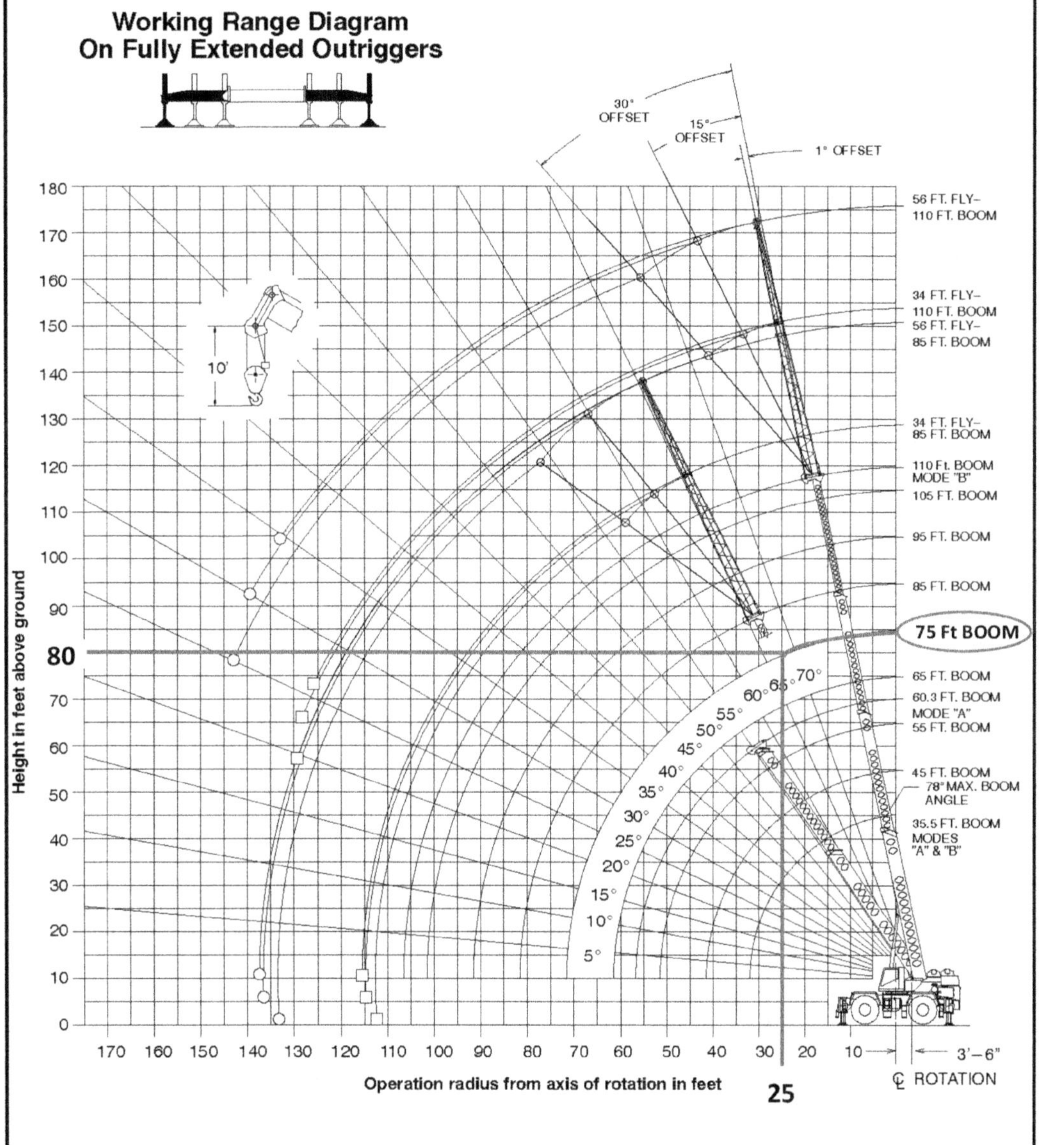

○ Denotes Main Boom + 56' Fly–Boom Mode "B"
□ Denotes Main Boom + 34' Fly–Boom Mode "B"
Note: Boom and fly geometry shown are for unloaded condition and crane standing level on firm supporting surface. Boom deflection, subsequent radius and boom angle change must be accounted for when applying load to hook.

14. Rated lifting capacities are based on correct reeving. Deduction must be made for excessive reeving. Any reeving over minimum required (See Wire Rope Strength) is considered excessive and must be accounted for when making lifts.
Use working rage diagram to estimate the extra feet of rope then

Boom Length = 75 ft
Radius = 25 ft
Boom Height is 80 ft X 2 Parts
Of Excess Reeving = 160 X 1 lb
Per foot = 160 lbs

WINCH PERFORMANCE

Winch Line Pulls			Drum Rope Capacity (ft)	
	Two Speed Winch			
Wire Rope Layer	Low Speed Available lb	High Speed Available lb	Layer	Total
1	16,266*	7,726	102	102
2	14,998*	7,124	111	213
3	13,914*	6,609	120	333
4	12,976*	6,164	128	461
5	12,156	5,774	137	598
6	11,434	5,431	145	743

* Reduce to 12,920 lb if using Type RB Rope.

WIRE ROPE STRENGTH

Maximum Lifting Capacities Based On Wire Rope Strength		
Parts of Line	3/4" Type RB	Notes
1*	12,920	Capacities shown are in pounds and working loads must not exceed the ratings on the capacity charts in the Crane Rating Manual.
2	25,840	Study Operator's Manual for wire rope inspection procedures.
3	38,760	*Use of swivel end with 1 part of line is not recommended.
4	51,680	
5	64,600	
6	77,520	
7	90,440	
8	103,360	
9	116,280	
10	129,200	

LBCE	DESCRIPTION
TYPE RB	18 X 19 Rotation Resistant – Extra Improved Plow Steel – Preformed Right Lay – Regular Lay, Swaged

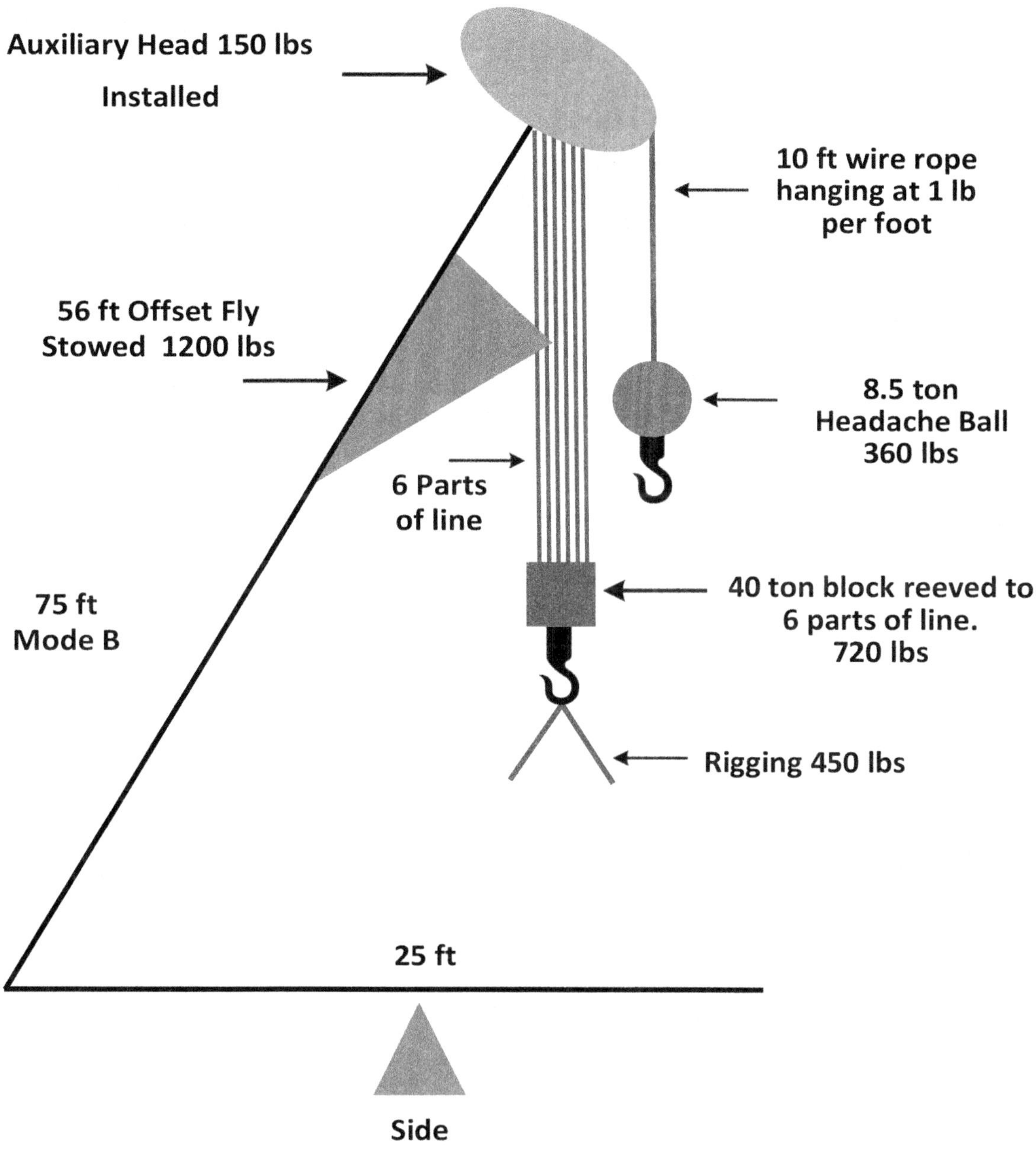
Auxiliary Head 150 lbs
Installed
10 ft wire rope
hanging at 1 lb
per foot
56 ft Offset Fly
Stowed 1200 lbs
8.5 ton
Headache Ball
360 lbs
6 Parts
of line
75 ft
Mode B
40 ton block reeved to
6 parts of line.
720 lbs
Rigging 450 lbs
25 ft
Side

Telescopic Boom Crane—Swing Cab (TLL)			
Outriggers	Fully Extended & Set	Extension/Jib	56 ft Offset Fly Erected
Counterweight	Full	Extension/ Jib Offset	30°
Tires	N/A	Block	60 Ton Hook Block Reeved to 4 Parts
Crawlers	N/A	Overhaul Ball	8.5 Overhaul Ball Hook Ball Hanging 15 ft
Main Boom	Length 85 ft (Mode B)	Rigging	210 lbs
Main Boom Angle	N/A	Wire Rope (Main Hoist)	LBCE Type RB 18 X 19
Radius	35 ft	Wire Rope (Auxiliary)	LBCE Type RB 18 X 19
Other	Auxiliary Head Installed	Other	Spreader Bar 450 lbs

5. You are making a lift over the front of the crane using the 60 ton Hook Block. What is your Net Capacity with the crane in this configuration

A. 15,145 lbs

B. 15,330 lbs

C. 15,540 lbs

D. 15,690 lbs

Gross Capacity	=	25,400 lbs
Auxiliary Head	-	150 lbs
56 ft Offset Fly Stowed	-	7800 lbs
60 ton Hook Block	-	1100 lbs
Excess Wire Rope	-	170 lbs
8.5 ton Hook Ball	-	360 lbs
Excess Wire Rope	-	15 lbs
Rigging	-	210 lbs
Spreader Bar	-	450 lbs
Net Capacity	=	15,145 lbs

Note: For line in use, when wire rope deductions are required, the rope length shall be measured from the boom tip to the ground.

65 Ft. To 85 Ft. Main Boom										
Load Radius In Feet	65 Ft.			75 Ft.			85 Ft.			Load Radius In Feet
	Loaded Boom Angle (Deg.)	360°	Over Front	Loaded Boom Angle (Deg.)	360°	Over Front	Loaded Boom Angle (Deg.)	360°	Over Front	
12	77.0	42,000	42,000							12
15	74.5	42,000	42,000	77.0	42,000	42,000				15
20	70.0	42,000	42,000	73.0	42,000	42,000	75.5	36,000	36,000	20
25	65.5	42,000	42,000	69.0	41,700	41,700	72.0	31,500	31,500	25
30	60.5	40,700	40,500	65.0	37,100	37,100	68.5	28,200	28,200	30
35	55.0	30,700	35,100	60.5	30,900	32,500	64.5	25,400	**25,400**	**35**
40	49.0	24,200	27,600	56.0	24,400	27,800	61.0	23,000	23,000	40
45	43.0	19,500	22,300	51.0	19,700	22,600	57.0	19,900	21,100	45
50	35.5	16,000	18,400	46.0	16,300	18,700	52.5	16,400	18,800	50
55	27.0	13,300	15,400	40.0	13,600	15,600	48.0	13,700	15,800	55
60	13.5	11,100	12,900	33.5	11,500	13,200	43.0	11,700	13,400	60
65				25.0	9,700	11,300	38.0	9,900	11,500	65
70				12.5	8,200	9,700	31.5	8,400	9,900	70
75							24.0	7,200	8,500	75
80							12.0	6,100	7,300	80
Min. Boom Angle/ Cap.	0°	8,000	8,000	0°	5,900	5,900	0°	4,300	4,300	Min. Boom Angle/ Cap.

Load Handling Equipment	**Weight (lb)**
Auxiliary Head Attached	150
60 Ton Hook Block (See Hook Block For Actual	1,100
40 Ton Hook Block (See Hook Block For Actual Weight)	720
8.5 Ton Hook Ball (See Hook Ball For Actual Weight)	360
Lifting From Main Boom With:	
22 Ft. Fly Tip Stowed On Boom Base	300
34 Ft. Offset Fly Stowed On Boom Base	900
34 Ft. Offset Fly Erected But Not Used	4,400
56 Ft. Offset Fly Stowed On Boom Base	1,200
56 Ft. Offset Fly Erected But Not Used	7,800

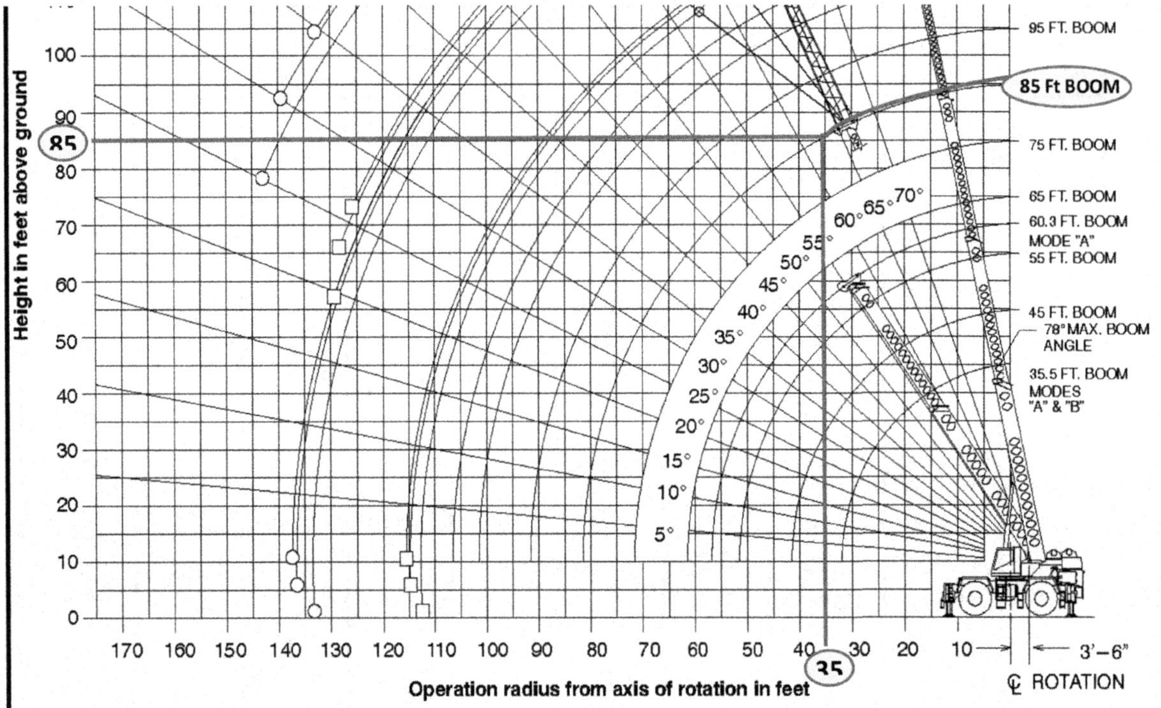

2 parts excess reeving X 85 ft = 170 X 11 lbs per ft = 170 deduction

○ Denotes Main Boom + 56' Fly—Boom Mode "B"

□ Denotes Main Boom + 34' Fly—Boom Mode "B"

Note: Boom and fly geometry shown are for unloaded condition and crane standing level on firm supporting surface. Boom deflection, subsequent radius and boom angle change must be accounted for when applying load to hook.

Maximum Lifting Capacities Based On Wire Rope Strength

Parts of Line	3/4" Type RB	Notes
1*	12,920	Capacities shown are in pounds and working loads must not exceed the ratings on the capacity charts in the Crane Rating Manual.
2	25,840	Study Operator's Manual for wire rope inspection procedures.
3	38,760	*Use of swivel end with 1 part of line is not recommended.
4	51,680	
5	64,600	
6	77,520	14. Rated lifting capacities are based on correct reeving. Deduction must be made for excessive reeving. Any reeving over minimum required (See Wire Rope Strength) is considered excessive and must be accounted for when making lifts. Use working range diagram to estimate the extra feet of rope then deduct 1 lb. for each extra foot of wire rope before attempting to lift the load.
7	90,440	
8	103,360	
9	116,280	
10	129,200	

LBCE	DESCRIPTION
TYPE RB	18 X 19 Rotation Resistant – Extra Improved Plow Steel – Preformed Right Lay – Regular Lay, Swaged

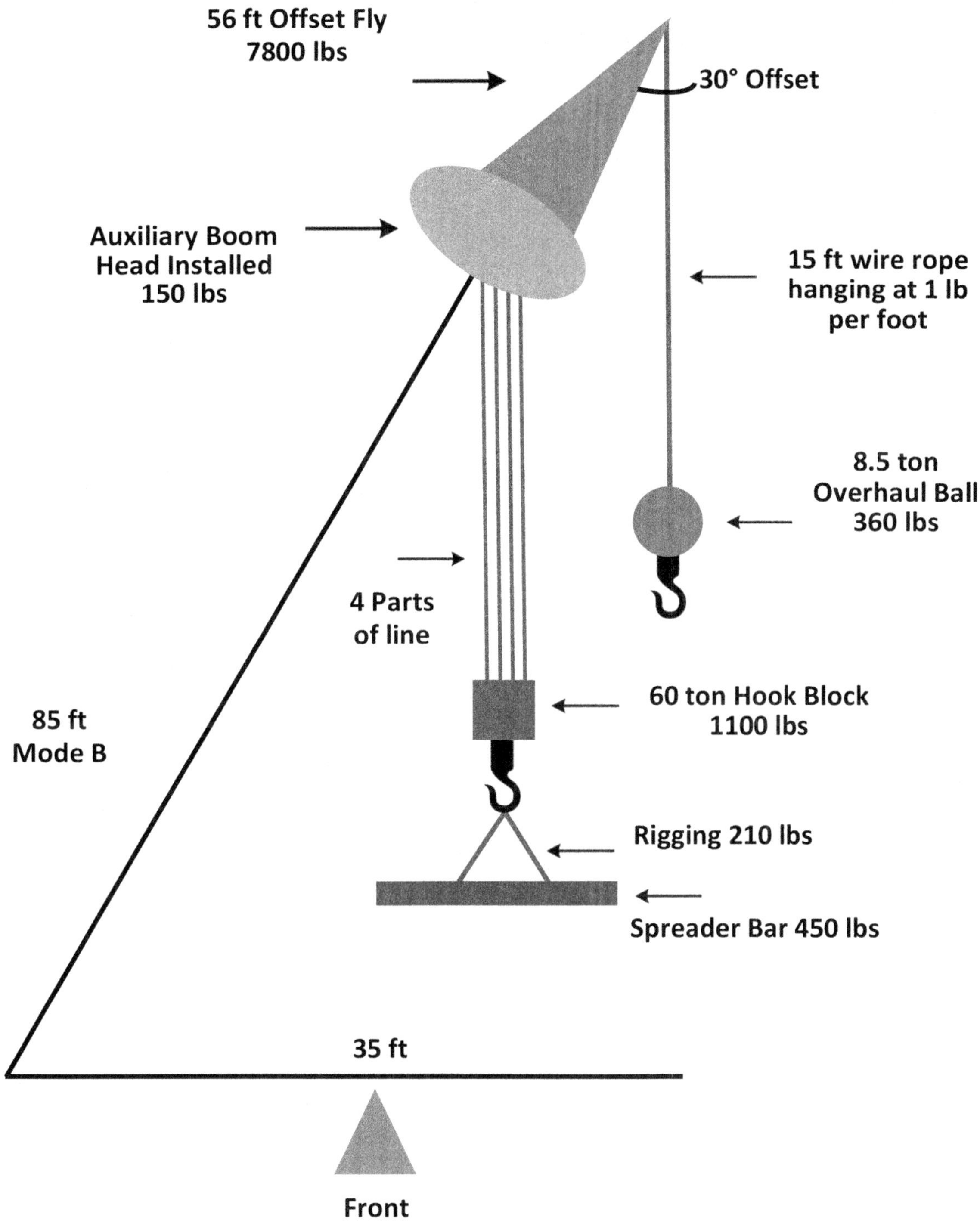
56 ft Offset Fly
7800 lbs
30° Offset
Auxiliary Boom
Head Installed
150 lbs
15 ft wire rope
hanging at 1 lb
per foot
8.5 ton
Overhaul Ball
360 lbs
4 Parts
of line
60 ton Hook Block
1100 lbs
85 ft
Mode B
Rigging 210 lbs
Spreader Bar 450 lbs
35 ft
Front

Telescopic Boom Crane—Swing Cab (TLL)			
Outriggers	Full Extended and Set	Extension/Jib	56 ft Offset Fly Erected
Counterweight	Full	Extension/ Jib Offset	1°
Tires	N/A	Block	40 Ton Hook Block 4 Parts of Line Hanging 10 ft
Crawlers	N/A	Overhaul Ball	8.5 Ton Overhaul Ball
Main Boom	Length 110 ft (Mode B)	Rigging	150 lbs
Main Boom Angle	N/A	Wire Rope (Main Hoist)	18 X 19 Rotation Resistant 1 lb per ft
Radius	100 ft	Wire Rope (Auxiliary)	18 X 19 Rotation Resistant 1 lb per ft
Other	Auxiliary Head Installed	Other	N/A

6. You are planning a lift over the side from the Offset Fly. Your crane is properly set up on firm level ground. What is your net capacity in this configuration

A. 880 lbs

B. 2120 lbs

C. 2280 lbs

D. 3080 lbs

Gross Capacity	=	3500 lbs
Auxiliary Head	-	150 lbs
40 ton Hook Block	-	720 lbs
4 Parts of Rope Hanging	-	40 lbs
8.5 ton Hook Ball	-	360 lbs
Rigging	-	150 lbs
Net Capacity	=	2280 lbs

Note: For line in use, when wire rope deductions are required, the rope length shall be measured from the boom tip to the ground.

110 ft Main Boom + 56 Ft. Offset Fly							
Load Radius In Feet	1° Offset		15° Offset		30° Offset		Load Radius In Feet
	Loaded Boom Angle (Deg.)	360°	Loaded Boom Angle (Deg.)	360°	Loaded Boom Angle (Deg.)	360°	
40	77.0	7,000					40
45	75.5	7,000					45
50	74.0	7,000					50
55	72.5	7,000	77.5	6,400			55
60	71.0	6,400	75.5	5,900			60
65	69.0	5,900	73.5	5,400	78.0*	5,000	65
70	67.0	5,400	71.5	5,000	76.0	4,600	70
75	65.0	5,000	70.0	4,600	74.0	4,300	75
80	63.0	4,600	68.0	4,300	72.0	4,000	80
85	61.5	4,300	66.0	4,000	70.0	3,800	85
90	59.5	4,000	64.0	3,700	68.0	3,500	90
95	57.0	3,700	61.5	3,500	66.0	3,300	95
100	55.0	3,500	59.5	3,300	63.5	3,100	100
105	53.0	3,300	57.5	3,100	61.5	2,900	105
110	50.5	3,100	55.0	2,900	59.0	2,800	110
115	48.5	2,900	53.0	2,700	56.5	2,600	115
120	46.0	2,600	50.5	2,600	54.0	2,500	120
125	43.0	2,300	47.5	2,400	51.0	2,300	125
130	40.5	1,900	45.0	2,200	48.0	2,100	130
135	37.5	1,600	42.0	1,900	45.0	1,900	135
140			38.5	1,500	41.5	1,700	140
145					37.0	1,400	145

⚠ WARNING

Do Not Lower 56 Ft. Offset Fly In Working Position Below 34.5 Degrees Unless Main Boom Length Is 89 Ft. Or Less, Since Loss Of Stability Will Occur Causing A Tipping Condition.

Load Handling Equipment	Weight (lb)
Auxiliary Head Attached	150
60 Ton Hook Block (See Hook Block For Actual Weight)	1,100
40 Ton Hook Block (See Hook Block For Actual Weight)	720
8.5 Ton Hook Ball (See Hook Ball For Actual Weight)	360

14. Rated lifting capacities are based on correct reeving. Deduction must be made for excessive reeving. Any reeving over minimum required (See Wire Rope Strength) is considered excessive and must be accounted for when making lifts. Use working range diagram to estimate the extra feet of rope then deduct 1 lb. for each extra foot of wire rope before attempting to lift the load.

40 ton Block is reeved to 4 parts of line that is all excessive.

4 Parts hanging 10 lft X 1 lb per foot = 40 lbs of wire rope deduction

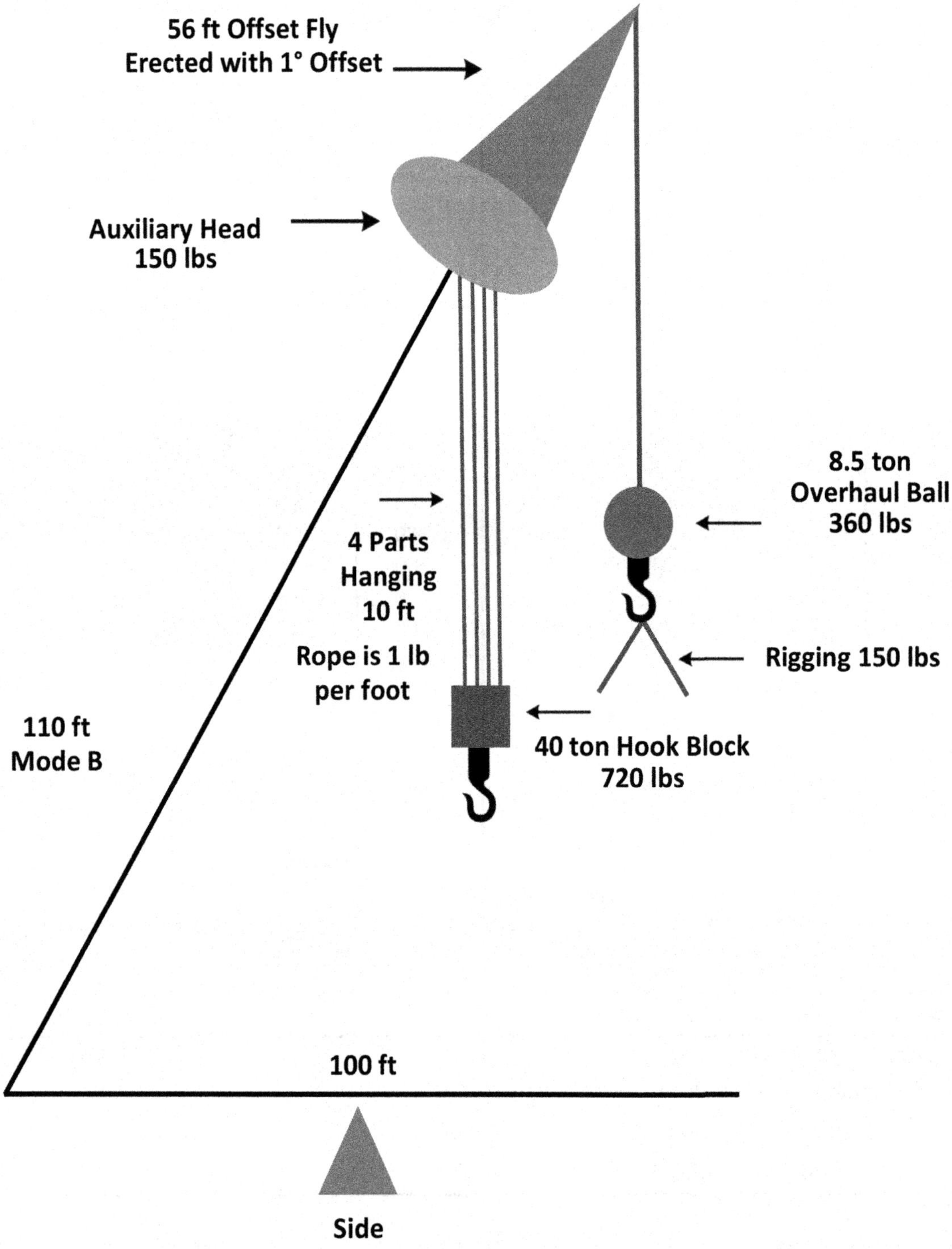
56 ft Offset Fly
Erected with 1° Offset
Auxiliary Head
150 lbs
8.5 ton
Overhaul Ball
360 lbs
4 Parts
Hanging
10 ft
Rope is 1 lb
per foot
Rigging 150 lbs
40 ton Hook Block
720 lbs
110 ft
Mode B
100 ft
Side

WORKING RANGE DIAGRAM

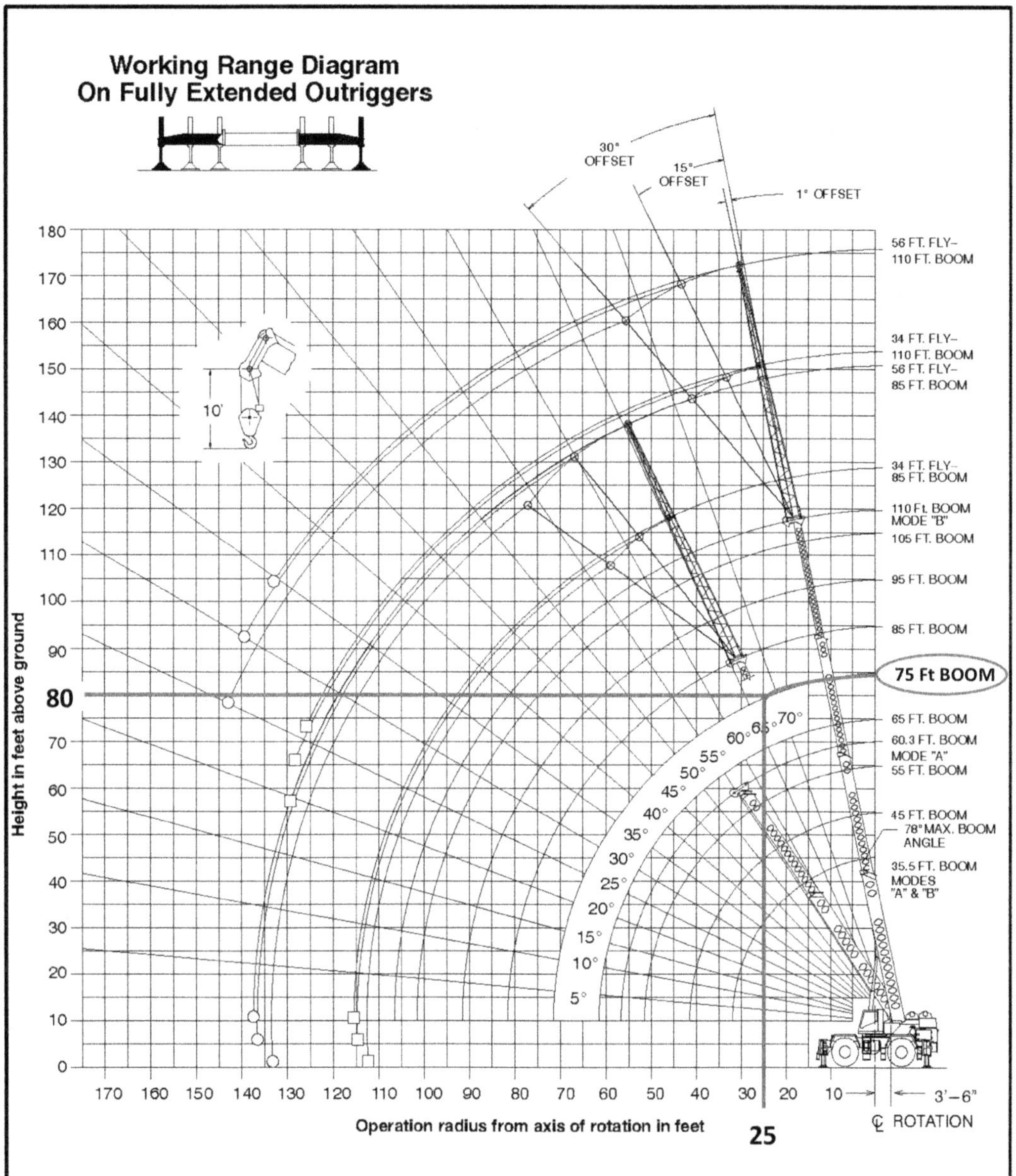

○ Denotes Main Boom + 56' Fly—Boom Mode "B"
□ Denotes Main Boom + 34' Fly—Boom Mode "B"

Note: Boom and fly geometry shown are for unloaded condition and crane standing level on firm supporting surface. Boom deflection, subsequent radius and boom angle change must be accounted for when applying load to hook.

14. Rated lifting capacities are based on correct reeving. Deduction must be made for excessive reeving. Any reeving over minimum required (See Wire Rope Strength) is considered excessive and must be accounted for when making lifts.
Use working rage diagram to estimate the extra feet of rope then

Boom Length = 75 ft
Radius = 25 ft
Boom Height is 80 ft X 10 Parts Of Excess Reeving
X 1 lb per foot = 800 lbs deduction

WINCH PERFORMANCE

Winch Line Pulls			Drum Rope Capacity (ft)	
	Two Speed Winch			
Wire Rope Layer	Low Speed Available lb	High Speed Available lb	Layer	Total
1	16,266*	7,726	102	102
2	14,998*	7,124	111	213
3	13,914*	6,609	120	333
4	12,976*	6,164	128	461
5	12,156	5,774	137	598
6	11,434	5,431	145	743

* Reduce to 12,920 lb if using Type RB Rope.

WIRE ROPE STRENGTH

Maximum Lifting Capacities Based On Wire Rope Strength		
Parts of Line	3/4" Type RB	Notes
1*	12,920	Capacities shown are in pounds and working loads must not exceed the ratings on the capacity charts in the Crane Rating Manual.
2	25,840	
3	38,760	Study Operator's Manual for wire rope inspection procedures.
4	51,680	*Use of swivel end with 1 part of line is not recommended.
5	64,600	
6	77,520	
7	90,440	
8	103,360	
9	116,280	
10	129,200	

LBCE	DESCRIPTION
TYPE RB	18 X 19 Rotation Resistant – Extra Improved Plow Steel – Preformed Right Lay – Regular Lay, Swaged

Lattice Boom Crane			
Outriggers	N/A	Extension/Jib	30 ft Erected 2,700 lbs
Counterweight	Crane Counter Wt. 179,100 lbs Car Body Counter Wt. 44,000 lbs	Extension/ Jib Offset	25°
Tires	N/A	Block	35 Ton Reeved 2 Parts 1,267 lbs
Crawlers	Crawlers Extended	Overhaul Ball	15 Ton (722 lbs) Hanging 15 feet from jib tip
Main Boom	230 ft	Rigging	160 lbs
Main Boom Angle	N/A	Wire Rope (Main Hoist)	1 in. 6 X 25 EIPS IWRC @ 2 lbs per ft
Radius	110 ft	Wire Rope (Auxiliary)	1 in. 6 X 25 EIPS IWRC @ 2 lbs per ft
Other	N/A	Other	N/A

1. Lifting on the main boom, the maximum radius at which you can place a net load of 10 tons over the side is

A. 125

B. 110

C. 130

D. 120

Net Capacity	=	20,000 lbs
30 ft. Jib	-	2700 lbs
Hook Block	-	1267 lbs
Overhaul Ball	-	722 lbs
Rigging	-	160 lbs
Wire Rope	-	882 lbs
Total Load	=	25,731 lbs

LIFTCRANE BOOM CAPACITIES
BOOM NO. 22EL
179,100 LB. CRANE COUNTERWEIGHT
44,000 LB. CARBODY COUNTERWEIGHT
360 DEGREE RATING

Lattice Boom Crane (LAT)

Meets ANSI B30.5 Requirements

Oper. Radius Feet	Boom Angle Degree	Boom Point Elevation Feet	Boom Capacity Crawlers Retracted Pounds	Boom Capacity Crawlers Extended Pounds
230 FT. Boom				
36	82.8	236.0	133,600 *	133,600 *
38	82.3	235.7	133,300 *	133,300 *
40	81.8	235.4	130,200	132,900 *
42	81.3	235.1	121,300	131,500 *
44	80.8	234.7	113,400	129,700 *
46	80.3	234.4	106,400	126,100
48	79.8	234.0	100,000	118,500
50	79.3	233.6	94,300	111,500
55	78.0	232.6	82,100	97,000
60	76.7	231.4	72,200	85,300
65	75.4	230.1	64,100	75,800
70	74.1	228.7	57,300	67,800
75	72.8	227.2	51,500	61,100
80	71.5	225.5	46,500	55,300
85	70.2	223.7	42,200	50,300
90	68.9	221.8	38,400	45,900
95	67.5	219.7	35,100	42,000
100	66.2	217.5	32,100	38,600
105	64.8	215.2	29,400	35,500
110	63.4	212.7	27,000	32,700
115	62.0	210.1	24,800	30,200
120	60.6	207.3	22,700	28,000
125	59.1	204.3	20,900	**25,900**
130	57.6	201.1	19,200	24,000
135	56.1	197.8	17,700	22,200
140	54.6	194.3	16,200	20,600
145	53.1	190.6	14,900	19,100
150	51.5	186.6	13,700	17,700
155	49.9	182.4	12,500	16,300
160	48.2	178.0	11,400	15,100
165	46.5	173.3	10,400	14,000
170	44.7	168.3	9,500	12,900
175	42.9	163.0	8,600	11,900
180	41.0	157.3	7,700	10,900
185	39.1	151.3	6,900	10,000
190	37.0	144.7	6,100	9,200
195	34.9	137.7	5,400	8,400
200	32.6	130.1	4,700	7,600
205	30.2	121.7	4,100	6,900
210	27.5	112.4		6,200
215	24.6	101.9		5,500
220	21.4	89.8		4,800
225	17.5	75.2		4,200

Oper. Radius Feet	Boom Angle Degree	Boom Point Elevation Feet	Boom Capacity Crawlers Retracted Pounds	Boom Capacity Crawlers Extended Pounds
240 FT. BOOM				
38	82.6	245.8	121,300 *	121,300 *
40	82.2	245.5	121,000 *	121,000 *
42	81.7	245.2	120,700 *	120,700 *
44	81.2	244.9	113,000	120,300 *
46	80.7	244.6	106,000	120,000 *
48	80.2	244.2	99,600	118,100
50	79.7	243.8	93,900	111,200
55	78.5	242.8	81,700	96,600
60	77.3	241.7	71,800	84,900
65	76.1	240.5	63,700	75,400
70	74.8	239.1	56,900	67,400
75	73.6	237.6	51,100	60,700
80	72.3	236.1	46,100	54,900
85	71.1	234.4	41,800	49,900
90	69.8	232.5	38,000	45,500
95	68.5	230.6	34,600	41,600
100	67.2	228.5	31,600	38,200
105	65.9	226.3	28,900	35,100
110	64.6	223.9	26,500	32,300
115	63.3	221.4	24,300	29,800
120	61.9	218.7	22,300	27,500
125	60.5	215.9	20,500	25,400
130	59.2	213.0	18,800	23,500
135	57.7	209.8	17,200	21,800
140	56.3	206.5	15,800	20,100
145	54.9	203.0	14,500	18,600
150	53.4	199.3	13,200	17,200
155	51.9	195.4	12,100	15,900
160	50.3	191.3	11,000	14,700
165	48.7	187.0	10,000	13,600
170	47.1	182.4	9,000	12,500
175	45.4	177.5	8,100	11,500
180	43.7	172.3	7,300	10,500
185	41.9	166.8	6,500	9,600
190	40.1	161.0	5,700	8,800
195	38.2	154.7	5,000	8,000
200	36.2	148.0	4,300	7,200
205	34.1	140.8		6,500
210	31.9	132.9		5,800
215	29.5	124.3		5,100
220	26.9	114.7		4,500

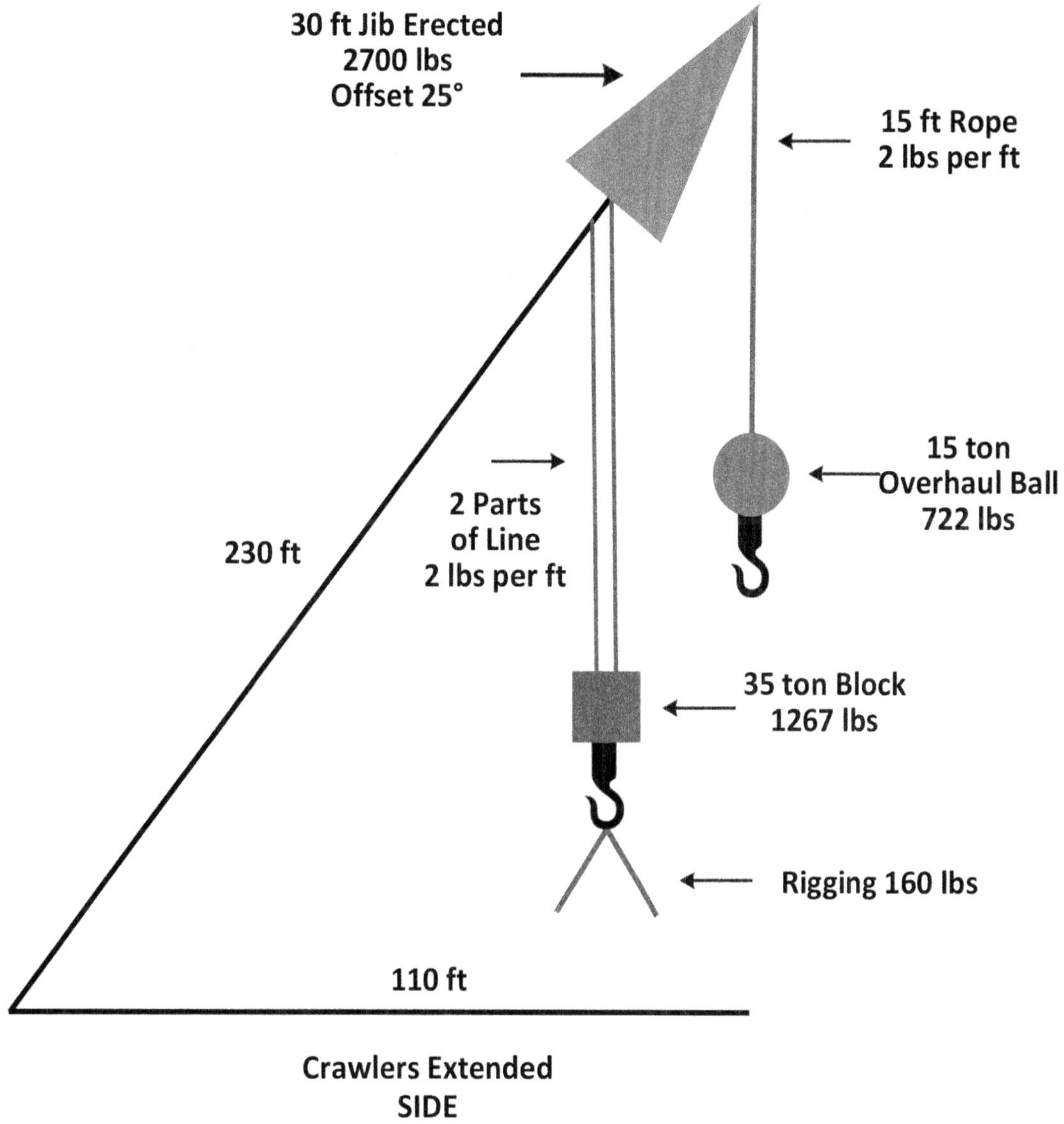
30 ft Jib Erected
2700 lbs
Offset 25°
15 ft Rope
2 lbs per ft
15 ton
Overhaul Ball
722 lbs
2 Parts
of Line
2 lbs per ft
230 ft
35 ton Block
1267 lbs
Rigging 160 lbs
110 ft
Crawlers Extended
SIDE

Lattice Boom Crane			
Outriggers	N/A	Extension/Jib	N/A
Counterweight	Crane Counter Wt. 179,100 lbs Car Body Counter Wt. 44,000 lbs	Extension/ Jib Offset	N/A
Tires	N/A	Block	70 Ton Reeved 8 Parts of Line 1,711 lbs
Crawlers	Crawlers Extended	Overhaul Ball	N/A
Main Boom	290 ft	Rigging	350 lbs
Main Boom Angle	63.8°	Wire Rope (Main Hoist)	1 in. 6 X 25 EIPS IWRC @ 2 lbs per ft
Radius	N/A	Wire Rope (Auxiliary)	1 in. 6 X 25 EIPS IWRC @ 2 lbs per ft
Other	Spreader Bar 550 lbs	Other	N/A

2. Your crane is set up on firm level ground with your crawlers extended. You are lifting over the rear of the crane. What is your net capacity in this configuration

A. 8,835 lbs

B. 9,185 lbs

C. 9,735 lbs

D. 14,012 lbs

Gross Capacity	=	20,900 lbs
Spreader Bar	-	550 lbs
70 Ton Block	-	1711 lbs
Wire Rope Boom Point is 267.3 ft X 8 parts of line X 2 lbs per ft	-	4277 lbs
Rigging	-	350 lbs
Net Capacity	=	14,012 lbs

Oper. Radius Feet	Boom Angle Degree	Boom Point Elevation Feet	Boom Capacity Crawlers Retracted Pounds	Boom Capacity Crawlers Extended Pounds
290 FT. Boom				
44	82.7	295.4	75,900 *	75,900 *
46	82.3	295.2	75,600 *	75,600 *
48	81.9	294.9	75,300 *	75,300 *
50	81.5	294.6	75,100 *	75,100 *
55	80.5	293.7	74,300 *	74,300 *
60	79.5	292.8	71,100	73,100 *
65	78.5	291.8	63,000	71,000 *
70	77.5	290.7	56,200	66,800
75	76.5	289.5	50,300	60,000
80	75.5	288.2	45,400	54,200
85	74.4	286.8	41,000	49,100
90	73.4	285.3	37,200	44,700
95	72.4	283.8	33,800	40,900
100	71.3	282.1	30,800	37,400
105	70.3	280.3	28,100	34,300
110	69.2	278.4	25,700	31,500
115	68.2	276.4	23,500	29,000
120	67.1	274.3	21,500	26,700
125	66.0	272.1	19,600	24,600
130	64.9	269.8	17,900	22,700
135	**63.8**	**267.3**	16,400	**20,900**
140	62.7	264.7	14,900	19,300
145	61.6	262.0	13,600	17,800
150	60.4	259.2	12,400	16,400
155	59.3	256.3	11,200	15,100
160	58.1	253.2	10,100	13,800
165	57.0	249.9	9,100	12,700
170	55.8	246.6	8,200	11,600
175	54.6	243.0	7,300	10,600
180	53.3	239.3	6,400	9,700
185	52.1	235.5	5,600	8,800
190	50.8	231.4	4,900	7,900
195	49.5	227.2	4,200	7,100
200	48.2	222.7		6,400
205	46.8	218.1		5,600
210	45.5	213.2		4,900
215	44.0	208.1		4,300

LIFTCRANE BOOM CAPACITIES
BOOM NO. 22EL
179,100 LB. CRANE COUNTERWEIGHT
44,000 LB. CARBODY COUNTERWEIGHT
360 DEGREE RATING

Lattice Boom Crane (LAT)

Meets ANSI B30.5 Requirements

Weight of all load block, hooks, weight ball, slings, hoist lines, etc., beneath boom and jib point sheaves, is considered part of jib load. Boom and jib are not to be lowered beyond radii where combined weights are greater than rated capacity. Where no capacity is shown, operation is not intended or approved.

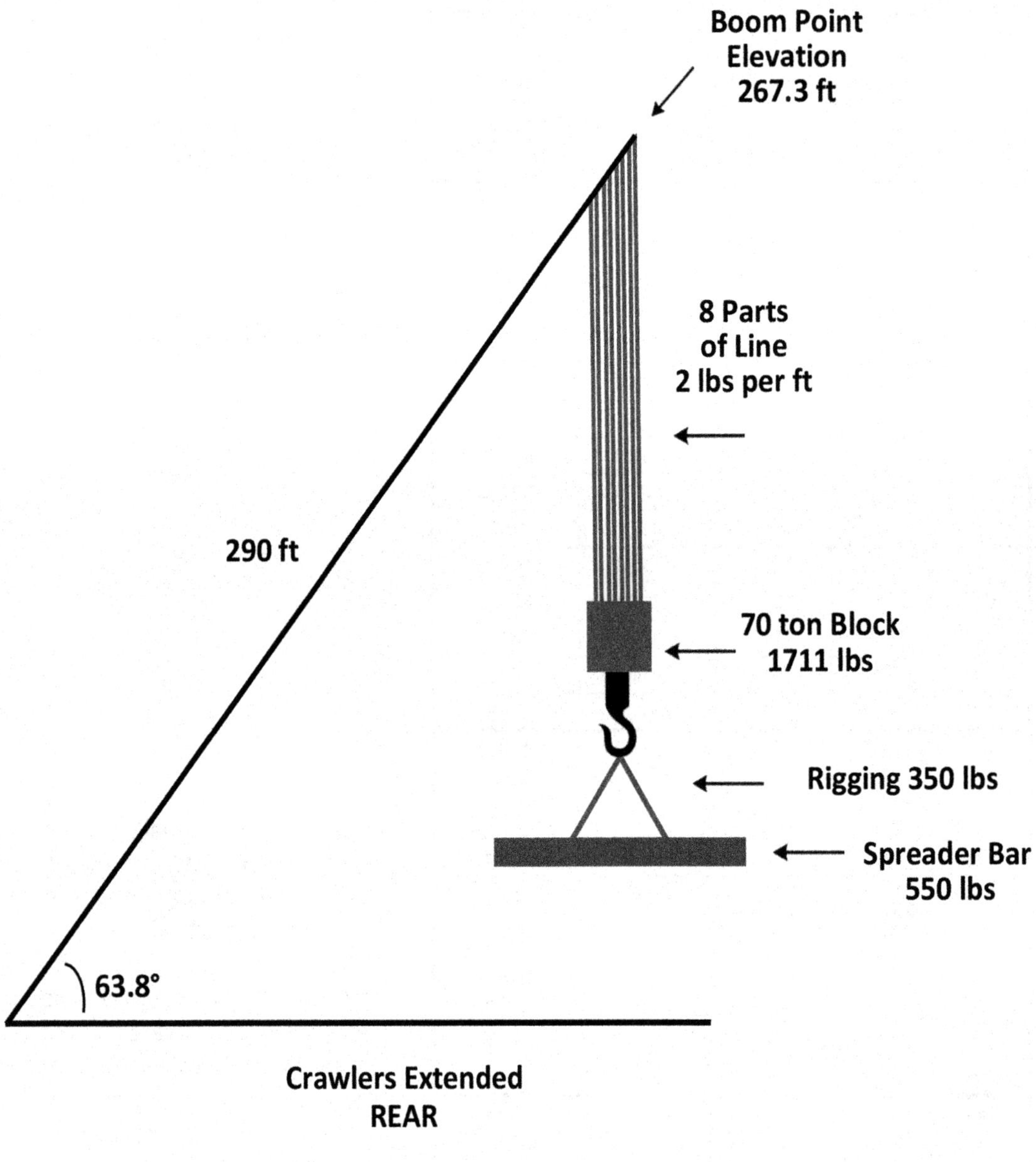
Boom Point
Elevation
267.3 ft
8 Parts
of Line
2 lbs per ft
290 ft
70 ton Block
1711 lbs
Rigging 350 lbs
Spreader Bar
550 lbs
63.8°
Crawlers Extended
REAR

Lattice Boom Crane			
Outriggers	N/A	Extension/Jib	N/A
Counterweight	Crane Counter Wt. 179,100 lbs Car Body Counter Wt. 44,000 lbs	Extension/ Jib Offset	N/A
Tires	N/A	Block	70 Ton Reeved 4 Parts 1,711 lbs
Crawlers	Crawlers Retracted	Overhaul Ball	N/A
Main Boom	180 ft	Rigging	195 lbs
Main Boom Angle	N/A	Wire Rope (Main Hoist)	1 in. 6 X 25 EIPS IWRC @ 2 lbs per ft
Radius	80 ft	Wire Rope (Auxiliary)	1 in. 6 X 25 EIPS IWRC @ 2 lbs per ft
Other	N/A	Other	N/A

3. What would your net capacity be when lifting with the main boom over the "side" of the crane

A. 45,220

B. 48,500

C. 48,305

D. 45,349

Gross Capacity	=	48,500 lbs
Hook Block	-	1711 lbs
Rigging	-	195 lbs
Wire Rope Boom Point is 171.7 ft Elevation X 4 = 686.8 lbs X 2 lbs per ft	-	1374 lbs
Net Capacity	=	45,220 lbs

LIFTCRANE BOOM CAPACITIES
BOOM NO. 22EL
179,100 LB. CRANE COUNTERWEIGHT
44,000 LB. CARBODY COUNTERWEIGHT
360 DEGREE RATING

Lattice Boom Crane (LAT)

Meets ANSI B30.5 Requirements

Oper. Radius Feet	Boom Angle Degree	Boom Point Elevation Feet	Boom Capacity Crawlers Retracted Pounds	Boom Capacity Crawlers Extended Pounds
170 FT. BOOM				
28	83.0	176.5		213,500 *
30	82.3	176.2	200,500	209,400 *
32	81.6	175.9	182,100	205,400 *
34	80.9	175.6	166,600	200,000
36	80.3	175.2	153,400	183,300
38	79.6	174.9	141,900	169,000
40	78.9	174.4	131,900	156,600
42	78.2	174.0	123,000	145,800
44	77.5	173.6	115,200	136,200
46	76.8	173.1	108,200	127,700
48	76.1	172.6	101,900	120,100
50	75.4	172.0	96,200	113,300
55	73.7	170.6	84,100	98,800
60	71.9	168.9	74,300	87,200
65	70.1	167.1	66,200	77,800
70	68.3	165.2	59,500	69,900
75	66.4	163.0	53,800	63,200
80	64.6	160.6	48,800	57,500
85	62.7	158.1	44,500	52,500
90	60.8	155.3	40,700	48,100
95	58.8	152.3	37,400	44,300
100	56.8	149.1	34,400	40,900
105	54.7	145.6	31,700	37,800
110	52.6	141.8	29,300	35,000
115	50.4	137.7	27,100	32,500
120	48.2	133.3	25,100	30,300
125	45.9	128.5	23,300	28,200
130	43.4	123.3	21,600	26,300
135	40.9	117.6	20,000	24,500
140	38.2	111.4	18,600	22,900
145	35.3	104.5	17,200	21,300
150	32.2	96.8	16,000	19,900
155	28.8	88.1	14,800	18,600
160	25.0	77.9	13,700	17,300
165	20.5	65.6	12,600	16,200

Weight of all load block, hooks, weight ball, slings, hoist lines, etc., beneath boom and jib point sheaves, is considered part of jib load. Boom and jib are not to be lowered beyond radii where combined weights are greater than rated capacity. Where no capacity is shown, operation is not intended or approved.

Oper. Radius Feet	Boom Angle Degree	Boom Point Elevation Feet	Boom Capacity Crawlers Retracted Pounds	Boom Capacity Crawlers Extended Pounds
180 FT. Boom				
30	82.7	186.3	199,200 *	199,200 *
32	82.1	186.0	181,900	195,500 *
34	81.4	185.7	166,400	191,800 *
36	80.8	185.4	153,100	183,100
38	80.2	185.0	141,600	168,800
40	79.5	184.6	131,600	156,400
42	78.9	184.2	122,700	145,500
44	78.2	183.8	114,900	136,000
46	77.6	183.4	107,900	127,500
48	76.9	182.9	101,600	119,800
50	76.2	182.4	95,900	113,000
55	74.6	181.0	83,700	98,500
60	72.9	179.5	74,000	86,900
65	71.2	177.8	65,900	77,500
70	69.5	175.9	59,200	69,500
75	67.8	173.9	53,400	62,900
80	66.1	**171.7**	**48.500**	57,100
85	64.3	169.3	44,200	52,100
90	62.5	166.8	40,400	47,800
95	60.7	164.0	37,000	43,900
100	58.9	161.0	34,100	40,500
105	57.0	157.8	31,400	37,500
110	55.0	154.3	29,000	34,700
115	53.0	150.6	26,800	32,200
120	51.0	146.6	24,800	29,900
125	48.9	142.3	22,900	27,800
130	46.7	137.6	21,300	25,900
135	44.5	132.6	19,700	24,200
140	42.1	127.1	18,300	22,500
145	39.7	121.2	16,900	21,000
150	37.1	114.7	15,700	19,600
155	34.3	107.6	14,500	18,300
160	31.3	99.6	13,400	17,100
165	28.0	90.6	12,400	15,900
170	24.3	80.0	11,400	14,800
175	19.9	67.3	10,500	13,800

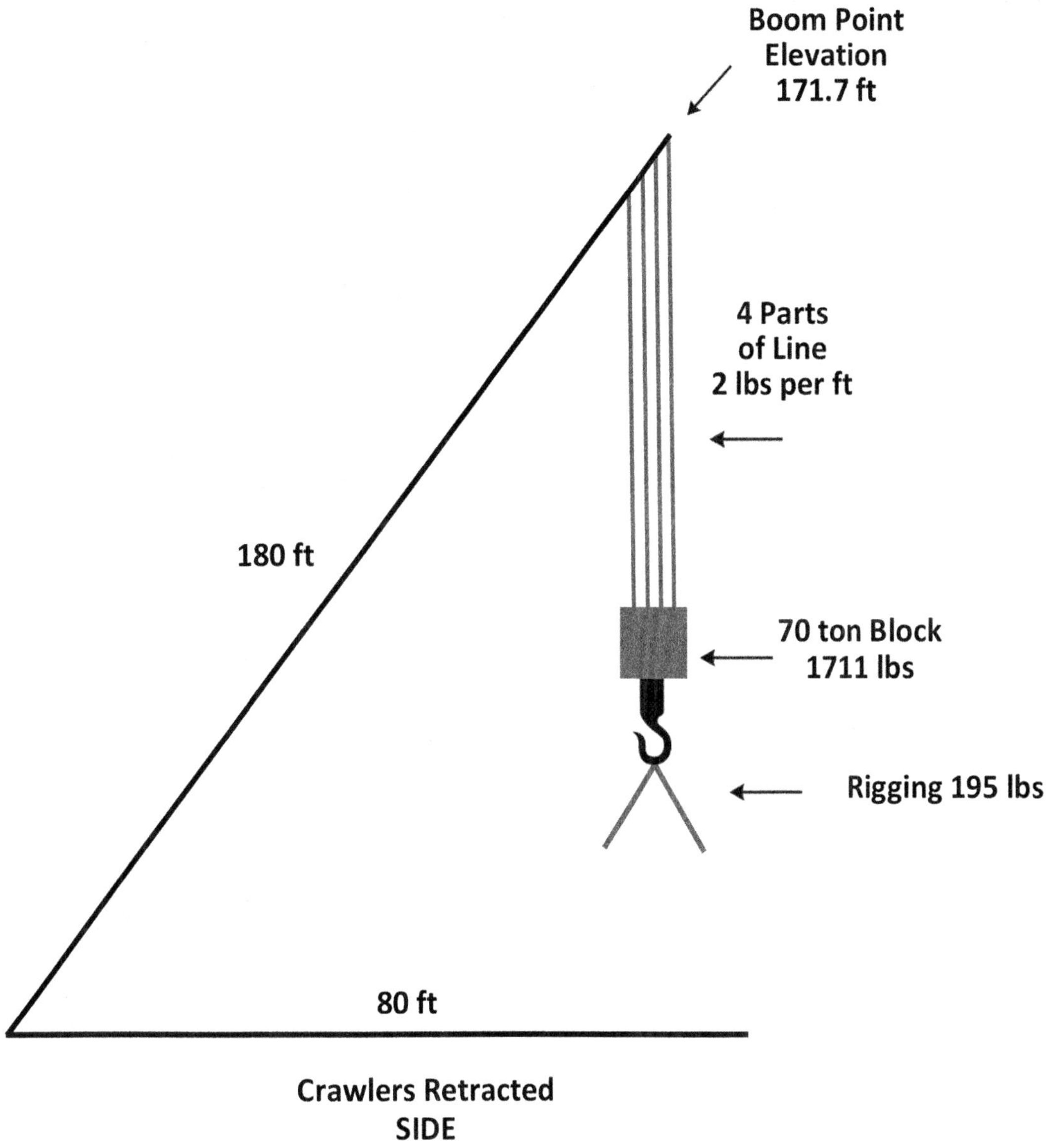
Boom Point
Elevation
171.7 ft
4 Parts
of Line
2 lbs per ft
180 ft
70 ton Block
1711 lbs
Rigging 195 lbs
80 ft
Crawlers Retracted
SIDE

Lattice Boom Crane			
Outriggers	N/A	Extension/Jib	30 ft Jib
Counterweight	179,000 lb Crane Counter Wt. 44,000 lbs Car Body Counterweight	Extension/ Jib Offset	5°
Tires	N/A	Block	70 Ton (1711 lbs) Reeved 6 Parts of Line Hanging 30 ft
Crawlers	Extended	Overhaul Ball	15 ton 722 lbs
Main Boom	260 ft	Rigging	225 lbs
Main Boom Angle	63°	Wire Rope (Main Hoist)	1 in. 6 X 25 EIPS IWRC @ 2 lbs per ft
Radius	N/A	Wire Rope (Auxiliary)	1 in. 6 X 25 EIPS IWRC @ 2 lbs per ft
Other	Spreader Bar 550 lbs	Other	N/A

4. You are planning a lift over the rear of your crane utilizing the 30 ft jib. What is your net capacity in this configuration

A. 16,402 lbs

B. 17,453 lbs

C. 17,846 lbs

D. 18,343 lbs

Gross Capacity	=	20,500 lbs
Spreader Bar	-	550 lbs
70 Ton Block	-	1711 lbs
6 Parts of Line Hanging 30 ft = 180 ft of rope X 2 lbs per ft	-	360 lbs
1.5 ton Overhaul Ball	-	722 lbs
265 ft Hoist Rope X 2 lbs per ft	-	530 lbs
Rigging	-	225 lbs
Net Capacity	=	16,402 lbs

30 Ft Jib

		5 DEGREE OFFSET			15 DEGREE OFFSET			25 DEGREE OFFSET			
BOOM LENGTH FEET	JIB OPER. RADIUS FEET	BOOM ANGLE DEGREE	JIB POINT ELEVA-TION FEET	JIB CAPACITY POUNDS	BOOM ANGLE DEGREE	JIB POINT ELEVA-TION FEET	JIB CAPACITY POUNDS	BOOM ANGLE DEGREE	JIB POINT ELEVA-TION FEET	JIB CAPACITY POUNDS	JIB OPER. RADIUS FEET
260	55	80.6	293.7	59,000*							55
	60	79.6	292.8	59,000*	80.6	291.9	59,000*				60
	65	78.6	291.8	59,000*	79.6	290.9	59,000*	80.6	288.9	50,400*	65
	70	77.5	290.7	59,000*	78.6	289.8	59,000*	79.6	287.8	49,200*	70
	75	76.5	289.5	59,000*	77.6	288.6	59,000*	78.6	286.6	48,100*	75
	80	75.5	288.2	55,400	76.5	287.3	56,800	77.6	285.3	47,000*	80
	85	74.5	286.8	50,400	75.5	285.9	51,600	76.5	283.9	46,000*	85
	90	73.5	285.3	46,000	74.5	284.4	47,100	75.5	282.4	45,000*	90
	100	71.4	282.1	38,600	72.4	281.1	39,600	73.4	279.1	40,700	100
	110	69.3	278.4	32,700	70.3	277.5	33,600	71.2	275.4	34,500	110
	120	67.1	274.3	27,900	68.1	273.3	28,700	69.1	271.2	29,500	120
	130	65.0	269.7	23,900	66.0	268.8	24,600	66.9	266.6	25,300	130
	140	62.8	264.7	20.500	63.7	263.7	21,100	64.7	261.6	21,800	140
	150	60.5	259.2	17,600	61.5	258.2	18,200	62.4	256.0	18,700	150
	160	58.2	253.2	15,000	59.2	252.1	15,600	60.0	249.9	16,100	160
	170	55.8	246.5	12,800	56.8	245.5	13,300	57.6	243.1	13,700	170
	180	53.4	239.3	10,900	54.3	238.2	11,300	55.1	235.8	11,700	180
	190	50.9	231.4	9,100	51.8	230.3	9,500	52.6	227.7	9,800	190
	200	48.2	222.7	7,600	49.2	221.5	7,900	49.9	218.9	8,200	200
	210	45.5	213.2	6,100	46.4	212.0	6,500	47.1	209.2	6,700	210
	220	42.6	202.7	4,900	43.5	201.4	5,100	44.2	198.5	5,400	220

Weight of all load block, hooks, weight ball, slings, hoist lines, etc., beneath boom and jib point sheaves, is considered part of jib load. Boom and jib are not to be lowered beyond radii where combined weights are greater than rated capacity. Where no capacity is shown, operation is not intended or approved.

Wire rope deduction include 6 parts of line hanging 30 ft from main boom point.

180 ft of rope X 2 lbs per ft = 360 lbs

Wire rope deductions include 265 ft of hoist line below jib point.

265 ft of rope X 2 lbs per ft = 530 lbs

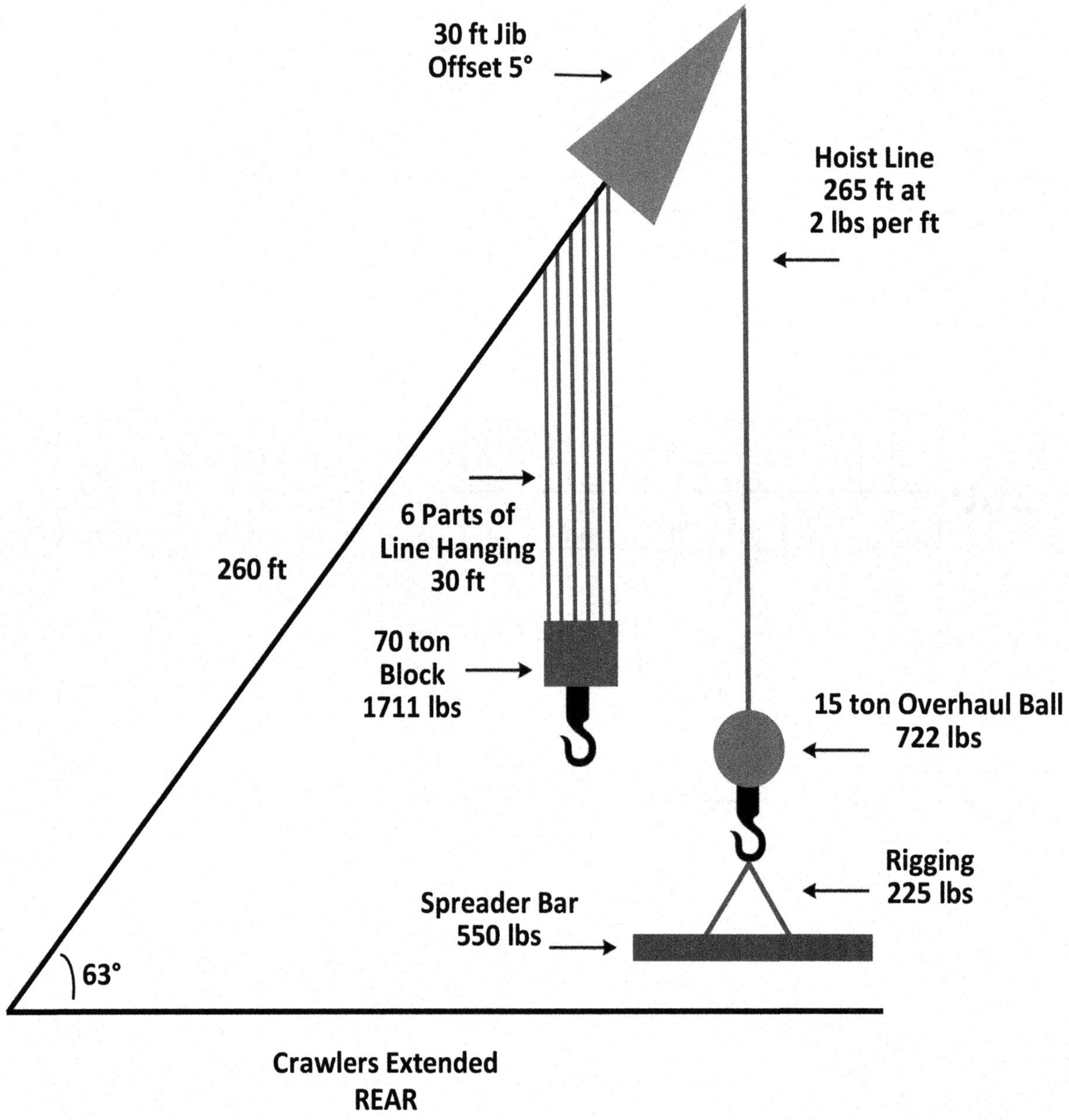
30 ft Jib
Offset 5°
Hoist Line
265 ft at
2 lbs per ft
6 Parts of
Line Hanging
30 ft
260 ft
70 ton
Block
1711 lbs
15 ton Overhaul Ball
722 lbs
Rigging
225 lbs
Spreader Bar
550 lbs
63°
Crawlers Extended
REAR

Lattice Boom Crane			
Outriggers	N/A	Extension/Jib	30 ft Jib
Counterweight	179,1000 lb Crane Counter Wt. 44,000 lbs Car Body Counterweight	Extension/ Jib Offset	25°
Tires	N/A	Block	70 Ton (267 lbs) Reeved 4 Parts of Line Hanging 20 ft
Crawlers	Extended	Overhaul Ball	15 ton 722 lbs
Main Boom	150 ft	Rigging	175 lbs
Main Boom Angle	N/A	Wire Rope (Main Hoist)	1 in. 6 X 25 EIPS IWRC @ 2 lbs per ft
Radius	83 ft	Wire Rope (Auxiliary)	1 in. 6 X 25 EIPS IWRC @ 2 lbs per ft
Other	N/A	Other	N/A

5. Lifting from the extension jib over the side, what is the net capacity of this crane in this configuration

A. 37,544

B. 37,212

C. 37,052

D. 39,200

Net Capacity	=	39,200 lbs
70 Ton Block	-	267 lbs
4 Parts of Line Hanging 20 ft = 80 ft of rope X 2 lbs per ft	-	160 lbs
Overhaul Ball	-	722 lbs
Hoist Line 166 ft X 2 lbs per ft	-	332 lbs
Rigging	-	175 lbs
Net Capacity	=	37,544 lbs

BOOM LENGTH FEET	JIB OPER. RADIUS FEET	5 DEGREE OFFSET			15 DEGREE OFFSET			25 DEGREE OFFSET			JIB OPER. RADIUS FEET
		BOOM ANGLE DEGREE	JIB POINT ELEVA-TION FEET	JIB CAPACITY POUNDS	BOOM ANGLE DEGREE	JIB POINT ELEVA-TION FEET	JIB CAPACITY POUNDS	BOOM ANGLE DEGREE	JIB POINT ELEVA-TION FEET	JIB CAPACITY POUNDS	
140	35	80.7	175.4	59,000*							35
	40	79.0	174.4	59,000*	80.7	173.6	59,000*				40
	45	77.2	173.3	59,000*	79.0	172.5	59,000*	80.8	170.6	50,600*	45
	50	75.5	172.0	59,000*	77.3	171.2	59,000*	79.0	169.3	48,600*	50
	55	73.7	170.5	59,000*	75.5	169.7	58,700*	77.2	167.9	46,700*	55
	60	72.0	168.9	59,000*	73.7	168.1	57,900*	75.4	166.2	45,000*	60
	65	70.2	167.1	59,000*	71.9	166.3	57,200*	73.6	164.4	43,400*	65
	70	68.4	165.1	59,000*	70.1	164.3	56,600*	71.8	162.4	42,000*	70
	75	66.5	163.0	59,000*	68.2	162.1	55,900*	69.9	160.2	40,700*	75
	80	64.7	160.6	58,500*	66.4	159.7	55,300*	68.0	157.8	39,400*	80
	85	62.8	158.1	54,900	64.5	157.2	54,800*	66.1	155.1	38,300*	85
	90	60.8	155.3	50,500	62.5	154.4	51,200	64.1	152.3	37,300*	90
	100	56.9	149.0	43,200	58.5	148.1	43,900	60.0	145.9	35,400*	100
	110	52.7	141.8	37,400	54.3	140.8	37,900	55.8	138.5	33,800*	110
	120	48.3	133.3	32,600	49.9	132.2	33,100	51.2	129.8	32,400*	120
	130	43.5	123.3	28,600	45.1	122.1	29,000	46.3	119.5	29,300	130
	140	38.3	111.4	25,200	39.8	110.1	25,500				140
	150	32.3	96.8	22,300	33.7	95.3	22,500				150
	160	25.1	77.9	19,700							160
150	40	79.6	184.6	59,000*							40
	45	78.0	183.6	59,000*	79.6	182.7	59,000*				45
	50	76.3	182.3	59,000*	78.0	181.5	59,000*	79.7	179.7	49,300*	50
	55	74.7	181.0	59,000*	76.3	180.1	59,000*	78.0	178.3	47,500*	55
	60	73.0	179.5	59,000*	74.7	178.6	58,300*	76.3	176.7	45,800*	60
	65	71.3	177.8	59,000*	73.0	176.9	57,600*	74.6	175.0	44,300*	65
	70	69.6	175.9	59,000*	71.3	175.1	57,000*	72.9	173.1	42,800*	70
	75	67.9	173.9	59,000*	69.5	173.0	56,400*	71.1	171.1	41,500*	75
	80	66.2	171.7	59,000*	67.8	170.8	55,800*	69.3	168.8	40,300*	80
	85	64.4	169.3	54,400	66.0	168.4	55,200*	67.6	166.4	39,200	85
	90	62.6	166.7	50,000	64.2	165.8	50,800	65.7	163.8	38,100*	90
	100	58.9	161.0	42,800	60.5	160.0	43,400	62.0	157.9	36,200*	100
	110	55.1	154.3	36,900	56.7	153.3	37,500	58.1	151.1	34,600*	110
	120	51.1	146.5	32,100	52.6	145.5	32,600	53.9	143.1	33,100	120
	130	46.8	137.6	28,100	48.3	136.5	28,600	49.6	133.9	28,900	130
	140	42.2	127.1	24,700	43.6	125.9	25,100	44.8	123.2	25,400	140
	150	37.1	114.7	21,800	38.5	113.4	22,100				150
	160	31.4	99.6	19,300	32.7	98.0	19,500				160
	170	24.4	80.0	17,000							170

Wire rope deduction include 6 parts of line hanging 30 ft from main boom point.

180 ft of rope X 2 lbs per ft = 360 lbs

Wire rope deductions include 265 ft of hoist line below jib point.

265 ft of rope X 2 lbs per ft = 530 lbs

Weight of all load block, hooks, weight ball, slings, hoist lines, etc., beneath boom and jib point sheaves, is considered part of jib load. Boom and jib are not to be lowered beyond radii where combined weights are greater than rated capacity. Where no capacity is shown, operation is not intended or approved.

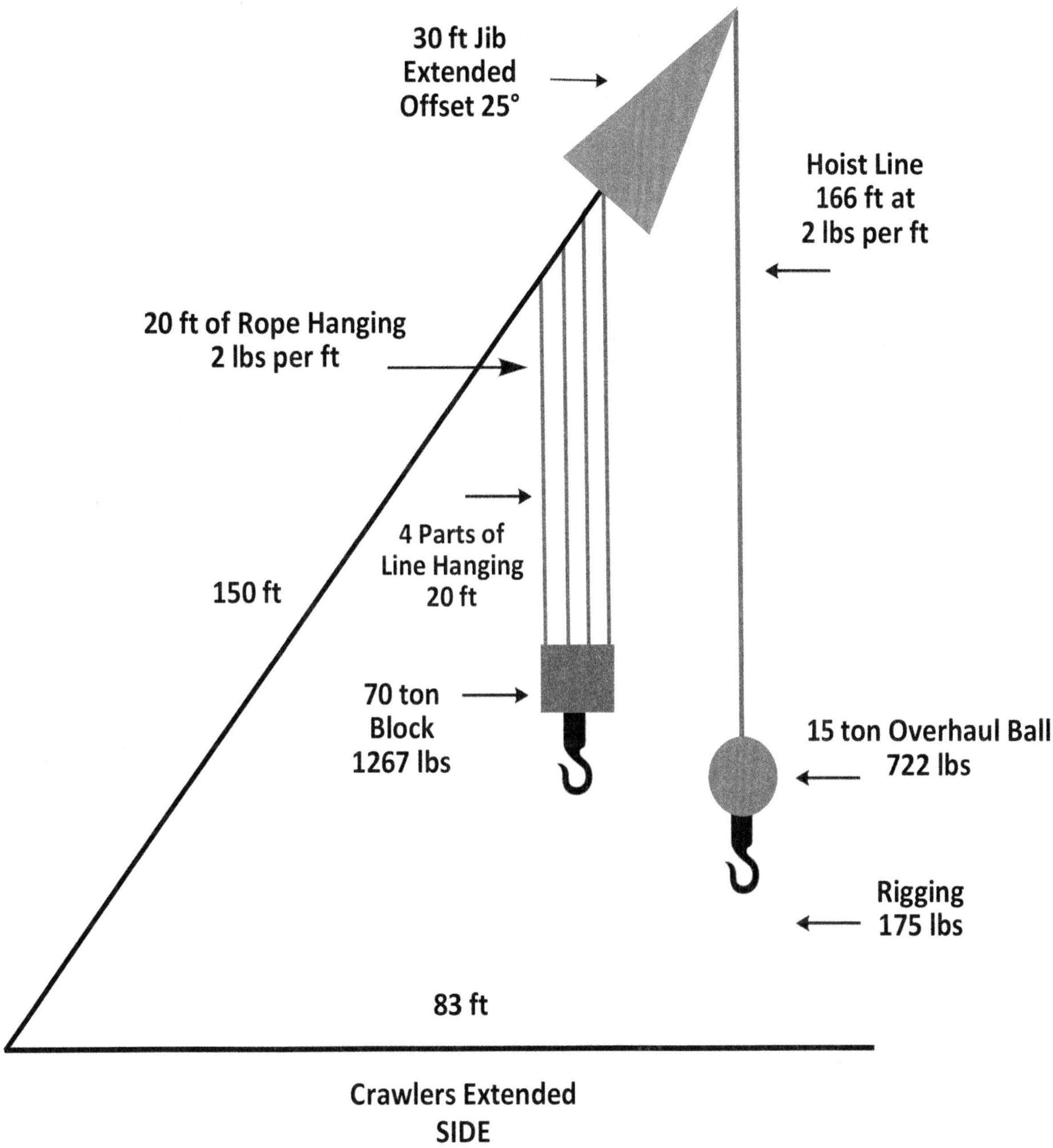
30 ft Jib
Extended
Offset 25°
Hoist Line
166 ft at
2 lbs per ft
20 ft of Rope Hanging
2 lbs per ft
4 Parts of
Line Hanging
20 ft
150 ft
70 ton
Block
1267 lbs
15 ton Overhaul Ball
722 lbs
Rigging
175 lbs
83 ft
Crawlers Extended
SIDE

Lattice Boom Crane			
Outriggers	N/A	Extension/Jib Erected	40 ft Jib
Counterweight	179,1000 lb Crane Counter Wt. 44,000 lbs Car Body Counterweight	Extension/ Jib Offset	15°
Tires	N/A	Block	70 Ton (1711 lbs) Reeved 8 Parts of Line Hanging 35 ft
Crawlers	Extended	Overhaul Ball	15 ton 722 lbs
Main Boom	120 ft	Rigging	445 lbs
Main Boom Angle	N/A	Wire Rope (Main Hoist)	1 in. 6 X 25 EIPS IWRC @ 2 lbs per ft
Radius	82 ft	Wire Rope (Auxiliary)	1 in. 6 X 25 EIPS IWRC @ 2 lbs per ft
Other	N/A	Other	Equalizer Beam 475 lbs

6. Your crane is set up on firm level ground with your crawlers extended. What is the net capacity of the crane in this configuration

A. 36,996 lbs

B. 37,441 lbs

C. 37,151 lbs

D. 35,997 lbs

Gross Capacity	=	40,200 lbs
70 Ton Block	-	1711 lbs
8 Parts of Line Hanging 35 ft = 280 ft of rope X 2 lbs per ft	-	560 lbs
15 Ton Overhaul Ball	-	722 lbs
Hoist Line 145 ft X 2 lbs per ft	-	290 lbs
Rigging	-	445 lbs
Equalizer Beam	-	475 lbs
Net Capacity	=	35,997 lbs

40 Ft Jib

BOOM LENGTH FEET	JIB OPER. RADIUS FEET	5 DEGREE OFFSET			15 DEGREE OFFSET			25 DEGREE OFFSET			JIB OPER. RADIUS FEET
		BOOM ANGLE DEGREE	JIB POINT ELEVA-TION FEET	JIB CAPACITY POUNDS	BOOM ANGLE DEGREE	JIB POINT ELEVA-TION FEET	JIB CAPACITY POUNDS	BOOM ANGLE DEGREE	JIB POINT ELEVA-TION FEET	JIB CAPACITY POUNDS	
120	35	80.7	165.2	53,500*							35
	40	78.8	164.1	52,200*							40
	45	77.0	162.9	50,900*	79.5	161.8	46,100*				45
	50	75.1	161.5	49,700*	77.6	160.4	45,200*	80.1	158.1	39,500*	50
	55	73.3	160.0	48,600*	75.7	158.9	44,300*	78.2	156.5	37,900*	55
	60	71.4	158.2	47,500*	73.8	157.1	43,500*	76.3	154.7	36,400*	60
	65	69.4	156.3	46,600*	71.9	155.2	42,800*	74.3	152.7	35,000*	65
	70	67.5	154.2	45,600*	69.9	153.0	42,100*	72.3	150.5	33,800*	70
	75	65.5	151.9	44,700*	67.9	150.7	41,400*	70.3	148.2	32,600*	75
	85	63.5	149.3	43,900*	65.9	148.1	40,800*	68.2	145.5	31,500*	80
	85	61.5	146.5	43,100*	63.8	145.3	40,200*	66.1	142.7	30,600*	85
	90	59.4	143.5	42,400*	61.7	142.3	39,700*	63.9	139.6	29,700*	90
	100	55.0	136.7	41,100*	57.3	135.4	37,200*	59.5	132.6	28,100*	100
	110	50.4	128.7	38,700	52.7	127.3	34,500*	54.7	124.3	26,700*	110
	120	45.5	119.2	33,900	47.7	117.7	32,300*	49.6	114.4	25,600*	120
	130	40.1	107.8	29,900	42.2	106.1	30,300	43.9	102.4	24,700*	130
	140	33.9	93.8	26,500	35.9	91.8	26,800				140
	150	26.4	75.5	23,500							

Weight of all load block, hooks, weight ball, slings, hoist lines, etc., beneath boom and jib point sheaves, is considered part of jib load. Boom and jib are not to be lowered beyond radii where combined weights are greater than rated capacity. Where no capacity is shown, operation is not intended or approved.

Wire rope deduction include 8 parts of line hanging 35 ft from main boom point.

280 ft of rope X 2 lbs per ft = 560 lbs

Wire rope deductions include 145 ft of hoist line below jib point.

145 ft of rope X 2 lbs per ft = 290 lbs

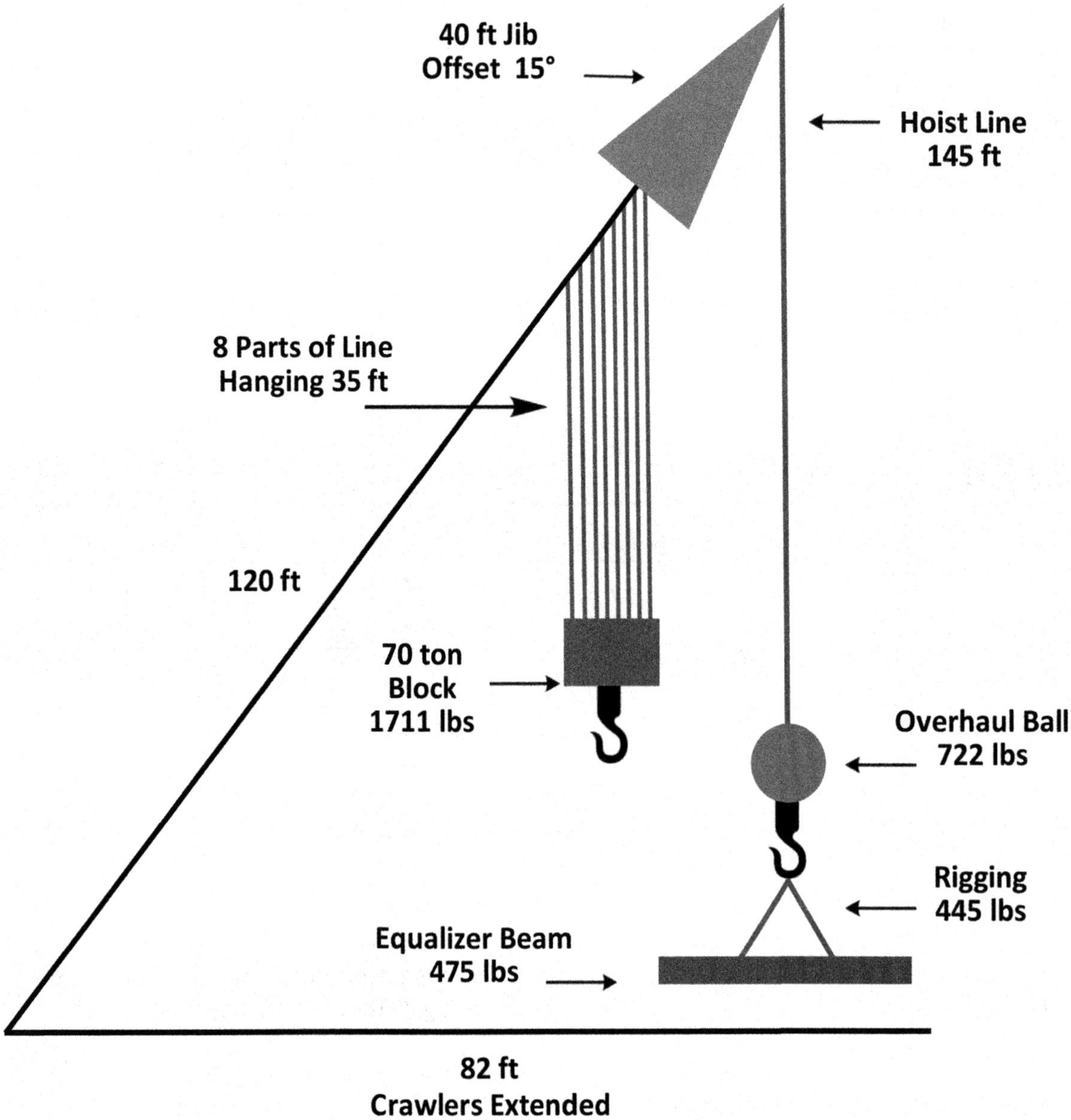
40 ft Jib
Offset 15°
Hoist Line
145 ft
8 Parts of Line
Hanging 35 ft
120 ft
70 ton
Block
1711 lbs
Overhaul Ball
722 lbs
Rigging
445 lbs
Equalizer Beam
475 lbs
82 ft
Crawlers Extended

Chapter 9

1. Tag lines shall be held by an experienced person, and when working in the proximity of energized electric lines
   c. should be nonconductive

2. If the load line of your crane contacts a power line, and there is no sign of any other danger such as fire, you must
   c. stay in the crane until the line is de-energized

3. The required clearance for normal voltage when operating a crane near high-voltage power lines over 200 to 350 kV is
   d. 20 feet

4. For cranes in transit with no load or boom lowered, the minimum required clearance up to 0.75 to 50 kV is
   b. 6 feet

5. The third most common cause of death among construction workers is
   a. contact with power lines

6. All overhead power lines should be considered energized unless
   d. an electrical utility authority verifies the lines are de-energized

7. When determining the safe working proximity to energized electric lines, it is also important to consider the
   c. sway or whipping caused by wind

8. The required clearance for normal voltage when operating a crane near high-voltage power lines over 350 to 500 kV is
   b. 25 feet

9. If the use of a tag line is necessary while making a lift with a crane, which of the following types offers better insulating properties than most other ropes when working around power lines?
   d. Polypropylene

10. You are setting up an all-terrain crane near an overhead power line that you are told is 850 kV. According to OSHA 1926.1408, what is the minimum clearance distance you must maintain between any part of your crane and the load you are lifting?
    b. 45 feet

Chapter 10

1. Running ropes shall be removed from service when there are
   b. six randomly distributed broken wires within one lay length or three broken wires in one strand in one lay

2. The frequency of wire rope lubrication will depend on
   c. the severity of duty and corrosive conditions

3. When replacing wire rope on your mobile crane, it is important to make sure that
   c. it is the original capacity and use

4. The design factor for running ropes that wind on drums or travel over sheaves shall not be less than
   b. 3.5

5. A wedge socket assembly must be assembled with the load line pulling in a straight line with the
   c. pinhole

6. The correct method for installing a new wire rope from a reel to the hoist drum is to
   b. unspool the rope from the top of the reel to the top of the hoist drum

7. The design factor for live or rotation-resistant ropes that travel over sheaves shall be
   b. 5.0 or greater

8. Rotation-resistant wire rope requires a higher safety factor because
   b. it is more likely to become damaged internally without being detected

9. According to the ASME B30.5 mobile crane safety standard, wire rope clips of the single-saddle U-bolt–type clip shall be
   c. drop forged steel

10. Wire rope clips attached with U-bolts shall have
    a. the U-bolt over the dead end of the rope and the live rope resting in the saddle

11. According to the ASME B30.5 mobile crane safety standard, the minimum that the drum flange shall extend over the top layer of rope at all times is
    b. ½ inch

12. According to the ASME B30.5 mobile crane safety standard, the winch drums shall provide a first-layer rope pitch diameter of not less than
    c. 18 times the nominal diameter of the rope used

13. According to the ASME B30.5 mobile crane safety standard, all hook and ball assemblies shall be labeled with their
    c. rated capacity and weight

14. According to the ASME B30.5 mobile crane safety standard, the design factor for boom pendants or standing ropes shall not be less than
    b. 3.0

15. According to the ASME B30.5 mobile crane safety standard, the design factor for live or running ropes that wind on drums or travel over sheaves shall not be less than
    a. 3.5

16. According to ASME B30.5 mobile crane safety standard, for supporting rated loads, rotation-resistant ropes shall have a design factor of
  c. 5 or greater

17. Wire rope inspection on your crane should
  a. occur daily

18. A rope lay is described as the
  a. axial distance along the rope in which one strand makes one complete turn around the rope

19. Standing or pendant ropes should be removed from service when there are
  c. three or more broken wires in one rope lay anywhere in the wire rope or two or more broken wires between end connections

20. According to the ASME B30.5 mobile crane safety standard, how many full wraps of rope shall remain on the drum when the hook is in the extreme low position and the boom is at maximum elevation and extension?
  b. Two full wraps

21. Wire rope inspections on your crane need not be conducted at equal calendar intervals and should be more frequent
  c. as the rope approaches the end of its life

22. You are installing rotation-resistant wire rope and a 7.5-ton overhaul ball on your 100-ton truck crane. In this configuration, it is permissible to use a swivel
  d. if the wire rope manufacturer approves

Chapter 11

1. The standard OSHA hand signal shown here is
  a. lower the boom and raise the load

2. The standard OSHA hand signal shown here is
  d. extend telescoping boom

3. The standard OSHA hand signal shown here is
  b. crawler crane travel, one track

4. The standard OSHA hand signal shown here is
  a. raise boom

5. The standard OSHA hand signal shown here is
  b. crawler crane travel, both tracks

6. The standard OSHA hand signal shown here is
  c. move slowly

7. The standard OSHA hand signal shown here is
d. swing

8. The standard OSHA hand signal shown here is
b. use auxiliary hoist

9. The standard OSHA hand signal shown here is
c. dog everything

10. The standard OSHA hand signal shown here is
d. hoist

11. The standard OSHA hand signal shown here is
a. raise the boom and lower the load

12. The standard OSHA hand signal shown here is
d. extend boom

13. The standard OSHA hand signal shown here is
d. emergency stop

14. The standard OSHA hand signal shown here is
c. use main hoist

15. The standard OSHA hand signal shown here is
d. retract boom

16. The standard OSHA hand signal shown here is
a. lower

17. The standard OSHA hand signal shown here is
d. emergency stop

18. The crane operator should move a load only after receiving a signal from
c. the designated signal person

19. When moving the crane, the following signal shall be used for stop:
d. One short audible signal

20. When moving the crane, the following signal shall be used for backing up:
b. three short audible signals

21. According to the ASME B30.5 crane safety standard, all directions given to the crane operator shall be given from
a. the operator's direction perspective

22. You are communicating during a lift by voice signals. According to ASME B305, each series of voice signals shall contain three elements. Which voice command should be given first?
d. Function and/or direction

Chapter 12

1. An important safety procedure to have in place when lifting personnel in a platform suspension system over water is to
   d. have a boat with rescue personnel available

2. When lifting personnel with a crane, the outriggers or stabilizers must
   d. be fully extended, blocked, and locked

3. When lifting personnel with a crane, the operator should remain in continuous sight or in communication, and hand signals to the operator should be in accordance with
   d. ASME B30.5

4. According to federal requirements, OSHA permits the hoisting of personnel by an approved lifting system when
   d. there is no less hazardous or practical way

5. When lifting personnel in an approved lifting system, the total weight of the loaded platform must not exceed
   d. 50% of the crane's charted capacity

6. Before being put into use, personnel platforms are to be loaded to 125% of their rated capacity and lifted and held for
   a. 5 minutes

7. According to the ASME B30.23 standard, when personnel lifts are conducted over water, personnel shall
   d. wear an approved floatation device

8. According to the ASME B30.23 safety standard, while in the platform, personnel fall protection devices with quick-release features shall be provided and required to be worn *except*
   c. when being lifted over water

9. When occupying an approved lifting device, personnel are not to be lifted in winds in excess of
   c. 20 mph

10. The personnel lifting system will have an identification plate that shall be provided and displayed to allow easy viewing from
    d. both the interior while hoisted and the exterior while not hoisted

Chapter 13

1. All lifting hooks should have which of the following to retain slings and lifting attachments?
   c. Safety latch

2. Lifting hooks shall be removed from service when the throat opening exceeds what percent?
   a. 5%

3. Crane hooks shall be equipped with a safety latch
   b. unless the application makes the use of a latch impractical

4. Safety latches shall close
   d. over the entire throat opening

5. Crane hooks are designed and manufactured to lift specific loads. Load hook block assemblies shall
   d. be labeled with rate capacity and weight

6. Crane hooks shall be removed from service when there is
   d. any visible bend or twist from the plane of the unbent hook

7. Crane hooks shall be removed from service
   d. if wear exceeds 10% of the original section dimension

8. Crane hooks having any deficiencies shall be removed from service unless
   b. a qualified person approves their continued use and initiates corrective actions

9. Crane hooks approved for continued use shall be subjected to
   a. periodic inspection

10. Crane hooks shall be removed from service when
    c. an increase in throat opening exceeds 5%

Chapter 14

1. During pick and carry operations with a rough-terrain crane, you should
   d. keep the crane boom as short as possible

2. Rough-terrain mobile cranes are designed to travel with a suspended load
   d. on firm, level ground capable of supporting the crane and the suspended load

3. Crane operators should plan a path of travel during pick and carry operations
   d. always; a path of travel should always be planned, and the manufacturer's rules and specifications should always be reviewed.

4. When is the crane operator required to de-rate the load chart during pick and carry operations?
   d. Some manufacturers may require you to de-rate the load chart, depending on the ground conditions or other special requirements.

5. You are planning a pick and carry operation on your construction project. Who will determine whether the crane is capable of on-rubber operations?
   d. It can be approved only if the manufacturer has approved it for pick and carry operations, and under those specifications using the on-rubber load chart.

6. When should crane operators consider the crane's fore-and-aft slope conditions when planning a pick and carry operation?
   c. Always; crane manufacturers may have important instructions or limitations that are often overlooked.

Chapter 15

1. The National Oceanic and Atmospheric Administration tells us that after hearing thunder or seeing lightning, we should go to a safe shelter and remain there for
   c. 30 minutes after hearing the last sound of thunder

2. The best shelters to go to during thunder and lightning storms are
   c. enclosed structures

3. Lightning will travel about one mile every
   a. 5 seconds

4. After a lightning strike near or on a jobsite where your crane is located, it is important to
   d. perform a complete and thorough inspection of the crane

5. Since a lightning strike can produce an extreme form of heat, what must always be replaced?
   d. All the wire rope

6. On average, how many people are killed in the United States every year from lightning?
   b 20–33 people

Chapter 16

1. The supervisor in charge of lattice crane boom assembly and disassembly operations is titled the
   d. assembly and disassembly director

2. When dismantling a lattice boom, which pins should be removed first?
   c. Both lower

3. During the assembly or dismantling of the lattice booms, the manufacturer's reference to "cantilever length" means the
   b. length of boom suspended beyond the last pendant hookup

4. When assembling lattice booms,
   d. the manufacturer's instructions must be followed

5. When assembling or dismantling lattice booms, working or walking under the boom
   d. must never be allowed unless the A/D director makes special provisions

Chapter 17

1. Any lift that involves more than one crane is considered to be a multiple-crane lift as well as a
   c. critical lift

2. How much of a crane's charted capacity should not be exceeded when making a tandem lift?
   c. 75%

3. According to OSHA, a critical lifting operation is present when
   d. any lifts involving more than one crane are considered critical lifts

4. When two or more cranes are used to lift a load, who is responsible for the operation?
   d. A person who meets the criteria for both a competent person and a qualified person, called the *lift director*

5. According to the ASME B30.5 safety standard, the person responsible for control of a multiple-crane lift is the
   d. project lift director

6. What is the exclusion zone used for during mobile crane lifts?
   d. It is a defined area where people are not allowed to go when particular work is being done to avoid the risk of being injured.

7. Which of the following can be considered a critical crane lift?
   c. Lifting of personnel

8. Particularly strict, no-nonsense attention must be given to all the cranes set up and to the alignment being used for multiple-crane lifts to avoid what type of pressure on the cranes' booms?
   d. Side loading

9. According to OSHA regulations and ASME safety standards, a critical lift means
   d. any lifting operation that involves increased risk factors

Chapter 18

1. The basic weight of 1 square foot of steel 1 inch thick is approximately 40 pounds. How much will 2 plates of steel each measuring 1½ inches × 3 feet × 6 feet weigh?

   2 (plates) × 1.5 inches (thick) × 3 feet × 6 feet × 40 pounds (per square foot) = 2,160 pounds

2. You have a 5-inch × 3-inch × ¼-inch angle flattened out to approximately an 8-inch × ¼-inch plate. The weight per square foot is 40 pounds. How many pounds would the flattened out plate weigh?

   40 pounds (per square foot) × 8 inches × 12 inches (1 foot) × 0.25 (¼) inch = 6.65 pounds per square foot

3. You are planning to lift a unit of spruce lumber that is 12 feet long, 3 feet high, and 4 feet wide. The weight per cubic foot of dry spruce is approximately 28 pounds. How much does this unit of lumber weigh?

   12 feet × 3 feet × 4 feet × 28 pounds (per cubic foot) = 4,032 pounds

4. A pipe 6 feet in diameter has a circumference of 6 feet × 3.14, or 18.84 feet. To compute the weight of this pipe, which is 30 feet long and made of ⅜-inch plate, mentally unroll it and flatten it out. This represents a plate that is 18.84 feet wide by 30 feet long by ⅜ inch thick. The weight per square foot is 40 pounds.
   a. How much would this 30 foot long pipe weigh?

      30 feet × 0.375 (⅜) inch × 40 pounds (per square foot) = 8,478 pounds

   b. The pipe has an end cap ⅜ inch thick and 6 feet in diameter (Figure 18.6), it would have a surface area of 3.14 × 6 feet × 6 feet divided by 4 = 28.26 square feet (we'll round it off to 28.3). How much would this 6-foot diameter pipe end cap weigh?

      28.3 square feet× 0.375 (⅜) inch × 40 pounds (per square foot) = 425 pounds

   c. What would be the combined weight of the pipe and the end cap?

      28.3 square feet × 0.375 (⅜) inch × 40 pounds (per square foot) = 8,903 pounds

Chapter 19

1. When lifting with two slings with a 60-degree angle to the load, what would the length of each sling be if the distance between the lift points is 12 feet?
   b. 12 feet

2. When using two slings to lift a load, at what angle to the load will the load on each sling equal the weight of the load?
   b. 30 degrees

3. According to the ASME B30.9 safety standard for slings, a critical lift plan must be developed whenever sling angles are less than
   b. 30 degrees

4. Who must approve a critical lift plan when low sling angles are used?
   c. a qualified person

5. As a sling angle decreases, the sling
   c. stress increases

6. A load tilted 12 degrees will see the stress on one of the sling legs increase
   d. 70 to 80%

Chapter 20

1. Lifting and rigging are two separate operations. Lifting involves which of the following?
   c. The physical movement of the load from one location to another

2. What is the most current and effective way to rate rigging hardware for weight limits?
   d. Working load limit (WLL)

3. Which system used for rating hardware limits has potential legal complications and has been replaced?
   c. Safe working load (SWL)

4. Which system used for rating hardware limits is the most precise in specifying working conditions and the importance of staying within the equipment's capacity limits?
   a. Working load limit (WLL)

5. When the signal person, rigger, or others need to communicate with the crane operator, the person must use approved hand signals in accordance with which safety standard?
   b. ASME B30.5

6. Chain shackles are designed to be used for
   b. straight line vertical pull lifts only

7. Twist shackles are one of the special types of lifting devices because their body has a
   d. 90-degree twist

8. Wire rope clips used for overhead lifting must be made of
   c. drop forged steel

9. Wire rope clips are popular because they can be installed in the field. Depending on the diameter of the wire rope, they provide how much efficiency of the rope's breaking strength?
   c. 80 to 90% efficiency

10. What regulation or safety standard refers to the correct use of wire rope clips?
    a. ASME B30.9

11. According to which safety standard do eye bolt markings and the identification requirement come from?
    d. ASME B30.26

12. Which type of eye bolt should never be used for overhead lifts?
    a. Bent eye bolts

13. Which type of link is used to group as many as 4 slings together?
    c. Master link subassembly

14. Which type of link serves best when attaching to crane hooks?
    b. Oblong master links

15. What type of rigging hardware is best for forming bridles?
c. Master rings

16. Which ASME safety standard provides valuable information regarding links and rings used for hoisting and rigging operations?
a. ASME B30.26

17. What specialized overhead lifting device provides more headroom?
a. Lifting beam

18. What specialized overhead lifting device listed here is best suited to handle loads that need to be supported all the way throughout its length?
d. Lifting beam

19. Both spreader bars and lifting beams are designed to keep lifting slings below the units at or near what angle?
a. 90 degree angle

20. The ASME safety standard that covers spreader bars and lifting beams used by cranes for hoisting loads is
b. ASME B30.20

21. According to the ASME safety standard for metal mesh slings, a component will be removed from service when
a. there is a lack of flexibility of the mesh fabric

22. The maximum load weight of a sling that can be used during hoisting and rigging activities is
b. its rated capacity

23. Identification tags attached to slings used in hoisting and rigging operations must include
d. the manufacturer's name

24. What is the maximum included angle for a shackle?
c. 120 degrees

25. Synthetic slings must be inspected
d. before each use

26. A chain sling should be inspected by a
d. competent person

27. The ASME safety standard states that the minimum allowed chain grade that shall be used for overhead lifting is
a. Grade 80

28. When a new chain is received from the manufacturer, it should undergo an
d. initial inspection

29. When slings used in hoisting and rigging operations can come into contact with loads with sharp edges, riggers must see to it that
    c. the slings are protected with pads or softeners to safeguard and shield them

30. According to the ASME B30.9 safety standards for slings, how many randomly distributed broken wires in one rope lay calls for the removal of a wire rope sling from service?
    c. 10 wires

Chapter 21

1. What ASME safety standard pertains to the use of come-alongs for pulling or lifting loads?
   a. ASME B30.21

2. What ASME safety standard pertains to the use of turnbuckles for tightening down loads?
   d. ASME B30.26

3. What is the typical safety factor for come-alongs, cable pullers, and hand winches when lifting loads?
   b. 2:1 to 3:1

4. What is the typical safety factor for come-alongs, cable pullers, and hand winches when pulling loads?
   d. 4:1 to 5:1

5. The OSHA safety standard that pertains to turnbuckles is
   c. 1926.251

6. Why is it important to inspect the come-along's cable, hooks, and chains before every use?
   c. To identify wear, deformation, kinks, or cracks that could lead to a sudden failure while under load

7. What ASME safety standard pertains to the safe use of come-alongs, lever hoists, and chain lever hoists?
   b. ASME B30.21

8. Manufacturers should design turnbuckles with a minimum safety factor of
   d. five times the working load limit

9. What ASME safety standard pertains to the use of jacks?
   a. ASME B30.1

10. Many mobile crane manufacturers require the crane operator to derate the load chart when conducting duty cycle operations. The amount needed to derate the chart is
    a. 20%

11. The primary OSHA regulation that is titled Floating Cranes, Derricks, and Cranes on Barges is
    d. 29 CFR 1926.1437

12. The primary ASME safety standard covering mobile land cranes on barges is
  a. ASME B30.8

13. The ASME standard that applies to clamshell work with a mobile crane is
  a. ASME B30.5

14. The ASME standard that applies to the clamshell bucket itself that is used for clamshell work is
  d. ASME B30.20

15. What is the hoisting task called when a vessel is lifted from a horizontal load to a vertical position, much like flipping a load from one orientation to another?
  a. Tailing

16. Temporary trunnions are specialized rigging attachments that are used to provide
  b. secure hoisting or lifting points on loads that have none

17. Why are conventional lattice boom crawler cranes widely chosen for dragline work?
  d. They have heavy-duty construction to withstand high dynamic stresses that come with duty cycle dragline operations.

18. Hydraulic pile drivers used in pile driving operations use a combination of static weight and high frequency. They also use which of the following to move piles into the ground?
  a. Thump

19. A top safety concern when operating land cranes on barges is
  c. maintaining stability

20. A mobile land crane when working on water cannot exceed a minimum wind speed of
  b. 20 miles per hour

21. When a mobile land crane is working on a barge, what type of indicator must be clearly visible to the crane operator based on the crane's rated capacity to ensure the barge deck remains out of the water?
  d. List or trim indicator

22. When inspecting a dragline or clamshell bucket, be sure the equipment is on level ground and kept away from any edges by at least
  a. 15 feet

Index

References to figures are in italics.

About the Author

DANNY THIEMENS is an influential figure and respected advocate in the crane safety, operations, and certification fields, boasting an impressive and transformative career spanning more than four decades. He is widely regarded not only for his technical expertise and field leadership but also for his visionary contributions to training, education, and regulatory reform within the hoisting and rigging industry. His name is synonymous with excellence, dedication, and a tireless commitment to the safety of workers and the advancement of professional standards.

Born into a working-class family rooted in the construction trades, Danny was introduced to heavy machinery at a young age. Operating equipment for his father's construction projects instilled in him a profound respect for machines and the lives they affect. This early exposure planted the seeds of a lifelong devotion to safety, discipline, and craftsmanship. It wasn't just a trade for Danny, it was a calling.

Pursuing this passion, Danny entered the ranks of the International Union of Operating Engineers (IUOE), completing a rigorous apprenticeship that would lay the foundation for his future as an industry leader. He distinguished himself early on, progressing from journeyman to supervisor with nationally recognized contractors such as Kiewit Pacific Company. Danny's ability to lead, teach, and inspire those around him became evident through each phase of his journey, and his reputation for fairness, safety, and no-nonsense leadership quickly took root.

In 1992, Danny was appointed Apprenticeship Director for IUOE Local 370. At the time, training facilities were limited to a small section of a community college campus. Recognizing that this environment was inadequate to meet the evolving demands of crane safety and operator education, Danny spearheaded a campaign to expand and transform the program. He identified a critical shortage of certified crane operators and envisioned a new standard of education—one grounded in safety, competency, and accessibility.

To realize this vision, Danny earned certification from the National Grant Writers Association and successfully secured funding to construct a new training facility. His efforts led to the establishment of the Western Operating Engineers Training Institute; a nationally accredited, state licensed school built on a 60-acre campus designed for comprehensive heavy equipment training. Apprentices themselves, under Danny's guidance, helped build the very grounds on which they would learn—a fitting metaphor for the legacy Danny was constructing.

One of Danny's most significant contributions was his early and active promotion of formal crane operator certification. In 1993, he organized and hosted a 10-day crane certification class with renowned crane educator Ronald John William Cowper—considered by many to be one of the most knowledgeable consultants and trainers in North America. Together, they led transformative training sessions that received acclaim from the entire hoisting and rigging community. Cowper became both a close friend and mentor to Danny, shaping his philosophy toward instruction and becoming a father figure and co-architect of the new era in crane safety.

Danny's leadership and vision were further demonstrated in his work with industry partners with his collaboration with local contractors, catalyzed momentum that ultimately led to the statewide requirement for the certification of crane operators. After accepting a staff position working for The National Commission for the Certification of Crane operators (NCCCO) in 2003, he was made the State and Federal Safety and Health Officer. Soon after he was appointed as NCCCO's Washington State Representative, contributing to the formation of the national crane operator certification standards and regulations that remain cornerstones of the industry to this day. His time with NCCCO reinforced his belief that the safety and success of crane operations begin with properly trained, qualified individuals.

Over the years, Danny's influence spread through an extensive network of respected industry professionals. Among them, Paul Parish of Lampson Crane; a revered international crane consultant specializing in bull rigging, critical heavy hoisting and multiple crane lifts—was a key collaborator and peer. Their shared passion for precision and professionalism helped shape the most advanced mobile crane training curriculum the hoisting and rigging industries have had thus far. Danny's knowledge, paired with Paul's, created a synergy that raised expectations across the profession.

After decades of reshaping the crane industry, Danny now leads Fulcrum Crane Training International. Though his wife and business partner, Kathy Thiemens, has passed, her legacy endures through the mission they built together. Danny has remained steadfast in carrying out their shared vision, and now dedicates himself to mentoring his only son, Dillon Thiemens. With an eye toward generational continuity, Danny maintains his dedication to the various trades that make both the mobile crane and hoisting and rigging industries with the same dedication being shared to the craftsman working those industries.

Danny's personal life is as dynamic as his professional one. An avid traveler and vintage car enthusiast, he finds joy in restoring classic hotrods, a pursuit that reflects the same meticulous care he brings to his instructional and operational practices. He is a devoted fan of the Spokane Chiefs, finding a sense of hometown pride and camaraderie at every game. Despite the losses he has faced, Danny's spirit remains unbroken—he continues to embody resilience and forward momentum.

He refuses to retire—not out of necessity, but from purpose. Danny Thiemens remains relentlessly devoted to the advancement of the crane and rigging industry. He speaks at conferences, advises contractors, guides apprentices, and works alongside regulators, all while leading Fulcrum with a clear and uncompromising vision. He is as much an architect of educational standards as he is a field technician, with safety as his main priority with every lift made.

His legacy is already cemented in classrooms, certification manuals, legislative records, and in the safe hands of operators across North America. But Danny's mission is not yet complete. Until every crane operator is properly trained, until every lift is performed with the utmost care, and until every worker returns home safely, Danny Thiemens will remain in the arena—teaching, leading, building.

Through relentless dedication and unmatched industry influence, he ensures that the future of crane safety and operator education will be stronger, smarter, safer, and keeping a safety-first mentality in each and every hoisting and rigging operation. Incomparable is his devotion to educating future generations aspiring to be the qualified, certified hoisting and rigging leaders of tomorrow.